Ligaments and Ligamentoplasties

Springer
Berlin
Heidelberg
New York
Barcelona
Budapest
Hong Kong
London
Milan
Paris
Santa Clara
Singapore
Tokyo

L'Hocine Yahia (Ed.)

Ligaments and Ligamentoplasties

With 132 Figures and 12 Tables

Springer

L'Hocine Yahia, Ph.D.
Biomechanics/Biomaterials Research Group
Biomedical Engineering Institute
Ecole Polytechnique of Montreal
P.O. Box 6079, Station "Centre-Ville"
Montreal, Quebec H3C 3A7, Canada

ISBN-13:978-3-642-64404-7 e-ISBN-13:978-3-642-60428-7
DOI: 10.1007/978-3-642-60428-7

Library of Congress Cataloging-in-Publication Data. Ligaments and ligamentoplasties/L'Hocine Yahia (ed.). p. cm. Includes bibliographical references and index. ISBN-13:978-3-642-64404-7 (softcover: alk. paper) 1. Ligament prostheses. 2. Cruciate ligaments. I. Yahia, Hocine, 1952- . [DNLM: 1. Anterior Cruciate Ligament – surgery. 2. Anterior Cruciate Ligament – injuries. 3. Ligaments – physiology. 4. Implants, Artificial. 5. Biocompatible Materials. 6. Polyethylene Terephthalates. 7. Biomechanics. WE 870 L7233 1996] RD561.L533 1997 617.4'72 – dc20 DNLM/ DLC for Library of Congress 96-17116

Softcover reprint of the hardcover 1st edition 1997

Cover design: Design & Production GmbH, Heidelberg

Typesetting: Scientific Publishing Services (P) Ltd, Madras

SPIN: 10487458 24/3135/SPS – 5 4 3 2 1 0 – Printed on acid-free paper

Dedication

Gilbert Drouin was born in Quebec, 4 May 1941. He received his Mechanical Engineering Degree in 1966 from the École Polytechnique of Montreal, his Master's in 1972 from Cornell University, and his Ph.D. in Biomedical Engineering in 1976 from the University of Virginia, USA. Because of his wide-reaching interest in Biomechanics and Biomaterials research and education, he created in 1975 with Roland Doré the first interdepartmental research group, the Biomaterials/Biomechanics Research Group (BBRG), at the École Polytechnique.

From 1978 to 1990, as a director of the BBRG, Professor Drouin was intensively involved in research on ligaments and ligamentoplasties. These activities began in 1975 when Dr. Carroll Laurin, an orthopedic surgeon from Royal Victoria Hospital, brought to the engineers the problem of replacing the natural ligament by an artificial one. Since that time ligament research has become the cornerstone of the BBRG's activities.

Among Professor Drouin's early contributions was the development of a new concept for design of prostheses for the ligament replacements (G. Drouin and R. Doré, Sixth Congress of Biomechanics, Copenhagen, July, 11–14, 1977). In those years he focused on understanding the composite nature of the cruciate ligaments and published a mathematical model based on ligament constituents (*Journal of Biomechanical Engineering*, 1980). In 1980 he organized the Second Congress of the Canadian Biomaterials Society, held in Montreal.

Since 1980 Professor Drouin has expanded his efforts toward the development of surgical tools and new methods for the mechanical evaluation of implants. Between 1980 and 1985 several patents were developed, particularly an original fatigue testing machine for ligament prostheses. He introduced the reverse engineering approach in the design of prosthetic ligaments and presented these new ideas at the Ninth International Society of Biomechanics Congress in 1985. During this period several members of his research group – P. Thiry, R. Doré, J.-M. Dorlot, P. Gely, and L'H. Yahia – contributed to the development of new concepts and approaches in the biomechanics and biomaterials field. Several clinicians participated in this research: Drs. P. Ranger, L. Theot, C.-H. Rivard, and G. Tremblay.

In parallel to these activities, Professor Drouin was the director of the Research Center of the Rehabilitation Institute of Montreal from 1981 to 1990. He contributed to many projects in rehabilitation engineering and patented a multifunctional hand prosthesis.

From 1986 to 1990 the involvement of M. Masson, J. De Guise, N. Newman, and M. Fallaha opened new areas, such as proprioception role of ligaments, three-dimensional reconstruction of the knee joint, surgical planning and assistance. In 1988 he published a critical review on the carbon fiber ligaments (CRC Press, vol. 4, 2; 1988).

Professor Drouin was a driving force behind the developments at both BBRG and the Rehabilitation Institute, and many of his contributions were outstanding.

Professor Drouin has been the recepient of many outstanding awards, including the presidency of the FRSQ-CRIQ committee (1992–1994) and the presidency of the Mechanical Engineering Committee of the CRSNG (1987–1989).

He has also served as member of the editorial board for several international journals, including the *Journal of Biomechanics*, *Journal of Biomedical Engineering*, and the *Journal of Orthopedic Research*. He has provided guidance to many professional societies. He is a member of eight societies.

Professor Drouin has served as a member of several NSERC committees, including Steacie and Gold Medal Committees and the Committee on Research Grants. He has also been group chair for several of the Council's engineering grant selection committees.

Professor Gilbert Drouin is now the Dean of Studies and Research at École Polytechnique and a Council Member of the Natural Sciences and Engineering Research Council of Canada (NSERC).

Along with his most distinguished academic achievements, Professor Drouin is best known for his friendliness and kindness. He is well liked and well respected by his students, followers, and colleagues. To this list it would be most appropriate to add: an eminent engineer, a pioneer in biomechanics, a distinguished scientist, and our teacher, leader, and friend, Prof. Gilbert Drouin.

L'Hocine Yahia

Preface

It was as early as 1959 that John Gort and Olav Rostrup (University of Alberta, Edmonton, Canada) introduced Teflon fabric (polytetrafluoroethylene), a substance previously used for blood vessel prostheses, for ligament replacements. This material unfortunately had unsuitable mechanical properties to fulfill the desired function. Since this early work there has been a growing interest in the treatment of the ever-increasing number of ligament injuries. Despite the substantial body of data and literature that has accumulated to date, our understanding of the structure, function, and biology of the cruciate ligaments has not reached the point that we are able to repair this complicated structure or to reconstruct it to its preinjury state. Replacement of injured cruciate ligaments of the knee by artificial devices therefore continues to be a real challenge. In fact, the life span of existing prostheses remains short. Autografts are currently popular options, but these require the disturbance of adjacent structures involved in the knee's function and do not reproduce the strength properties of the original structure. Allografts do not entail resection of structures, but they do not reproduce mechanical properties better than autografts, and sterilization without deterioration of function is difficult and controversial. There is therefore no ideal replacement for ligaments to date. This has led to the current need to reconsider the approach and to develop a new generation of implants.

Although a number of books are available that provide information about ligamentoplasties, we believe that no single volume emphasizes sufficiently the progress that has been made outside of North America. Despite all the research activity during the past two decades the many textbooks on this subject are relatively uniform in their focus and perspective. The present text thus differs from the classical view by highlighting the neurological function of ligaments, fatigue testing in more physiological conditions, biomimetic designs, European surgical techniques, and new approaches deriving from tissue engineering.

We have tried to make this book as pragmatic as possible. The approach is to include knowledge which, in the FDA school of thought, was marginal. This book therefore takes a more international approach, and the contributions are closer to the SICOT/SIROT philosophy – thanks to the Swedish, German, Israeli, French, Italian, Brazilian, and Canadian authors.

This volume does not attempt to address every aspect of cruciate ligament replacement but concentrates on proposing a new philosophy where the neurological function of the ligaments is put forward. The editor presents here the views of several experts, and it is interesting to note in fact that almost all contributions reach the same conclusion – that the neurological role of ligaments is at least as important, indeed if not more important, than their mechanical function. Basically, this text is devoted to summarizing what is genuinely new in this field: basic knowledge such as ligament function, proprioception, dynamic stability, and surgical techniques.

This book is organized into three sections. The first examines some fundamental aspects of natural ligaments. The second deals with artificial ligaments. The history of artificial ligaments is presented, and the materials and designs employed are covered. Their mechanical performance and lifetimes are discussed, some surgical techniques and a fixation technique are presented, focusing upon clinical performance and biofunctionality. New directions are presented in the third section to expand our present understanding and to suggest future approaches.

With the diversity of contributors to this book, the reader will at times detect similarities and at others no doubt divergences of opinions among the authors with respect to some scientific issues, including surgical techniques. However, our goal is not to present a uniform and incontestable philosophy but to show the diversity of interpretation that is possible and to open discussion of the problem of ligaments and ligamentoplasties.

The editor would like to acknowledge the help and support of many individuals in the preparation of this book. First and foremost are my colleagues of the BBRG at the École Polytechnique of Montreal, Gilbert Drouin, Roland Doré, André Bazergui, Pierre Thiry, Nicolas Newman, Jacques de Guise, Nicolas Duval, Charles-Hilaire Rivard, and Jean-Marie Dorlot, who provide, as pioneers, a truly outstanding environment for biomechanics and biomaterials research. Next, my students and research associates Nicola Hagemeister, Eric DesRosiers, Stefane Methot, Maryam Tabrizian, Nancy St-Onge, Pascale Pigeon, Arnold Mitnitski, Francis Boureault, and Souad Rhalmi provided helpful comments. The several authors have undertaken a difficult task with enthusiasm, and I have done little as editor to change their original manuscripts.

This volume has been made possible through the work and support of the members of the BBRG at the École Polytechnique of Montreal. The participation in the editor's meetings by the BBRG members resulted in an exchange of multidisciplinary ideas. The insightful discussions that followed contributed positive input to the final book organization and content.

I would like to thank all the contributing authors for their time in meeting the manuscript deadlines and bringing their expertise to this project.

I would also like to thank my research assistant, Nicola Hagemeister, for her untiring efforts, diligence in contacting authors, patience in compiling the manuscripts, expertise in word-processing skills, and German touch (translation and organization).

The initial discussions which eventually led to this book began during the SIROT meeting in Seoul in 1993. I would like to thank Dr. Charles-Hilaire Rivard, the current President of the SIROT for his support and collaboration during the past 15 years.

I would also like to extend my gratitude to Dr. Nicholas Newman for his stimulating discussions on the proprioception of the knee joint ligaments.

The text emerged from all of these individuals' contributions.

Finally, I would like to thank the staff at Springer-Verlag for their patience and guidance in the preparation of this text.

L'Hocine Yahia

Contents

Contributors

Auger, F.A.
Laboratoire d'Organogénèse Expérimentale, Hôpital du Saint Sacrement, 1050 Chemin St-Foy, Québec, Québec G1S 4L8, Canada

Aunallah, J.
Department of Orthopedic Surgery and Traumatology, Western Galilee Hospital, P.O. Box 21, Nahariya 22100, Israel

Boss, J.H., MD
Department of Pathology, Bnai-Zion Medical Center, 47 Golomb St., P.O. Box 4940, Haifa 31048, Israel

Caron, C.
Laboratoire d'Organogénèse Expérimentale, Hôpital du Saint Sacrement, 1050 Chemin St-Foy, Québec, Québec G1S 4L8, Canada

Chaput, C., PhD
Chemical Engineering Department, Centre de Recherche Appliquée sur les Polymères, École Polytechnique, C.P. 6079, Succursale "Centre ville", Montréal, Québec H3C 3A7, Canada

de Guise, J.A., Prof.
École Supérieure de Technologie, Université du Québec, 4759 avenue Henri-Julien, Montréal, Québec H2T 2C8, Canada

Denti, M., MD
University of Milan, Orthopedic Clinic, Milan, Italy

DesRosiers, E.A., PhD
Apotex Research Inc., CNRC, Edifice M54, 1200 Chemin de Montréal, Ottawa, Ottawa K1A 0R6, Canada

Drouin, G., Prof.
Biomechanics and Biomaterials Research Group (BBRG), École Polytechnique, C.P. 6079, Succursale "Centre ville", Montréal, Québec, H3C 3A7, Canada

Duval, N., MD, MScA
Service d'Orthopédie, Porte G 4132, Hôpital Notre-Dame,
1560 Sherbrooke Est, Montréal, Québec H2L 4M1, Canada

Fabbriciani, C., MD
Universita Cattolica del Sacro Cuore, Instituto di Clinica Orthopedica,
Facolta di Medicina e Chirurgia "Agostino Gemelli", Franscesco Vitro 1,
00168 Roma-Largo, Italy

Fernandes, J.C., MD, MScA
Serviço de Ortopedia, Hospital de Clinicas da Universida Federal do
Parana, 200 Curitiba, Brazil

Franzese, S.
Universita Cattolica del Sacro Cuore, Instituto di Clinica Orthopedica,
Facolta di Medicina e Chirurgia "Agostino Gemelli", Franscesco Vitro 1,
00168 Roma-Largo, Italy

Germain, L.
Département de Chirurgie, Faculté de Médecine, Université Laval,
Québec, Québec G1K 7P4, Canada

Goulet, F.
Laboratoire d'Organogénèse Expérimentale, Hôpital du Saint Sacrement,
1050 Chemin St-Foy, Québec, Québec G1S 4L8, Canada

Hagemeister, N., MScA
Biomechanics and Biomaterials Research Group (BBRG),
École Polytechnique, C.P. 6079, Succursale "Centre ville", Montréal,
Québec H3C 3A7, Canada

Jírová, J.
Institute of Theoretical and Applied Mechanics of the Academy
of Sciences of the Czech Republic, Prosecka 74, 19000 Prague 9,
Czech Republic

Johansson, H., PhD
Division of Work Physiology, National Institute of Working Life,
P.O. Box 7654, 90713 Umeå, Sweden

Kafka, V., DSc, Eng.
Institute of Theoretical and Applied Mechanics of the Academy
of Sciences of the Czech Republic, Prosecka 74, 19000 Prague 9,
Czech Republic

Krudwig, W.K., MD
Department of Traumatology, von Hoerde'sches Marienhospital,
Von-Droste-Straße 14, 59597 Erwitte, Germany

Laboureau, J.P., MD
Centre de Chirurgie Orthopédique et Sportive,
44, boulevard Henri Bazin, 21300 Chenove, France

Larouche, S., Ing.
Biomedical Engineering Institute, École Polytechnique, C.P. 6079,
Succursale "Centre ville", Montréal, Québec H2C 3A7, Canada

Marnat-Perrichet, F., MD
Centre de Chirurgie Orthopédique et Sportive,
44, boulevard Henri Bazin, 21300 Chenove, France

Mendes, D.G.
Department of Pathology, Bnai-Zion Medical Center, 47 Golomb St.,
P.O. Box 4940, Haifa 31048, Israel

Méthot, S.
Centre de Recherche Pédiatrique, Sainte-Justine's Hospital,
3175 Côte St Catherine, Montréal, Québec H3T 1C5, Canada

Newman, N.
Orthopedica Surgery, Hôtel Dieu Hospital, 384 rue St Urbain, Montréal,
Québec H2W 1T8, Canada

Normand, A.
Laboratoire d'Organogénèse Expérimentale, Hôpital du Saint Sacrement,
1050 Chemin St-Foy, Québec, Québec G1S 4L8, Canada

Rancourt, D.
Laboratoire d'Organogénèse Expérimentale, Hôpital du Saint Sacrement,
1050 Chemin St-Foy, Québec, Québec G1S 4L8, Canada

Ranger, P., MD, MScA
Hôpital du Sacré-Coeur de Montréal, Montréal, Québec, Canada

Rhalmi, S., MScA
Centre de Recherche Pédiatrique, Sainte-Justine's Hospital,
3175 Côte St Catherine, Québec, Québec H3T 1C5, Canada

Rivard, C.H., MD
Centre de Recherche Pédiatrique, Sainte-Justine's Hospital,
3175 Côte St Catherine, Montréal, Québec H3T 1C5, Canada

Sati, M., PhD
Biomedical Engineering Institute, École Polytechnique, C.P. 6079,
Succursale "Centre ville", Montréal, Québec H2C 3A7, Canada

Schäferh, R.
Frauenhofer Institut für Werkstoffmechanik, Freiburg, Germany

Schiavone, Panni A., MD
Universita Cattolica del Sacro Cuore, Instituto di Clinica Orthopedica, Facolta di Medicina e Chirurgia "Agostino Gemelli", Franscesco Vitro 1, 00168 Roma-Largo, Italy

Selmani, A., Prof.
Chemical Engineering Department, Centre de Recherche Appliquée sur les Polymères, École Polytechnique, C.P. 6079, Succursale "Centre ville", Montréal, Québec H3C 3A7, Canada

Shajrawi, I.
Department of Pathology, Bnai-Zion Medical Center, 47 Golomb St., P.O. Box 4940, Haifa 31048, Israel

Sjölander, P., PhD
Division of Work Physiology, National Institute of Working Life, P.O. Box 7654, 90713 Umeå, Sweden

Solomon, H.
Department of Orthopedic Surgery and Traumatology, Bnai-Zion Medical Center, 47 Golomb St., P.O. Box 4940, Haifa 31048, Israel

Soltesz, U.
Frauenhofer Institut für Werkstoffmechanik, Freiburg, Germany

Soudry, M., MD
Department of Orthopedic Surgery and Traumatology, Rabin Medical Center, Campus Beilinson, 49100 Petah-Tikva, Israel

St-Onge, N.
Biomedical Engineering Institute, École Polytechnique, C.P. 6079, Succursale "Centre ville", Montréal, Québec H2C 3A7, Canada

Ulrich, D.
Frauenhofer Institut für Werkstoffmechanik, Freiburg, Germany

Witzel, U., Dr. Ing.
Forschungsinstitut für Biomechanik, Institut für Konstruktionstechnik, Ruhr-Universität Bochum, 44780 Bochum, Germany

Yahia, L'H., PhD, Prof.
Biomechanics and Biomaterials Research Group (BBRG), École Polytechnique, C.P. 6079, Succursale "Centre ville", Montréal, Québec H3C 3A7, Canada

Introduction: Evaluation of Cruciate Ligament Prostheses – Critique of Current Concepts and Procedures

L'H. Yahia, N. Hagemeister, G. Drouin, C.H. Rivard, S. Rhalmi,
and N. Newman

Introduction

Many studies have been published on ligament replacements over the past century, and the various research groups have voiced very conflicting opinions. To date none of the available reconstruction procedures has demonstrated adequate long-term behavior. While some of the techniques show promising results, many questions remain concerning material, structure, and reconstruction technique. Surgeons disagree even regarding the type of procedure to be used (autograft or synthetic ligament).

An understanding of the failures and successes of previous grafts and a study of the promising surgical techniques and prosthetic designs permits a number of hypotheses to be elaborated concerning the ligament replacement. These hypotheses should lead to further studies in this field, in the hope of leading to solutions to the problem. The importance of the cruciate ligaments of the knee joint is well understood. Failure of the anterior cruciate ligament (ACL) causes instability and excessive joint mobility that, if left untreated, leads to osteoarthritis. Since the beginning of this century interest has focused increasingly on knee ligament reconstructions. The basic alternatives for replacing the ACL are autografts, allografts, and synthetic ligaments.

Ligaments are now recognized to play both mechanical and neurological functions in the biomechanics of the knee (see Chaps. I.1 and I.2). Ligaments as a control system of a healthy knee joint supported by the muscular system are responsible for the rolling and gliding motion of the femoral condyles on the tibial plateau. The presence of mechanoreceptors in the human cruciate ligaments has been confirmed only in recent years. The first histological demonstration of mechanoreceptors in the human ACL was reported by Schultz et al. in 1984 [37]. In 1986 Zimny et al. [46] identified by using a modification of Gairn's gold chloride stain, the presence of Ruffini's end organs, Golgi tendon organs, and Pacini corpuscles. In 1992 we confirmed the presence of Ruffini's and Pacini's corpuscles and free nerve endings in canine cruciate ligaments [42, 43].

The development of prosthetic materials to replace ligaments has stimulated interest in the contribution of mechanoreceptors activity in the cruciate ligaments to the stability in the knee joint. The specificity of the gold chloride

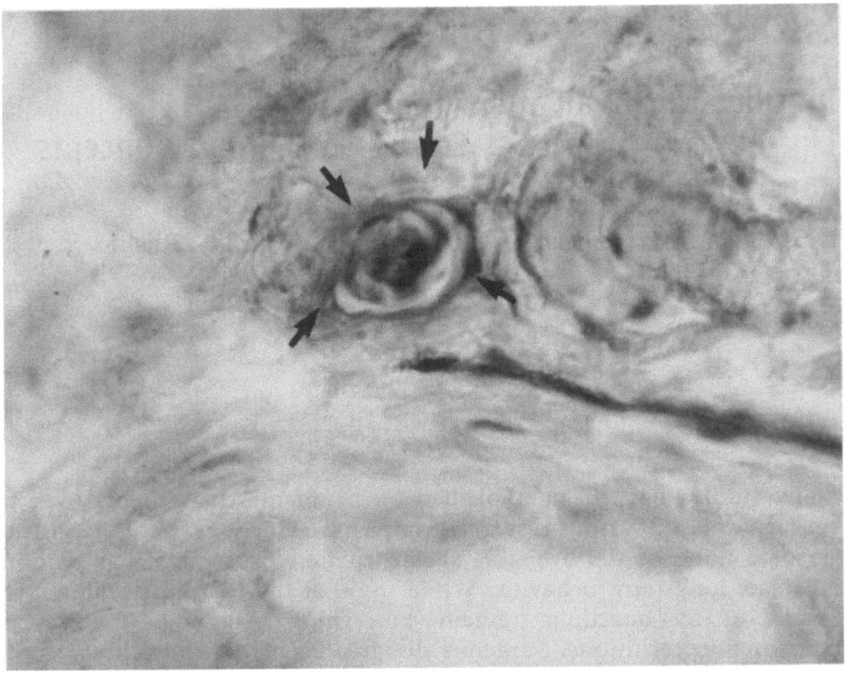

Fig. 1. Neurofilament protein immunoreactive Ruffini nerve ending in human anterior cruciate ligament (*arrowhead*). ×390

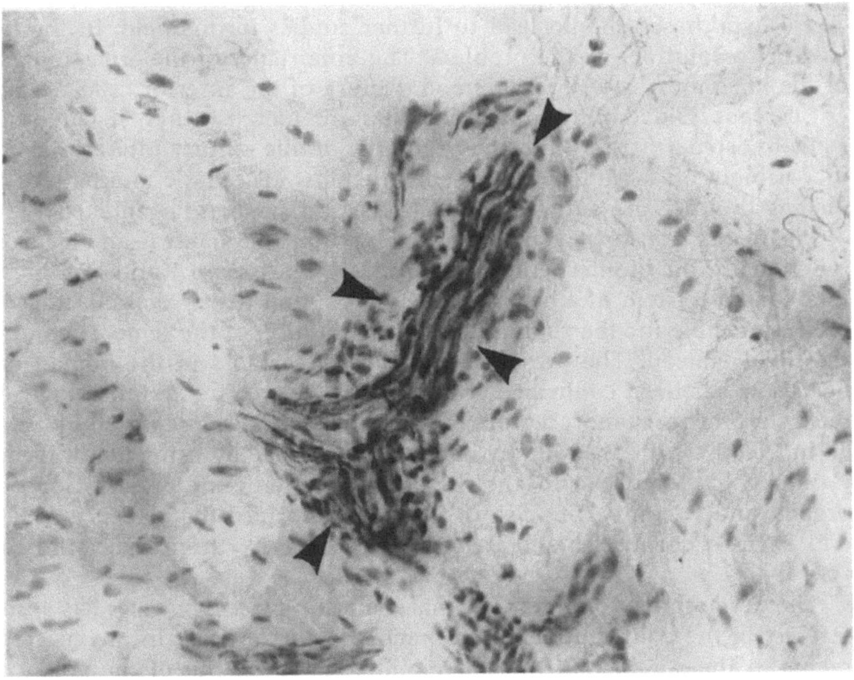

Fig. 2. Pacinian's corpuscle immunoreactivity to S-100 protein in human ACL showing the inner core and the lamellar system (*arrows*). ×125

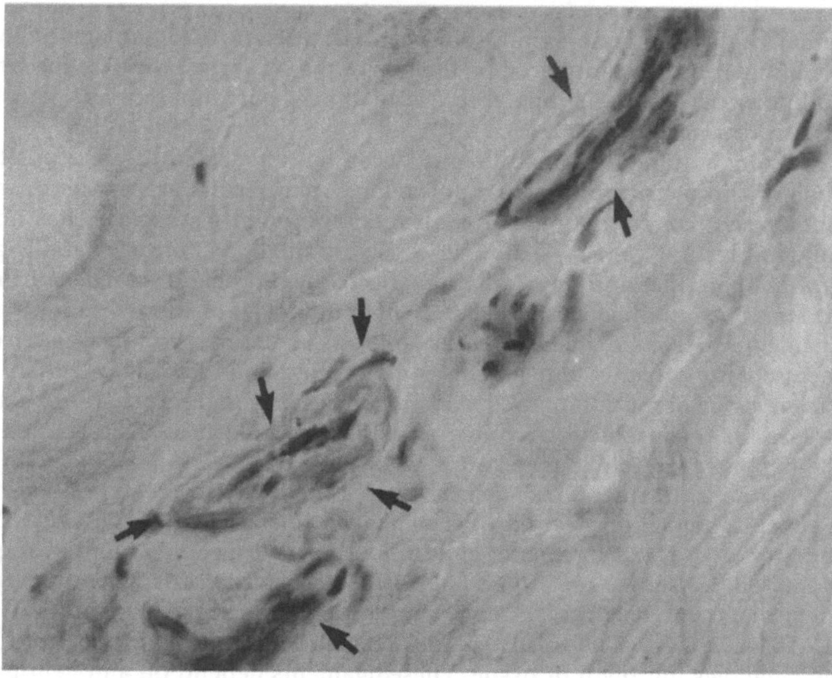

Fig. 3. S-100 immunoreactive peri-insertion nerves and Pacinian-like corpuscle in human anterior cruciate ligament (*arrows*). ×125

technique was demonstrated by immunohistochemical methods [34]. Using this highly specific method, we recently demonstrated the presence of the same mechanoreceptors in human cruciate ligaments [35] (Figs. 1–3). It is therefore accepted now that the knee joint ligaments contain Ruffini, Pacinian, and Golgi tendonlike organs, and free-nerve endings with varying capabilities of providing the central nervous system with information about movement and position as well as noxious events [21, 41]. The mechanoreceptors in the knee joint ligaments are suggested to play a key role in the joint stiffness and the functional joint stability [39].

Most of the presently used treatments for injured ligaments are based on the view that the ligaments primarily act as static, mechanical stabilizers of the joint. Consequently a lot of efforts have been devoted to find, for instance, substitution materials and structures which have mechanical properties comparable to those of the natural ligaments. However, the clinical results following ligament reconstruction show the need to restore both neuroproprioceptive and mechanical functions of the cruciate ligaments (see Chaps. I.1, I.5). In fact, impaired proprioception of the joint has been demonstrated in patients with a severely damaged or ruptured ACL [3].

Problems associated with the use of autografts are mainly the necrosis which occurs during the early postoperative period during revascularization and the loss of normal structures which contribute to the normal function of the knee joint. The use of bone-tendon-bone grafts causes further damage of the neuromuscular balance impaired by the loss of the ACL, including its receptors. In addition, the still existing receptors of the ligamentous stumps can be destroyed by large tunnels.

Although autografts and allografts have been reported to yield positive results in short-term follow-up studies, long-term results are not as promising. The present concern about the transmission of viral disease also has not been fully resolved. It appears that this risk is very small with proper handling of the graft, but sterilization by irradiation affects the physical properties of the graft and has been implicated in the failure of grafts [45]. For these reasons synthetic ligaments presented themselves to be used for the cruciate ligament replacements. The prosthetic ligament is biomechanically stronger which allows earlier range of motion of the knee, weight-bearing, and return to full activities. Its use for reconstruction avoids moreover proprioceptive disturbance due to the harvest of autologous tissues.

Prosthetic ligaments are generally classified into three types: permanent ligaments, stents, or scaffolds. Permanent ligaments include the Gore-Tex and the Stryker-Dacron ligaments and are designed to provide high strength and increased resistance to fatigue failure. The second type is a temporary prosthetic ligament used as a stent such as the Ligament Augmentation Device (LAD) or Trevira. The third type of prosthetic ligament is the scaffold device, which allows ingrowth of tissue. These ligaments depend on autogenous tissue ingrowth. The collagen tissue may grow into the existing graft, as in the Leeds-Keio ligament and the Ligament Advanced Reinforcement System (LARS).

Failure Analysis of the FDA-Approved Devices

Many synthetic ligaments have been introduced for human clinical trials, including the carbon fiber, Kennedy LAD, Leeds-Keio, Gore-Tex, and Stryker Dacron ligaments (see Chap. II.2). Gore-Tex, Stryker, and LAD have been approved by the FDA for use as a replacement for the ACL but with restrictions. The FDA gave approval in 1986 for the general use of Gore-Tex (expanded polytetrafluoroethylene; Gore, Flagstaff, Arizona) and in 1989 for Stryker-Dacron artificial ligaments (polyester fabric; Meadox Medical, Oakland, N.J., USA) but restricted their use to patients with failed intraarticular reconstructions. The Kennedy LAD (diamond-braided polypropylene, 3M, St. Paul, Minn., USA) was also approved in 1986 to augment the autogenous graft used in the Marshall-Macintosh procedure, i.e., the quadriceps tendon-prepatellar tissue-patellar tendon autograft [38]. In 1991 the LAD was also approved as an augmentation for the iliotibial band, the semitendinous tendon, and portions of the patellar tendon used for ACL replacement.

The Gore-Tex ligament was predicted to have a life to rupture of 4×10^9 cycles and less than 4% elongation at 3×10^8 cycles [5]. Assuming an average of 4.2×10^6 cycles per year, the ligament has a life expectancy of 950 years prior to rupture and 70 years prior to 4% elongation [13]. This has not been confirmed in actual clinical application, and the ligament has failed both because of rupture and elongation. Despite the encouraging preliminary results of Gore-Tex ligament reconstruction, the results of many 4- to 5-year follow-up studies have demonstrated an apparent deterioration in both subjective and objective results [30].

The Stryker-Dacron ligament was initially designed to be used as an augmentation device with the iliotibial band in ACL reconstructions, but the high strength of the ligament has led to its use as a permanent prosthesis. Fatigue testing showed the ability to withstand 134 000 fatigue cycles at 1730 N. This amount of force would presumably disrupt a normal ACL after a single cycle. The estimated in vivo life expectancy from the fatigue testing data is 8.5 years. Recent studies demonstrate a significant failure rate at 5 years.

The LAD prosthesis is a permanent, nonabsorbable augmentation device made of woven polypropylene. In vitro testing demonstrates a cyclic elongation of 4% after 1 million cycles between 50 and 500 N, resulting in a 9% strength loss. Bending fatigue parameters after 10 million cycles of 150–300 N, with flexion between 15° and 40°, yielded 4.5% cyclic elongation and 23% strength loss [30].

Many clinical studies have demonstrated early success, but to our knowledge no long-term investigations have shown convincingly that these constructs remain intact, or that they replace the function of the ACL effectively. Very limited use of these prostheses in the United States has been disappointing. The life-span of the existing prostheses remains short. Fatigue, fretting, and wear have resulted in an unacceptably high rate of failure. As a consequence enthusiasm for the use of synthetic ligaments has been tempered by poor long-term results, and replacement of the cruciate ligaments seems impossible with today' s techniques. The FDA has adopted a low profile regarding this high rate of failure, and the bone-patellar tendon-bone appears as the "gold standard" solution and the best alternative to synthetic ligaments in North America. However, it is more appropriate to consider this solution as the "best available to date" [12].

Outside the United States several polyester prosthetic ligaments have been developed without the control of the FDA. However, recently the synthetic cruciate ligaments have also been subjected to some limitations in France. The implanting of artificial ligament prostheses is permitted as either stents or replacements, but severe restrictions have been imposed by the French health care system [26], although no scientific argument has been given. One of the better known investigational polyester prosthesis in United Sates is the Leeds-Keio implant, while in Canada LARS and Trevira Hochfest are gaining in popularity. In the United Kingdom the Leeds-Keio ligament was presented as a scaffold to ease the growth of a neoligament, but up to 1988 the clinical data did not provide the expected success [32]. Clinical trials for the Leeds-Keio prosthesis began in the United States in 1990. To our knowledged it has not yet been approved.

Another polyester ACL prosthesis, the Ligastic implant, has been introduced mainly in France. The particularity of this artificial ligament is its knitted structure, in which all the fibers are oriented longitudinally at their maximal length [7]. This first generation was very sensitive to the surgical technique, and a high rate of failure was reported recently by Cazenave et al. [8]. This is surprising and contradicts the previous results published by the same authors. Cazenave et al. [8] did not provide any scientific explanation for this difference.

Based on ligastic experience, a new design using mechanically and biologically selected polyester has been introduced recently by LARS [25]. This has two different parts: (a) The intra-osseous part is made of longitudinal fibers united by a transverse knitted structure which, unlike a woven structure, prevents elongation over time. (b) The intraarticular part is constructed only of longitudinal and parallel fibers which are pretwisted. This architecture is based on a biomimetical approach and showed a higher performance compared to the ligastic first generation [12, 23].

In Germany the Trevira Hochfest, a woven prosthetic ligament made of polyethylene terephthalate (PET), has been used since 1986. Clinical results revealed an incidence of prosthesis tearing or loosening of 5.15% after an implantation time of up to 4 years [22]. In case of salvage procedures about 18% ruptures after 4 years follow-up have been stated. No synovial reaction has been observed [27].

The surgical techniques using LARS and Trevira both try to preserve the mechanoreceptors present in the ruptured ACL and to minimize any further sensory damage. The Trevira ligament has bony tunnels of small diameter, which avoids excessive tissue harvest, and the LARS is inserted by preserving the stumps of the broken ligament. Surgeons believe that this technique preserves the mechanoreceptors which are present in the stumps (Laboureau, personal communication). These could therefore pass on information to the central nervous system on the position of the joint surfaces. In fact, damages to the innervation of the ACL leads to inadequate motor control of the joint and can give rise to disturbances in the synchronization of movements. This loss of control may result in uneven joint loading, leading to repetitive microtrauma and could be involved in premature fatigue of artificial ligaments. In the case of the Leeds-Keio ligament the harvest of the remnant of the original ligament may explain the absence of mechanoreceptors in the ligament stated by Denti et al. [11]. As regards the LAD, which is an augmentation of an autologous structure for the repair of the ACL, Ruffini, Pacini, and free nerve endings have been found. In this case the synthetic augmentation device is completely surrounded with autologous structures.

The following clinical conclusion has been stated by Metak et al. [31]: "The evidence of nervous structures in cruciate ligament stumps up to 7 years after the injury renders the resection of these stumps obsolete. Regardless of the time of operative care, cruciate ligament stumps should be integrated in the reconstruction." However, it is not evident whether the preservation of the stumps has an influence on the life time of artificial ligaments. Further research on retrieved ligaments is therefore needed to confirm this hypothesis.

Unforeseen Failure of Prosthetic Ligaments

The poor clinical results observed with the ligaments developed in the United States under FDA control are surprising since in vitro evaluation of the implants has been performed following the FDA guidelines. Why, then, do artificial ligaments fail?

Considering that most of the devices involved have been subjected to rigorous premarket testing in compliance with the current guidelines, there arise a number of questions. Were the evaluations properly conducted? Was the simulation of product usage conditions inadequate? Or is there something fundamentally inadequate in the current approach?

To begin to answer to these questions one must compare current practices in biomechanical and biocompatibility testings with the state of the art in analytical science. In doing so the assumptions inherent in the current recommended tests need to be identified and their limitations questioned.

Since these synthetic ligaments have been approved by the FDA, they are presumed to have fulfilled the FDA biomechanical requirements. In our opinion, the concepts used in the old designs and the procedures used for their evaluation are inadequate for the cruciate ligaments [42, 45]. In 1984 we were the first laboratory to suggest that the FDA guidelines are not appropriate (see below) [19]. Among the weaknesses raised, the loading mode (combined loading versus single load; displacement-controlled versus force-controled), the tunnel angulation, and the lengthening were identified as very critical for the prostheses life span [14].

However, the FDA rejected our innovative approach without discussion. In 1988 we had difficulty in publishing our work showing the twisting of the collagen fibers in the natural ACL and PCL [41]. Prejudice against artificial ligaments is also caused by problems frequently associated with the development of any new product or method. The results of previous generation ligaments were poor because of a lack of technical expertise. In essence, synthetic ligament reconstruction offers the surgeon a logical solution by creating no new damage to the knee [12].

To elucidate the reasons of the failure we must also understand the success. Konrad Lorenz, the renowned philosopher, physician, and ethnologist, has noted that much more can be learned from trial and success than from trial and error [28]. By studying only random clinical (retrieved "failed") implants, Black, the "father" of modern biomaterials, advised that we are making the same mistake as certain penologists, who, wishing to comprehend crime, study only *failed* criminals, those convicted and incarcerated [4]. It is therefore interesting to look at European devices, which showed better clinical results.

Mechanical Testing of Prosthetic Ligament: The State of the Art

Current Practices: Requirements According to the FDA Guidelines

Standard test methods are needed to predict the clinical performance of new biomedical devices; however, there are many problems associated with the current standards. This is made evident by the recent clinical failures of ligament prostheses. The test standards for the ligament prostheses are based on specific forces; clinical failures can be due to many factors, including the surface finish and the surgical technique. A device assessment should provide a life time prediction for the device, similar to the manner in which airplanes have been designed for decades; materials, components, and system testing is required to determine device performance.

Given the extensive recommendations offered by the FDA guidelines and the impressive body of scientific experience on which they are based, there is a strong temptation to follow the guidelines without actively considering their possible limitations or how the individual device being tested may require additional, more extensive testing. However, recent and well-publicized failures of medical devices (ligament prostheses) suggest that these recommendations must be subjected to constant questioning, and that, depending on the device and its application, testing protocols may need to go beyond the current guidelines to ensure the safety of the device.

It is clear that most of the in vitro testings do not reproduce the in vivo loading of the ligament prostheses. Moreover, the biological variability among the individuals and the large number of different surgical techniques result in varying in vivo loading of the prostheses. This knowledge led us to develop tools for the in vitro testing of artificial ligaments in more physiological conditions (torsion, tension, flexion testing machine, TTF [16]) and to adapt these tests to each procedure (see Chap. II.3).

Displacement-Controlled Versus Load-Controlled Procedure

A ligament prosthesis is loaded by a three-dimensional displacement and rotation of its bony attachments. Usually, a load controlled procedure is chosen, loading the ligament between predetermined values F_{min} and F_{max}. The load-elongation curves, irreversible changes in shape and the lifetimes of different prostheses can be compared and conclusions for the service lives under physiological conditions are drawn. However, the minimum load state of the prosthesis is not given by a lower load as in the standard test but by the position in the knee when it is unloaded. Recently a new version of our fatigue machine allowing load and position controlled cyclic testing has begun to be developed. The elongation of the prosthesis is load controlled in the upper loading period, and the machine can be switched to position control when during the unloading phase a certain level is been reached. Thus at zero load the actuator moves to the same position in all cycles. The consequence is that the residual elongation of the ligament which could be detected in the purely load controlled test causes buckling in this arrangement. This buckling occurs in a strongly concentrated area, and the prosthesis can be damaged there. As a consequence the cycling test itself or a tear test after cycling, reveals a distinct weakening of the prosthesis in this region. To avoid the influence of the residual elongation on the fatigue strength by buckling and on the destabilization of the joint, it is therefore suggested to include this test into routine methods. This test would in fact provide a better idea of the fatigue life in vivo of the prosthesis.

Fatigue Concept: Combined Load Versus Single Load

The fatigue strength of ligament prostheses should be tested under physiological load conditions, i.e., test under normal gait conditions. To date no knee joint simulator exists. The most difficult problem of the in vitro fatigue test of ligament prosthesis is the lack of knowledge regarding the biomechanics, i.e., the in vivo load conditions. Fatigue-abrasion which occurs in vivo certainly has the most deleterious effect on implant survival. Nevertheless, its pattern is difficult to reproduce in vitro.

Ensuring the biomechanical reliability of ligament prostheses over several years' service time requires laboratory tests that allow prediction of the mechanical long time behavior. Methods are suggested in the FDA guidelines for tensile and bending fatigue testings [17]; nothing is mentioned regarding torsion tests, or fatigue under realistic conditions, i.e., combined loading.

Most of the ligament prostheses approved by the FDA have been tested following their guidelines. A comparison of this approach to the state of the art suggests a significant shortcoming: these tests provide data only for non-physiological conditions. That is, current practices do not determine the fatigue resistance under service loading conditions such as the combined tension-torsion-flexion loadings to which the ACL is subjected. Consequently ligaments that have been tested to 40 million cycles in the laboratory rupture after 6 months in the human knee [18]. Using the TTF developed in our laboratory, we have shown that the fatigue life time of ACL prostheses is decreased dramatically compared to standard tensile fatigue testings used by most investigators [14].

To our knowledge, among the various manufacturers, only LARS has performed similar fatigue tests which are beyond the protocols and provided fatigue lifetime data by using combined tension-flexion-torsion tests [25]. By cycling at 2 Hz frequency under 110 N tension, 30° flexion, 10° torsion, the LARS prosthesis resisted to 22 millions of cycles without complete rupture and with 50% of their initial resistance (Laboureau, personal communication). The permanent lengthening was found to be 5.1% corresponding to a laxity of about 1.5 mm. This is a first attempt to test a prosthesis under such conditions. However, these data are hypothetical, and tools are needed to measure the range of parameters in vivo (torsion, flexion, and tension). Further efforts must therefore be made to ensure that the loads and displacements are near to those occurring in vivo (adapt the test to the surgical procedure and the patient's anatomy).

Boundary Conditions

It should be recognized that data on forces in the ACL during specific activities are needed to aid in the design of prostheses and to establish proper reconstructive techniques. The direct measurement of ligament forces at the knee in vivo is impracticable. As a result most investigators use the data estimated by Chen and Black [9]. In daily life the ACL is thought to be subjected to loads which vary from 70 N in climbing stairs, 210 N in walking, 485 N in descending an incline, and 630 N in jogging. These characteristics appear to leave a significant margin of safety for ACL protection with routine activities. The validity of these data is now questioned by the high rate of failure of ACL prostheses. In addition, loads for athletic activities not yet well known are significantly greater.

Service-Loading Conditions

The mechanical function of the ligament is to limit movement between femur and tibia. From this arises the concept that ligaments are loaded by bone displacement which demonstrate a three-dimensional movement. The corresponding force levels imposed by three-dimensional kinematics on ACL prostheses depend on their mechanical properties; i.e., their stiffness. However, most authors have used the tensile forces estimated for natural ligaments by Chen and Black [9] while artificial ligaments are much stiffer [10]. As a result, for the same range of movement, the prostheses undergo much higher stresses in service. In addition, the surgical technique determines the angle of torsion and bending and the strains and stresses to which the synthetic ligament is submitted. These stresses are different from those existing in natural ligaments on which mechanical evaluations are often based.

Influence of the Direction of the Drill Holes

The relative motion of the ligament insertions (loaded by displacement) give the boundary conditions for the deformations of the ligament when no other interaction of the ligament with other ligaments or with bones occur. Most authors consider only a single mode of deformation, i.e., the longitudinal deformation. The torsional deformation of the cruciate ligaments, which has been previously addressed in a qualitative sense by Palmer [33] and in a quantitative sense by our laboratory [15, 19], is often overlooked. The reported global torsions from Gely et al. [19] for six different femoral and seven different tibial anchoring channels for the prosthetic ACL ranged from 36° to 86°. The surgical technique is then important for obtaining some initial torsion in the artificial ligament when the knee is in extension so that all fibers are arranged properly for the functional joint positions when the maximum strength of the artificial ligament is required. During every knee flexion the prosthesis is bent at the drill hole entrances. At the same time the prosthesis is subjected to torsional stress. The changes of the bending angles at the drill hole entrances of tibia and femur and the torsion angle of the intraarticular part of the prosthesis depend on the direction of the drill-canals.

We have investigated the influence of different localizations of these drill hole outlets on the torsion of ligament prosthesis and on the bending angle of the prosthesis in relation to the axis of the drill hole [15, 19]. The torsion angle depends on the combination of lateral femoral and anteromedial tibial drill hole entrance. At the femur the bending of the prosthesis in the anterior direction predominates; at the tibia the ligament is bent mostly posteriorly. The bending angle of the prostheses is generally greater at the femur than at the tibia. Stress concentrations occur at the outlets of the drill canals at the tibial plateau and at the femoral condyle. With every flexion of the knee the prosthesis is bent around the edge of the drill hole under a simultaneous torque. This combined bending-torque stress is superposed by a tensile force, which depends on the flexion angle. The resulting stress depends on the properties of the prostheses itself, the implantation technique and its integration inside the joint.

Claes et al. [10] have shown that the bending radius has a very important influence on the fatigue of ligament prostheses. To achieve long-term endurance of the prostheses this requires a large bending radius at the drill tunnel entrance. However, the realization of this requirement is limited by the anatomical circumstances inside the joint. Fatigue tests using larger bending radii do not correspond to the clinical conditions and reveal excessively high fatigue results.

Effect of the Method of Implantation

The method of implantation also has an important effect on the loading conditions to which the prosthetic ligament is submitted. Some important aspects are highlighted here.

As discussed above, the preservation of the innervation in the joint (by minimizing the bony tunnels and avoiding the harvest of autologous structures) has an influence on the loads on the prosthesis. The tightness of the reconstruction (pretension) depends usually on the surgeon's judgment. The optimum amount of pretensioning load is still subject for debate. It may vary between 20 and 70 N according to the literature. Witzel show that a pretensioning of 70 N results in tolerable loosening of the prosthesis for the larger knee flexions (Chap. II.5). It is important to note that small changes in the tightness of an implant cause significant differences in stability and range of movement [1]. The optimum tightness allows the knee to be extended fully, but with a "springy" feeling caused by a high implant tension, since weight bearing compresses the joint and relax the implant a little. The implant, whether isometric or not, should not be under tension continuously since this predisposes it to creep.

Isometry, when applied to an ACL reconstruction, is defined as following: the distance between the femoral and tibial attachments of the reconstruction remains constant as the knee is moved in flexion/extension [1]. This definition makes no reference to the position of the reconstruction. The definition of isometry means that the femoral attachment is constrained to travel approximately in an arc centered at the tibial attachment, when viewed in the sagittal plan. Because of the three-dimensional nature of the ACL the femoral attachment should be described as moving in a spherical locus centered at the tibial attachment, relative to the tibia. Isometry should be measured by using a suture passed along the ligament fibers, linked to a displacement transducer or isometer. It is interesting to note that isometry depends on the range of movement of the knee.

Hefzy et al. [20] have shown how an increase in the range of movement over which isometry is desired reduces the attachment area available with a given length change tolerance. Cazenave and Laboureau [7] took radiographs of pins in five cadaveric ACL-damaged knee joints at 0° and 90° knee flexion. They concluded that the center of the circular arc of the posterior part of the lateral femoral condyle, connected with the anatomical center of the tibial attachment, was nearly isometric, finding a length change less than 2 mm between those two positions. This does not mean that there was no length change at the other positions of knee flexion. Isometry is normally aspired to because this type of reconstruction avoids excessive length changes of the prosthesis (permanent stretching) during knee function.

Cruciate ligament tissue fails if it is extended of approximately 20%. A review of the literature shows that the ACL has a mean reported length of 32 mm; therefore an elongation of 7 mm (22%) could be expected to cause rupture. A more significant limit, however, is the viscoelastic nature of the tissue; this means that cyclic elongation of only 6% has been reported to cause permanent stretching of grafts, and that represents an elongation of only 2 mm. Since the extension to failure of the natural ligament corresponds to a force of 1725–2200 N, it seems likely that the normal motion forces are approximately up to 600 N. At a load of 500 N some artificial implants (the Gore-Tex band and the Trevira) stretched of 2% to 3%. In this case with the implant much stiffer than the natural ligament, the length change is only about 1 mm.

A rigid ACL replacement would impose knee motion that would not be physiological. Thus, rather than pursuing isometry by checking on reconstructions, a more logical approach would be to know how to recreate the behavior of the ACL-intact knee. It seems logical to suggest that what is needed is an approach that allows the surgeon to identify the isometric spot that would have been found when the knee was intact and then to base the reconstruction on that location. Even with this approach, normality could be restored only if the graft were a reasonable facsimile of the mechanical properties of the natural ACL [2]. Witzel et al. show in fact that a nearly isometric behavior is obtained when the tunnel is placed dorsomedially of the native insertion point (Chap. II.5).

Concluding Remarks

Need for Standard Procedure Revision

It is an important part of ASTM procedure that all standards be periodically revised or reviewed and reapproved to keep them up to date so that important advances are not impeded or ignored. Standards are not meant to be written in stone but to be revisable and adjustable, keeping up with the advances that are inevitable in science and surgical specialties. It is vital that new materials and devices be allowed to take their place in our armamentarium for providing care for our patients [24].

New Generation of Designs

New biomimetical designs have recently been introduced to reduce abrasion. This approach is based on the reverse engineering concept [45]. The natural ACL fans out from the center making the femoral and tibial attachments larger than the midsection. The fascicles of the ACL spiral from the femoral to the tibial attachments (counterclockwise for the left knee and clockwise for the right) [6, 41]. The amount of twist varies in different parts of the ligament. At about 70° flexion, the bundles spiral about 170°–200° for canine ACL. The maximum rotation observed by Butler et al. [6] was 114°–122°. Total anatomical twist was then calculated to be approximately 180°.

The presence of a ligament torque during axial loading has implications in the design and selection of a ligament substitute. Based on this natural design, LARS and Trevira Hochfest developed a design to minimize the torsion stresses. In the case of the LARS ligament the design and architecture of the fibers are arranged in parallel to avoid the interfibers abrasion that occurs with woven or braided structures. In each cases the intraarticular fibrous part of the prosthesis is implanted under a prerotation to the left or to the right similar to natural ligaments. This structure avoids shearing forces between the fibers during the combined tension, torsion, and flexion and minimizes the torsion angle. These polyethylene terephtalate devices allow tissue ingrowth in the intraarticular part. This ingrowth of connective tissue into the structure of the prosthesis is a great biomechanical advantage [10]. Soft tissue between the ligament fibers acts as a viscoelastic element and protects the ligament fibers against friction relative to the drill canal as well as against friction among the fibers itself. The friction of fibers among themselves contributes to wear of ligament prostheses.

The natural ligaments originate from relatively large areas of bone; therefore it is impossible for most of the fibers to be isometric; for example, those anterior to the axis must stretch during flexion [2]. Multibundled implant structures provide a closer approach to normal knee behavior, with their structure acting in a manner analogous to the fiber bundles.

A prosthesis implanted isometrically does not change its length during flexion and performs no relative movements in relation to drill canal and deflection points. Isometric implantation would be useful for endurance and stress of a prosthesis since a constant tensile loads would be generated during pure flexion movement and wear at the insertion spots could be minimized. A look at natural ligaments shows that this aim of isometric implantation can never be achieved. Even natural ligaments are subjected to strain of up to 4% during pure flexion, corresponding to a change in length of up to 1.2 mm for an average length of 30 mm for the ACL [10].

Loads in Prostheses: A Redundant Biomechanical Problem

We have seen that a redundant problem related to synthetic ligaments is to know precisely the loads, which are applied in service conditions on the prostheses. These forces depend on the surgical procedure, the material, and the design of the prosthesis.

We have recently developed tools to predict prosthetic ligament deformations occurring in vivo (Chap. II.3). This program is based on our previous work to analyze prosthetic ligament deformation for optimal insertion orientation. Software for measuring and visualizing personalized knee kinematics in vivo has been developed. The system consists of a mechanical system, equipped with magnetic position tracking sensors, which clamps onto the knee noninvasively to obtain accurate in vivo knee kinematics. Personalized knee geometry is obtained using three-dimensional medical imagery and incorporated into an interactive three-dimensional graphics environment. This system has been found to be sufficiently accurate to calculate prosthetic ligament deformations in vivo [36]. These deformations which are more realistic can then be introduced in the TTF fatigue testing system.

Acknowledgements. This work was supported by NSERC and FCAR grants. We wish to thank Dr.J.P. Laboureau for his comments.

References

1. Amis AA (1989) Anterior cruciate ligament replacement. Knee stability and the effects of implants. J Bone Joint Surg Br 71:819–824
2. Amis AA, Zavras TD (1995) Isometricity and graft placement during anterior cruciate ligament reconstruction. Knee 2 (1):5–17
3. Barrack RL, Skinner HB, Buckley SL (1989) Proprioception in the anterior cruciate deficient knee. Am J Sport Med 17:1–6
4. Black J (1994) The last jungle - learning about biomaterials from clinical experience. Proceedings of the Scientific/Regulatory Interface Symposium. Montreal, October 6–7, pp 39–40
5. Bolton CW, Bruchman WC (1985) The Gore-Tex expanded polytetrafluoroethylene prosthetic ligament. An in vitro and in vivo evaluation. Clin Orthop 196:202–213
6. Butler DL, Stouffer DC (1983) Tension-torsion characteristics of the canine anterior cruciate ligament. II. Experimental Observations. J Biomech Eng 105:160–165
7. Cazenave A, Laboureau JP (1990) Isometric reconstruction of the anterior cruciate ligament. Pre- and per-operative determination of the femoral isometric point. Fr J Orthop Surg 4:255–259
8. Cazenave A, Baert D, Ridoux PE (1994) Reconstruction du ligament croisé antérieur avec un ligament artificial (Ligastic). Résultats cliniques de 72 patients. Rev Chir Orthop 80:413–419
9. Chen EH, Black J (1980) Materials design analysis of the prosthetic anterior cruciate ligament. J Biomed Mater Res 14:567–586
10. Claes LE, Durselen L (1990) Biomechanical considerations in the design and use of ligament prostheses. In: Williams DF (ed) Current perspectives on implantable devices, vol 2. JAI, London, pp 1–63

11. Denti M, Monteleone M, Berardi A, Schiavone Panni (1994) Anterior cruciate ligament mechanoreceptors. Histological studies on lesions and reconstruction. Clin Orthop Relat Res 308:29–32
12. Dericks G (1995) Ligament advanced reinforcement system anterior cruciate ligament reconstruction. Op Tech Sports Med 3(3):187–205
13. DiGiovine NM, Shields CL (1991) Ligaments in ACL reconstructions. A review. Am J Knee Surg 4(1):42–48
14. Drouin G (1985) Prosthetic ligaments: a review. In Winter DA, Norman RW, Walls RP, Hayes KC, Patla AE (eds) Biomechanics, IX-A. International series on biomechanics, vol 5A, Human Kinetics, Chamoign, pp 207–211
15. Drouin G, Yahia L'H (1987) Effect of fixation on the loading of ligament. In: Claes LC, Durselen L (eds) Biomechanics of the human knee ligaments. Proceedings of the workshop of the European Society of Biomechanics. University of Ulm, Ulm, Germany, p 38
16. Drouin G, Masson, M, Yahia L'H (1991) In vitro fatigue testing of prosthetic ligaments: a new concept. Biomed Mat Eng 1:159–165
17. Food and Drug Administration (1987) Guidance document for the preparation of investigational device exemptions and premarket approval applications for intra-articular prosthetic knee ligament devices. FDA, 1 September 1, pp 1–49 (revised 18 February 18 1993)
18. Friedman MJ (1991) Prosthetic anterior cruciate ligaments. Clin Sports Med 10:499–513
19. \rho\lambdaGely P, Drouin G, Thiry PS, Tremblay GR (1984) Torsion and bending imposed on a new ACL prosthesis during knee flexion: an evaluation method. J Biomech Eng 106:285–294
20. Hefzy MS, Grood ES, Noyes FR (1989) Factors affecting the region of most isometric femoral attachments. II. The anterior cruciate ligament. Am J Sports Med 17:208–16
21. Johansson H, Sjölinder P, Sojka P (1991) Receptors in the kne joint ligaments and their role in the biomechanics of the joint. Crit Rev Biomed Eng 18(5):341–368
22. Kock HJ, Stürmer KM, Letch R (1991) Biologische Gewebereaktion nach alloplastischem Ersatz des forderen Kreuzbandes durch PET-Band (Trevira hochfest). Unfallchirurgie 94:594–602
23. Laboureau JP (1995) Two-bundles posterior cruciate ligament reconstruction: technique and results. Op Tech Sports Med 3(3):206–221
24. Laing PG (1994) World standards for surgical implants: an American perspective. Biomaterials 15(6):403–407
25. LARS (1994) Ligament advanced reinforcement system. Technical report. LARS, Arc-sur-Tille
26. Quotidien du médecin (1995) Les 147 références médicales de 1995. Quotidien du médecin, no 5561, 27 January
27. Letch R, Stürmer KM, Kock HJ, Wissing H, Schit-Neuerburg KP (1994) Der Ersatz des forderen Kreuzbandes durch ein PET-Band (Trevira hochfest) als "Salvage Procedure" bei chronisch instabilen voroperierten Kniegelenken. Unfallchirurgie 20(6):293–301
28. Lorenz K (1977) Behind the mirror: a search for a natural history of human knowledge. Chaucer, Bungay
29. Masson M, Yahia L'H, Drouin G (1988) In vitro evaluation of ACL prosthesis life: a mechanical testing machine using simultaneous tension, bending and torsion. Transactions of the 34th annual meeting of the Orthopedic Research Society. 1–4 February 1988, Atlanta, p 193
30. McCarthy DM, Tolin BS, Schwendeman L, Friedman MJ, Woo SL-Y (1993) Prosthetic replacement for the anterior cruciate ligament. In: Jackson DW, Arnoczky SP, Frank CB, Woo SL-Y, Simon TM (eds) The anterior cruciate ligament: current concepts and future concepts. Raven, New York, pp 343–356
31. Metak G, Scherer M.A, Herfeldt K, Aumann M, Blümel G (1995) Nervous structures in the anterior cruciate ligament (ACL): resection or reconstruction of the ligament stumps when using grafts. Transactions of the European Orthopedic Society. 2–3 July, Munich, p 10

32. Moyen B, Lerat JL (1994) Artificial ligaments for anterior cruciate replacement. J Bone Joint Surg Br 76 (2):173–175
33. Palmer I (1938) On the injuries to the ligaments of the knee joint. Acta Chir Scand Suppl 53
34. Rhalmi S, Yahia LH, Newman N, Isler M (1993) Immunohistochemical study of nerves in lumbar spine ligaments. Spine 18 (2):264–267
35. Rivard CH, Yahia LH, Rhalmi S, Newman N (1993) Immunohistochemical demonstration of sensory nerve fibers and endings in human anterior cruciate ligaments. 39th Annual Meeting, Orthopedic Research Society, February 15–18, San Francisco, Calif.
36. Sati M, de Guise JA, Drouin G (1996) Computer assisted planning and guidance system for the knee. J Image-Guided Surg (submitted)
37. Schultz RA, Miller DG, Kerr CS, Micheli L (1984) Mechanoreceptors in human cruciate ligaments. A histological study. J Bone Joint Surg Am 66:1072–1076
38. Silver FH, Tria AJ, Zawadsky JP, Dunn MG (1991) Anterior cruciate ligament replacement: a review. J Long-Term Effect Med Implants 1(2):135–154
39. Sjölander P (1989) A sensory role for the cruciate ligaments. Regulation of joint stability via reflexes onto the γ-muscle spindle system. Umea University Medical Dissertations, new series no 245:1–48
40. Stouffer DC, Butler DL, Kim H (1983) Tension-torsion characteristics of the canine anterior cruciate ligament. I. Theoretical framework. J Biomech Eng 105:154–159
41. Yahia L'H, Drouin G (1989) Microscopical investigation of canine anterior cruciate ligament and patellar tendon: collagen fascilce morphology and architecture. J Orthop Res 7:243–251
42. Yahia LH, Newman N (1991) Mechanoreceptors in the canine cruciate ligaments. Anat Anz Jena 173:233–238
43. Yahia LH, Newman N, St-George M (1992) Innervation of the canine cruciate ligaments. A neurohistological study. Anat Histol Embryol 21 (1):1–8
44. Yahia L'H, Drouin G, Zukor D (1993) The irradiation effect on the initial mechanical properties of meniscal grafts. Biomed Mater Eng 3(4):211–221
45. Yahia L'H, Hagemeister N, Drouin G, Sati M, Rivard C-H (1994) Conceptual design of prosthetic anterior cruciate ligaments: the need for a biomimetical approach. Biomimetics 2(4):309–330
46. Zimny ML, Schutte M, Dabezies E (1986) Mechanoreceptors in the human anterior cruciate ligament. Anat Rec 214:204–209

Part I
Fundamental Aspects of the Ligaments

I.1 Passive and Dynamic Stability of the Knee Joint

N. Duval and N. St-Onge

Introduction

The ultimate goal of a knee ligament reconstruction is to provide the injured knee with a normal function. It should correct the instability and reestablish the normal knee mechanics in order to prevent the articular degeneration process. The ligaments of the knee act as passive restraints to abnormal knee motion. They also provide some dynamic stability by stimulating the neuromuscular system through their sensory innervation.

The first part of this chapter reviews thecurrent understanding of the normal behavior of ligaments of the knee, with emphasis on the anterior cruciate ligament (ACL). The second part of the chapter discusses the problems created by rupture of the ACL and their significance to the stability of the knee.

Functional Anatomy of the Knee

The knee is a complex joint (Fig. 1). The shape of the femoral condyles and tibial plateaus is important in determining knee ligament tension, but the articular cartilage surfaces, because of their extremely low coefficient of friction, are unable to resist translation. The ligaments, tensed by the compressive loading of the articular cartilage, determine the blend of gliding and sliding motion of the knee joint.

The two femoral condyles are asymmetric: the lateral condyle is slightly wider, and the medial condyle provides a longer surface for articulation with the tibia. The weight-bearing surface of the medial femoral condyle extends farther anteriorly than does that of the lateral femoral condyle. The medial condyle contains an additional annular sector of articular surface that extends nearly to the midline, whereas the weight-bearing surface of the lateral condyle ends in a depression called the sulcus terminalis. When the knee is fully extended, the lateral tibial plateau impinges at the sulcus terminalis, but because of the longer articular surface of the medial femoral condyle, the gliding continues in the medial compartment and accommodates the normal 10°–15° of external tibial rotation that accompanies terminal knee extension.

The tibial plateaus are also asymmetric: the medial tibial plateau is concave and the lateral tibial plateau is convex. Welsh [100] emphasized the clinical

Fig. 1. The human knee

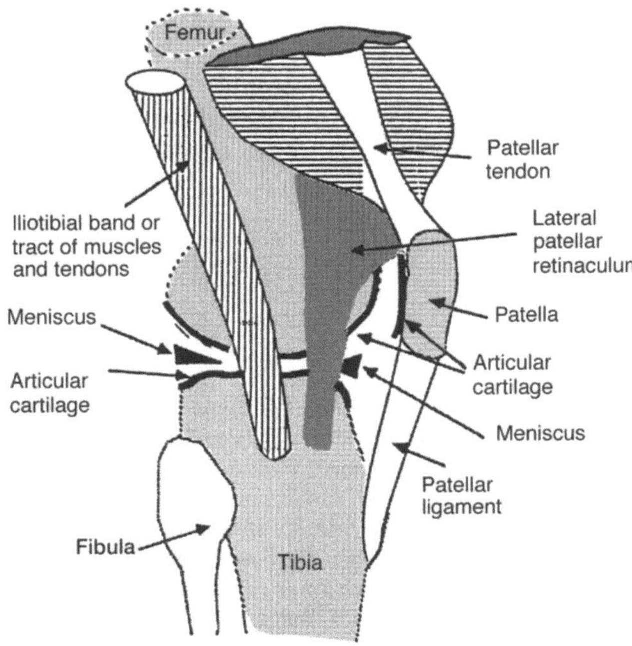

significance of the convex profile of the lateral tibial plateau which enables withdrawal of the lateral meniscus to prevent its impaction in flexion. The tibial plateaus are separated by an intercondylar region containing two tibial spines: the anteromedial and posterolateral spines. These spines prevent medial and lateral subluxation of the tibia and form a central pivot for internal and external rotation.

The menisci improve congruence between the femoral condyles and the tibial plateau. They act as weight-bearing structures to distribute compressive forces across the knee joint and aid in passive stability of the knee. Shoemaker and Markolf [86] demonstrated that the compressed menisci do resist anterior tibial subluxation after section of the ACL in cadaver knees loaded at 20° of flexion. Two mechanisms can explain that resistance to anterior tibial displacement: first, the menisci act as a physical barrier, and second, they distract the articular surfaces, tensing the ligaments and capsular tissues.

The ACL courses from the medial aspect of the lateral femoral condyle to an area just anterior and slightly lateral to the anteromedial tibial spine (Fig. 2). Its position makes the ACL a primary stabilizer of anterior displacement of the tibia as well as a secondary stabilizer of adduction and internal rotation. Kennedy et al. [51] found the average length of the ACL to be 39 mm, Girgis et al. [34] found 38 mm, Odensten and Gillquist [72] found 31 mm, and Noyes et al. [68] found 26.9 mm. Its multifascicular structure is a continuum that has been commonly simplified as consisting of two bands (anteromedial and posterolateral) by some authors including Arnoczky [5]. This structure is

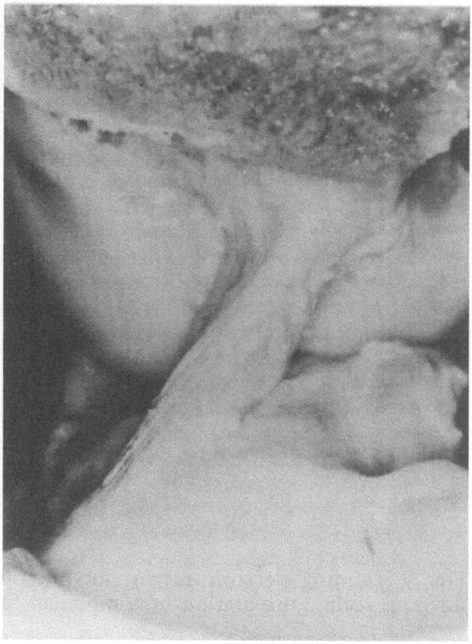

Fig. 2. Medial view of the ACL after the medial femoral condyle was removed with an osteotome. (Reproduced from [72] with permission)

particularly evident when the knee is flexed to 90°, and one can see the fibers make a spiral structure, the most anterior fibers on the tibia arising from a high, posterior, and medial origin and the most posterior fibers arising from a lower, more anterior, and more lateral origin of the medial aspect of the lateral femoral condyle. According to Norwood and Cross [66] the ACL is made of three bands while Odensten and Gillquist [72] and Welsh [100] made no anatomic subdivisions of the ACL, offering a concept of functionally different portions of the ACL without obvious different anatomic portions.

Microstructural Organization and Innervation of Ligaments

As all ligaments, the ACL consists of collagen, elastin, polysaccharides, glycolipids, glycoproteins, water, and cells [1]. Collagen constitutes 72%–80% of the dry weight of the ligament and is responsible for its tensile strength. Water, a major component of the ground substance, constitutes 71%–80% of the total weight of the ligament. Water and proteoglycans confer viscoelastic properties to the ligament.

Danylchuk et al. [22] observed the complex network of interlacing collagen fibrils forming collagen fibers (Fig. 3). The direction of the majority of fibers is parallel to the long axis of the ligament, and a large number of fibers form the subfascicular unit. A thin band of loose connective tissue, the endotenon, surrounds the subfasciculi. Between 3 and 20 subfasciculi bind together to form the fasciculus, surrounded by a denser connective tissue, the epitenon. The fasciculi join to form the ligament, surrounded by the thick paratenon.

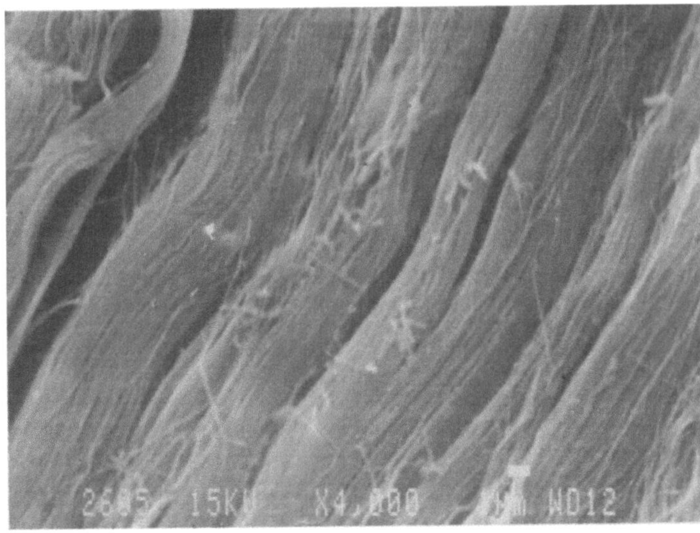

Fig. 3. Scanning electron microscope photomicrograph of human anterior cruciate collagen fibers, revealing orientation of component fibrils (×4000)

Johansson et al. [42] and Madey et al. [57] summarized the information published on mechanoreceptors and the innervation of the knee. Neural tissues compose 1%–2.5% of the total volume of the ACL [82, 105]. Three main branches carry afferent fibers from all the ligaments of the knee: the lateral articular nerve (LAN), the medial articular nerve (MAN) and the posterior articular nerve (PAN). The cruciate ligaments and the collateral ligaments are innervated by the PAN [5, 27, 29, 49, 76]. The medial collateral ligament (MCL) is also innervated by the MAN, and the lateral collateral ligament (LCL) by the LAN. Although the ACL is innervated mainly by the PAN, it also receives, but to a lesser degree, innervation from the MAN and from the LAN [35].

Four types of receptors are identified in the ligaments of the knee: Ruffini endings, Golgi tendon organlike endings, Pacinian corpuscules, and free nerve endings. These receptors are present in the cat [3, 13, 15, 29, 89, 91], dog [20], and humans [14, 38, 49, 81, 82, 104, 105]. In a cat model, all the receptors are located near the surface of the ligament and close to the tibial and femoral bony insertions [89]. In the dog, the majority of the receptors are found in the middle of the ligaments, well beneath the external synovial sheath [27]. In humans, most are found close to the bony insertions and only a few in the midsubstance [38, 49, 81, 82, 104, 105]. Free nerve endings are nociceptors [78, 100, 104]. They are activated by abnormal mechanical deformations of the ligament [3, 15, 29]. The three other types of receptors are mechanoreceptors. All of them are present in cat and human knee ligaments. Ruffini endings are active at intermediate angles [91]. Golgi tendon organlike endings have a high threshold of activation, making them probably active at extreme angles [29, 100, 104]. They are activated by the joint movement, but their impulses subside once a new position is maintained [13, 29, 91]. Pacinian corpuscules are also dynamic mechanoreceptors that are inactivated by immobilization of the joint and by displacement at a constant speed [29, 104].

Biomechanics of the Anterior Cruciate Ligament

Graf [37] presented the ACL as the primary restraint to anterior displacement of the tibia and emphasized that its function is much more than that of a simple check rein. With the posterior cruciate ligament (PCL), it forms the central pivot guiding the normal knee kinematics.

Frankel et al. [28] and later Müller [64] and Gerber and Matter [33] studied the instant-center of knee rotation. When the knee flexes, there is always combined gliding and rolling motions between the articular surfaces of the femur and tibia. The contact point moves posteriorly on the tibia indicating rolling motion, and the contact points between the femur and tibia are more spaced on the femur than on the tibia, indicating gliding motion. The ratio of rolling to gliding varies from 1:2 in early flexion to 1:4 in late flexion. The instant center of knee rotation constantly changes throughout knee motion and depends on the position of the knee. Factors influencing the location of the centrode include the shape of the femoral condyle, ligamentous restraint, and muscle tension. It always lies within the femoral contours and implies parallel movement of the joint surfaces for every knee movement. This instant center method neglects a functionally significant but quantitatively slight rotation of the knee. During the terminal extension of the knee, a 15° automatic external rotation of the tibia occurs. As noted above, the shape of the femoral condyle combined to cruciate ligaments tension affect this movement.

Kennedy et al. [51], Girgis et al. [34], Crowinshield et al. [21], Arms et al. [4] and Beynnon et al. [12] all concluded that tension of the ACL is least between 30° and 60° of knee flexion. The ACL is taut in full extension and between 70° and 90° of flexion. The tension within the anteromedial and the posterolateral fibers of the ligament varies according to the position of the knee. Viewed simplistically, the anteromedial band is tight in flexion, and the posterolateral band is tight in extension. The anteromedial band resists the anterior translation of the tibia of the knee flexed at 90° while the posterolateral band does the same on the fully extended knee. However, Rosenberg and Rasmussen [80], using an in vivo arthroscopic method of assessing the tension of the ACL, were unable to document anatomical subdivisions of the ACL or to support the concept of two reciprocally functioning ACL bands. Kennedy et al. [51] documented an increased tension of the ACL in all positions of flexion when the tibia is kept in internal rotation. They also noted an increased tension in combined abduction and external rotation. Arms et al. [4] reported passive varus, passive internal rotation, and to a lesser degree passive valgus, to increase ACL strain. Passive external rotation increased strain only during the first few degrees of flexion. The anteromedial capsule was an important aspect of strain protection of the ACL.

Noyes et al. [68] first described the nonlinear mechanical behavior of the ACL. At rest, the collagen fiber appearance shows a relaxed wavy arrangement.

The initial application of a load stresses only the fibers parallel to the direction of loading and progressively recruits the nonparallel fibers until they all become oriented in the direction of loading. This recruitment occurs early with only a slight increase in load, causes little anterior tibial translation, and corresponds to the realignment zone of the load/deformation curve (Fig. 4). When all the fibers become oriented, the linear portion of the curve begins and reflects the elastic behavior of the ligament (the ability to return to the initial resting length when the load is removed). This portion continues until it reaches a yield point beyond which microfailure of fibers occurs and the ligament loses its elastic properties (plastic deformation) reflected by the irregular zone of the curve. The end point is the complete rupture of the ligament. The area underneath the curve represents the energy absorbed by the ligament and the slope of the linear portion represents the stiffness of the ligament. Under cyclic loading the yield point decreases by increasing compliance of the ligament. This property results in a reduction of the steepness of the slope of the load/deformation curve.

Noyes et al. [68] demonstrated that bone-ACL-bone preparations failed at a higher load and at a greater elongation, and absorbed significantly more energy, at a fast rate of deformation than at a slow rate. At a slow rate, the bone insertion area of the tibia was the weakest component as shown by the frequency of tibial avulsion failures. At a fast rate, which more closely approximated physiological loading conditions, the ligament and tibial osseous component were balanced as to strength properties, and there was an increased frequency of ligamentous failure.

Kennedy et al. [50] determined the ultimate failure strength of the ACL (48.2 kg at 12.5 cm/min, 63.8 kg at 50 cm/min) and of the tibial collateral ligament to be roughly equal, while the PCL was almost twice as strong. These results were the same for both strain rates. All ligaments showed viscoelastic properties, absorbing more energy and requiring more force to rupture as the

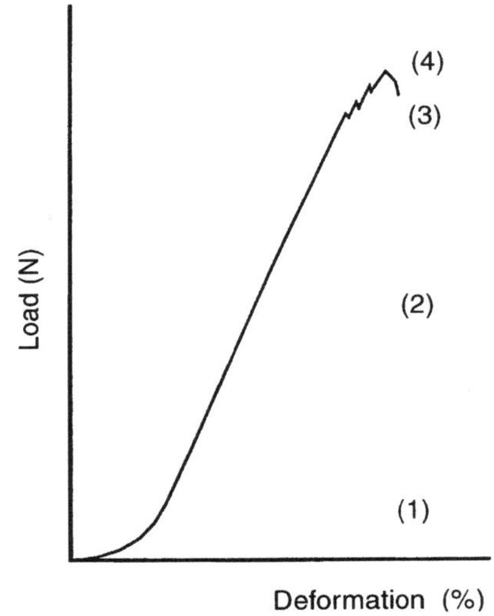

Fig. 4. Load/deformation curve of a ligament. *1*, Realignment zone; *2*, elastic deformation zone; *3*, plastic deformation zone; *4*, breaking zone

loading rate was increased. Noyes and Grood [69] measured the ultimate strength of the ACL to be 1730 N. Rupture of the ACL was usually midsubstance with microscopically ruptured collagen fiber bundles dispersed throughout the body of the ligament. There were nonuniform lesions along the ligament length with some degree of osseous deformation. According to Butler et al. [16], this higher force to failure reflects the younger age of the donors (16–26 years), the very high strain rates employed, and the elongation measurement techniques used to reflect the response of the entire femur-ACL-tibia unit. This may explain the differences between the results of these studies.

Noyes and Grood [69] reported a 44% strain to maximum stress and a 60% strain to failure. Kennedy et al. [50] reported a 20%–30% strain to failure. Tremblay et al. [98] reported a 25% elongation at yield point using a 47-kg force at a 5 cm/min rate of elongation. They noted that the ACL elongates by 14% of its resting length during normal flexion and extension of the knee. After 20 000 cycles with 14% deformation within 30 h, the ligament returned to its normal resting length, proving its truly elastic behavior.

Numerous factors may alter the mechanical properties of a ligament. The dimension, structure, strain rate, donor age, systemic function, vascular supply, mobility status, and previous steroid administration all deserve attention. Noyes [67] showed that immobility can decrease the normal strength of a ligament up to 60% and significantly increase ligament compliance.

Neuromuscular Stability of the Knee

Three categories of experiments have been made in order to understand the relationship between ligament stimulation and the neuromuscular system. The activity of the ligaments afferences, of the muscles and of the Ia afferences were recorded during these experiments.

Pope et al. [73, 75] and Krauspe et al. [52] recorded the activity of the PAN under mechanical stimulation of the ACL in cat. They showed that the PAN can be activated by pressure, tension, and knee extension, internal and external rotation (Fig. 5). Similar studies were carried out on the MCL in cat and in rabbit. Andrew and Dodt [3] showed a relationship between MCL loading and MAN activity. Stener [95] observed an increase in PAN activity under MCL stimulation.

Pope et al. [74, 77] and Solomonow et al. [94] studied the relationship between mechanical stimulation of the ACL and electromyographic (EMG) modifications of the hamstrings and quadriceps in cat. Solomonow et al. [94] were able to show an increase in hamstrings activity and an inhibition of quadriceps activity (Fig. 6) while Pope et al. [74, 77] did not record such changes. The difference between these results could be a reflection of the different position of the knee during the experiments as the amplitude of the H reflex, and perhaps the sensitivity of mechanoreceptors, are modified by the position of the articulation [23]. Similarly, Ekholm et al. [26] observed an increase in EMG activity of the hamstrings when a stress was applied on the MCL, while Anderson and Stener [2] did not find any change.

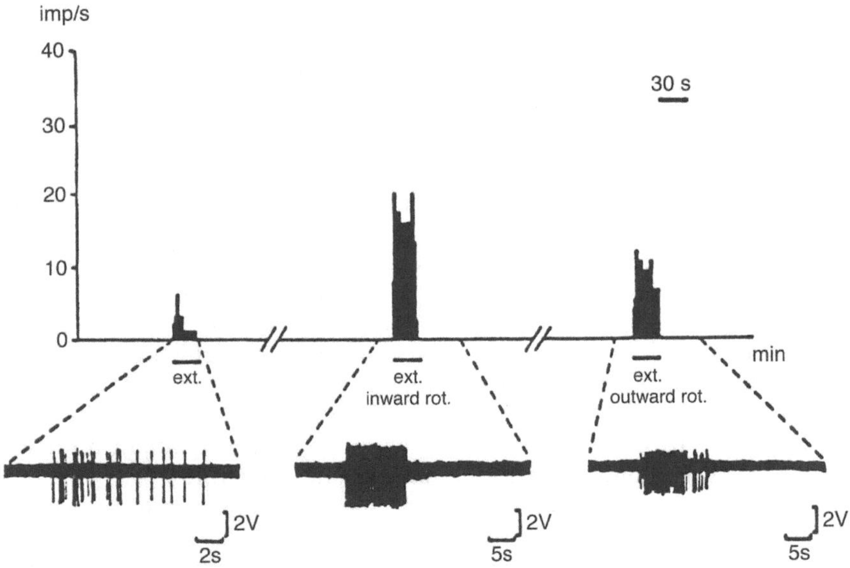

Fig. 5. An afferent unit of the anterior cruciate ligament with a conduction velocity of 28 m/s. The upper panels in the graphs display peristimulus time histograms, which show the number of action potentials in bins of 1-s duration. The action potentials were counted with the use of a window discriminator, which allows separation of these signals from the background noise. *Traces below*, the original action potentials of the fiber as monitored on the oscilloscope. (From [52] with permission)

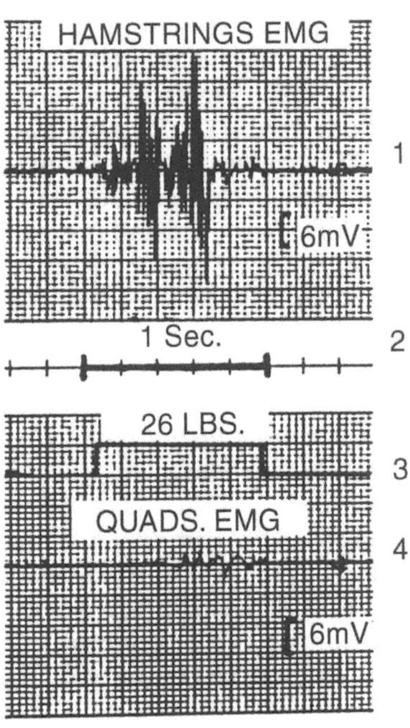

Fig. 6. The substantial increase in electromyographic activity of the cat's hamstrings (*trace 1*) over a 1 second duration (*trace 2*) of direct load application (*trace 3*) to the anterior cruciate ligament. The quadriceps electromyographic activity (*trace 4*) exhibits short initial low-level activity and then becomes inhibited for the duration of the ligament's loading. (From [94] with permission)

Johansson et al. [44] electrically stimulated the PAN in cat and reported an excitation of flexor γ-motoneurones and an inhibition of extensor γ-motoneurones. As the PAN carries afferences of many articular structures, it is not clear that their results are related to the stimulation of ligament afferences. However, Johansson et al. [41, 43], Sojka et al. [92, 93], and Sjölander et al. [88] reported an effect of mechanical stimulation of the ACL, PCL, and collateral ligaments on fusimotor neurones of the gastrocnemius and hamstrings in cat.

Muscles are important in voluntary movement. Their contribution to joint stability is also important. They can increase the rigidity of a joint, replace the action of ligaments in certain situations, and protect the ligaments. Markolf et al. [59, 60] and O'Connor et al. [71] observed an increase in knee rigidity and a decrease in knee laxity in anteroposterior, medial-lateral, varus-valgus, and rotation displacements, under a tibiofemoral contact force (Fig. 7). In vivo, body weight and muscles of the knee are responsible for that contact force. Therefore, cocontraction of antagonist muscles could be an important articular protection mechanism. Cocontraction of quadriceps and hamstrings muscles has been shown during normal walking [17, 18, 63], cycling [39] and slow extension of the knee [7, 24, 94].

Many authors including Arms et al. [4], Renström et al. [79], Yasuda and Sasaki [101, 102], Draganich and Vahey [25], Zavatsky and O'Connor [103], and Bagger et al. [6] studied the effect of quadriceps and hamstrings activation

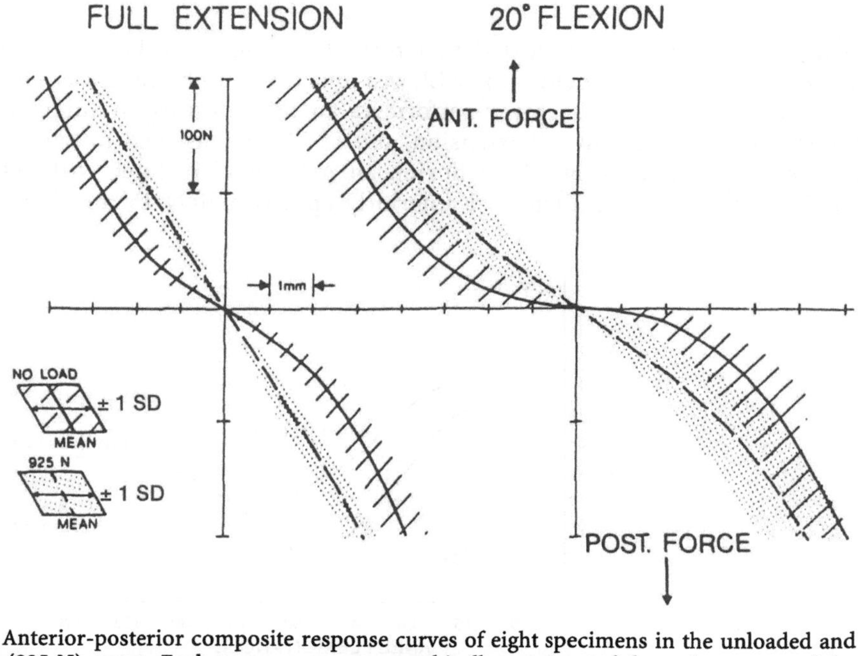

Fig. 7. Anterior-posterior composite response curves of eight specimens in the unloaded and loaded (925 N) states. Each mean curve was graphically constructed from the means of the stiffnesses and laxities of all knees tested. *Shaded regions*, 1-SD envelopes for these mean curves. (From [59] with permission)

on human tibial force, tibial displacement, and ACL strain. Isometric contraction of the quadriceps increased anterior tibial force [101] and ACL strain [4, 79] from 0° to 45° of knee flexion. Beyond 60°, isometric contraction decreased ACL strain. This decrease was significant at flexion angles of 105° and 120°. These results are in agreement with those of Zavatsky and O'Connor [103] who observed an anterior tibial force at small flexion angles and a posterior force at larger angles during isometric quadriceps contraction. Isometric contraction of the hamstrings produced a posterior force on the tibia at any angle of flexion [101]. Bagger et al. [6] observed a decrease in anterior displacement when the hamstrings were active during laxity measurements. Under simultaneous contraction of both muscles, an anterior tibial force was only present when knee flexion was less than 7° [102], and ACL strain was less than when only the quadriceps were contracted between 30° and 90° of flexion according to Renström et al. [79] (Fig. 8) and between 10° and 20° of flexion and at 90° of flexion according to Draganich and Vahey [25]. Beynnon et al. [12] documented in vivo a significantly greater strain during active range of motion as compared to passive range of motion at flexion angles between 10° and 40°, but no difference at greater flexion angles. Nisell et al. [65] reported important stresses arising in the ACL when the knee extended more than 60° during isokinetic extension exercises.

Biomechanics of the Knee with a Ruptured Anterior Cruciate Ligament

Warren and Levy [99] reported that medial menisectomy had no effect on anterior tibial displacement if the ACL was intact. When the ACL was sectioned first and a medial menisectomy performed, an 18% increase in anterior tibial translation was noted in full extension, and a 58% increase was noted at 90° of flexion. When loading at 20° of flexion an ACL-deficient knee with bilateral menisectomy, the tibia displaced anteriorly approximately 15 mm before the

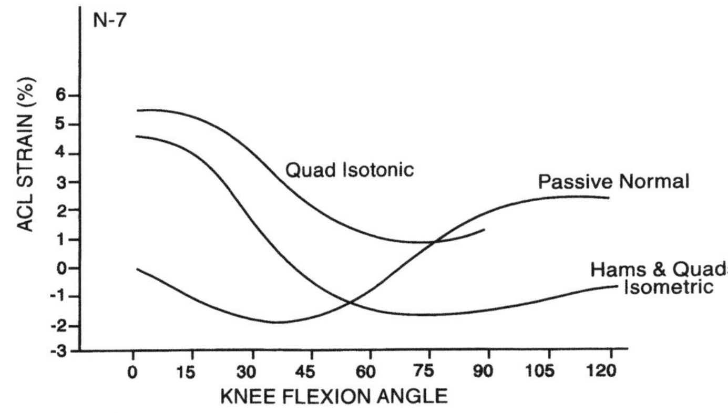

Fig. 8. Mean passive normal, simulated quadriceps isotonic, and simulated simultaneous quadriceps and hamstrings isometric strain patterns. (From [79] with permission)

soft tissues began to carry load and prevent the knee from dislocating [86]. However, Hughston and Barrett [40] found the meniscus, not the ACL, to be the anatomical structure that has the greater effect on the quality of anterolateral stability based on a long-term review of 154 operated knees.

Girgis et al. [34] severed the ACL and documented an hyperextension of the knee, an increased anterior drawer sign in the extended and flexed positions, an increased in internal tibial rotation of 8° in extension and 2° in flexion, and an increased external tibial rotation of 12° and 6°, respectively in flexion and extension.

Matsumoto and Seedhom [61] sectioned the ACL and under a valgus torque identified an increased internal rotation and sudden reduction between 20° and 40° of flexion (pivot shift phenomenon). Under external rotatory torque, no change in tibial rotation was observed. Under internal rotatory torque, the internal rotation slightly increased above normal between full extension and 30° of flexion. They concluded that an ACL-deficient knee develops the pivot shift because the axis of tibial rotation becomes located about the MCL. The pivot shift, as described by Galway and MacIntosh [31] and Galway et al. [30], is both a clinical phenomenon that gives rise to the complaint of giving way, and a physical sign that can be elicited upon examination of the injured knee. It is characterized by anterior subluxation of the tibia on the femur, although the instability dominates in the lateral compartment, and the spontaneous reduction of the subluxation during knee flexion.

Tamea and Henning [96] analyzed changes in instant centers of motion of ACL-deficient knee during internal rotation and valgus stress to duplicate the pivot shift maneuver. From 0° to 20° of flexion, the instant centers were located too near the joint line (Fig. 9) and displaced anteriorly. During the shift phase (20°–45°), the instant centers were located posteriorly and proximally on the femur. From 50° to 90°, the instant center progression was near normal. Gerber and Matter [33] reported on ACL-deficient knees a sudden shift of the centrode downward and forward between 20° and 40° of flexion and a shift back to its normal position before 90° of flexion.

Langrana and Bronfeld [54] found that between flexion angles of 30° and 50°, the fiber lengths of all four knee ligaments were minimum. At 25° flexion angle, 87.8% of the load of anterior tibial translation was resisted by the ACL. After ACL rupture, the MCL resisted 46.8%, the PCL resisted 29.6% and the LCL resisted 23.6% of the load. Noyes et al. [70] documented similar results between 30° and 90° of flexion, with the ACL resisting 85% of the load, and the other ligaments 15% of the load.

Neuromuscular Behavior of the Knee with an Injured Anterior Cruciate Ligament

Barrett [9], Barrack et al. [8], Skinner and Barrack [90], and Corrigan et al. [20] have compared proprioception in normal and injured knee of the same subjects, and in normal and injured subjects. They all reported a decrease in joint position sense and movement detection for the injured knees. The muscular

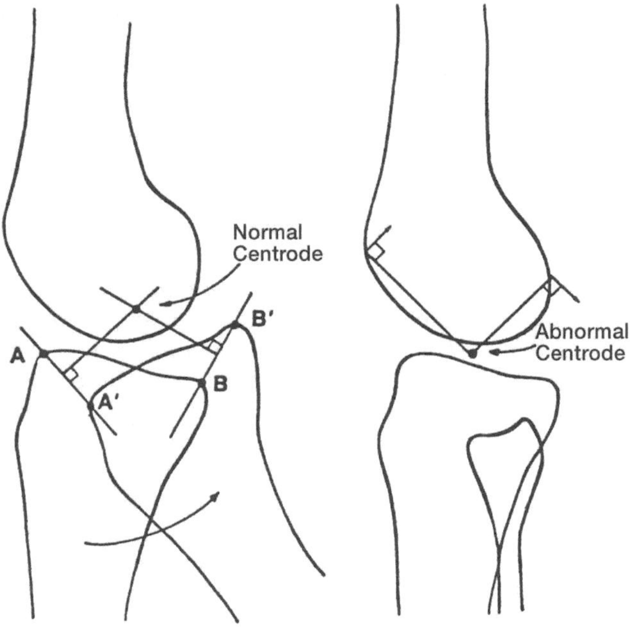

Fig. 9. *Left*, method of determining the centrode of an arc of motion in a normal knee; *right*, an abnormal centrode located at the joint line representing pure rolling motion. (From [96] with permission)

reflex latency after an ACL rupture has been studied by Beard et al. [10, 11]. They observed an increase in the latency of the hamstrings contraction in the injured knee following an anterior displacement of the tibia. There was a significant correlation between the latency of the reflex and the degree of instability reported by the subjects. There was no correlation between the measured objective passive laxity with an arthrometer (KT-1000) and the degree of subjective instability.

A tear of the ACL modifies the kinematics of active knee extension [45, 46] and flexion [48]. Jonsson et al. [46] reported an increased anterior tibial displacement while Kärrholm et al. [48] and Jonsson and Kärrholm [45] observed a smaller anterior displacement for injured knees. Kinematic studies during walking and running gave contradictory results. According to Shiavi et al. [84], the injured knee presented less flexion and more adduction during the stance phase of walking, while Tibone et al. [97], Marans et al. [58] and Kuster et al. [53] did not find any knee flexion difference during walking, running, or climbing stairs. Marans et al. [58] reported a greater anterior tibial displacement in injured knee during the swing phase (Fig. 10). One-leg jump studies also provided contradictory data: McNair and Marshall [62] found no difference between pathological and normal knees, while Gauffin and Tropp [32] found an increase in hip and knee flexion angle at touchdown, an increase in the maximal knee flexion angle, and a decrease in knee angular velocity of the injured leg.

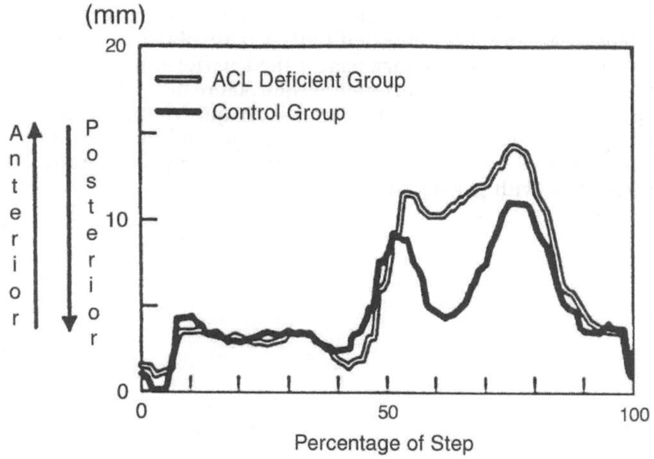

Fig. 10. Graphic comparison of the mean single-step patterns of anterior/posterior translation between the normal (control) and ACL ligament deficient groups. (From [58] with permission)

The activation of hamstrings and quadriceps while actively extending the knee is different in ACL deficiency [94]. In normal knee, the hamstrings activity is constant, and in the injured knee the activity increases between 37° and 46° of flexion (Fig. 11). This increase corresponds to a decrease in the quadriceps activity. Grabiner and Weiker [36] reported a higher hamstrings activity level during isometric knee extension of the injured knee. Most EMG studies during walking and running describe varous patterns of muscular activity between the normal and injured knees [19, 47, 53, 55, 56, 85, 87, 97]. Most of these studies reported greater hamstrings activity for injured knees. However, the results are more conflicting for gastrocnemius and quadriceps. Shiavi et al. [83] studied a pivot motion during walking. During stance phase, EMG activity of the quadriceps and gastrocnemius was lower, and the activity of hamstrings higher. During swing phase, the activity of all three muscle groups was lower. Branch et al. [14] studied a pivot motion during running. They found similar results during stance phase but during swing phase the activity of hamstrings was higher in injured knees. EMG study of the one-leg jump reported by Gauffin and Tropp [32] showed a lower EMG activity of the quadriceps in injured knees but a similar study by McNair and Marshall [62] did not show any change.

Conclusion

The ligaments of the knee give a passive and a dynamic stability to the articulation. The dynamic stability is made possible by mechanoreceptors that stimulate the neuromuscular system. A ligament injury alters this proprioceptive function.

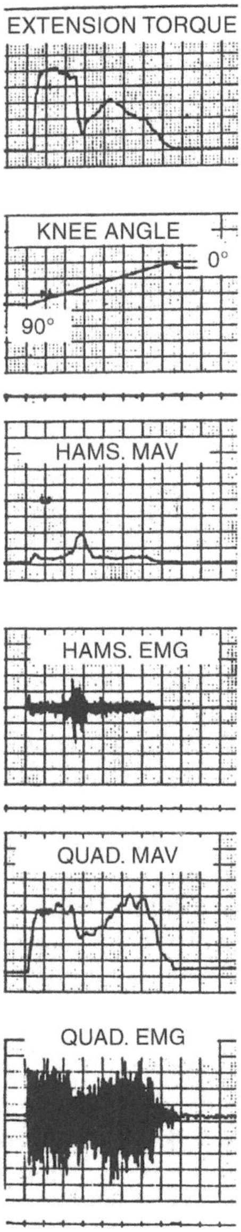

Fig. 11. Extension torque, knee angle, hamstrings, and quadriceps mean absolute value and EMG activity obtained from a patient with a midsubstance tear of the anterior cruciate ligament. Note the large subluxation torque failure near 42°, which appears simultaneously with decrease in quadriceps electromyographic activity/mean absolute value and increase in hamstrings electromyographic activity/mean absolute value. (From [94] with permission)

Presently there are no universal tools to measure passive and dynamic knee stability. The variability of the methods and results showed in this chapter is a reflection of this situation. Our understanding of the "knee disease" caused by the rupture of the ACL is still very limited. A three-dimensional passive knee stability testing system and a dynamic knee stability testing system must be developed to tailor the knee treatment to the patient's needs.

The reconstruction of the unstable knee is a challenging problem for the orthopedic surgeon. More clinical, biomechanical, and neuromuscular documentation is necessary to understand the knee joint, and to restore the injured knee the best possible function.

Acknowledgments. The authors thank Cyril Chaput for his help in this work.

References

1. Akeson W, Amiel D, Mechanic GL et al (1982) Ligament biology and biomechanics. AAOS Symposium on sports medicine: the knee. Mosby, St. Louis
2. Anderson S, Stener B (1959) Experimental evaluation of the hypothesis of ligamento-muscular protective reflexes. II. A study in cat using the medial collateral ligament of the knee joint. Acta Physiol Scand 48 [Suppl] 166:27–49
3. Andrew BL, Dodt E (1953) The development of sensory endings at the knee joint of the cat. Acta Physiol Scand 28:287–296
4. Arms SW, Pope MH, Johnson RJ, Fischer RA, Arvidsson I, Eriksson E (1984) The biomechanics of anterior cruciate ligament rehabilitation and reconstruction. Am J Sports Med 12:8–18
5. Arnoczky SP (1983) Anatomy of the anterior cruciate ligament. Clin Orthop 172:19–25
6. Bagger J, Ravn J, Lavard P, Blyme P, Sørensen C (1992) Effect of functional bracing, quadriceps and hamstrings on anterior tibial translation in anterior cruciate ligament insufficiency: a preliminary study. J Rehabil Res Dev 29:9–12
7. Baratta R, Solomonow M, Zhou BH, Letson ED, Chuinard R, D'Ambrosia R (1988) Muscular coactivation. The role of the antagonistic musculature in maintaining knee stability. Am J Sports Med 16:113–122
8. Barrack RL, Skinner HB, Buckley SL (1989) Proprioception in the anterior cruciate deficient knee. Am J Sports Med 17:1–6
9. Barrett DS (1991) Proprioception and function after anterior cruciate reconstruction. J Bone Joint Surg Br 73:833–837
10. Beard DJ, Kyberd PJ, Fergusson CM, Dodd CA (1993) Proprioception after rupture of the anterior cruciate ligament. An objective indication of the need for surgery? J Bone Joint Surg Br 75:311–315
11. Beard DJ, Kyberd PJ, O'Connor JJ, Fergusson CM, Dodd CAF (1994) Reflex hamstring contraction latency in anterior cruciate ligament deficiency. J Orthop Res 12:219–228
12. Beynnon B, Howe JG, Pope MH, Johnson RJ, Fleming BC (1992) The measurement of anterior cruciate ligament strain in vivo. Int Orthop 16:1–12
13. Boyd IA (1954) The histological structure of the receptors in the knee joint of the cat correlated with their physiological response. J Physiol (Lond) 124:476–488
14. Branch TP, Hunter R, Donath M (1989) Dynamic EMG analysis of anterior cruciate deficient legs with and without bracing during cutting. Am J Sports Med 17:35–41
15. Burgess PR, Clark FJ (1969) Characteristics of knee joint receptors in the cat. J Physiol (Lond) 203:317–335
16. Butler DL, Grood ES, Noyes FR, Sodd AN (1985) On the interpretation of our anterior cruciate ligament data. Clin Orthop 196:26–34
17. arlsöö S, Fohlin L, Skoglund G (1973) Studies of co-contraction of knee muscles. In: Desmedt E(ed) New development in EMG and clinical neurophysiology. Karger, Basel, pp 648–655
18. Carlsöö S, Nordstrand A (1968) The coordination of the knee-muscles in some voluntary movements and in the gait in cases with and without knee joint injuries. Acta Chir Scand 134:423–426

19. Ciccotti MG, Kerlan RK, Perry J, Pink M (1994) An electromyographic analysis of the knee during functional activities. I. The anterior cruciate ligament-deficient and -reconstructed profiles. Am J Sports Med 22:651–658
20. Corrigan JP, Cashman WF, Brady MP (1992) Proprioception in the cruciate deficient knee. J Bone Joint Surg Br 74:247–250
21. Crowinshield R, Pope MH, Johnson RJ (1976) An analytic model of the knee. J Biomech 9:397–405
22. Danylchuk KD, Finlay JB, Krcek JP (1978) Microstructural organization of human and bovine cruciate ligaments. Clin Orthop 131:294–298
23. Davies T, Lader M (1983) Effects of ankle joint angle on reflexes and motor threshold intensities of soleus in man. J Physiol (Lond) 339:22p
24. Draganich LF, Jaeger RJ, Kralj AR (1989) Coactivation of the hamstrings and quadriceps during extension of the knee. J Bone Joint Surg Am 71:1075–1081
25. Draganich LF, Vahey JW (1990) An in vitro study of the anterior cruciate ligament strain induced by quadriceps and hamstrings forces. J Orthop Res 8:57–63
26. Ekholm J, Eklund G, Skoglund S (1960) On the reflex effects from the knee joint of the cat. Acta Physiol Scand 50:167–175
27. Ferrell WR (1980) The adequacy of stretch receptors in the cat knee joint for signaling joint angle throughout a full range of movement. J Physiol (Lond) 299:85–99
28. Frankel VH, Burstein AH, Brooks DB (1971) Biomechanics of internal derangement of the knee. J Bone Joint Surg Am 53:945–962
29. Freeman MAR, Wyke B (1967) The innervation of the knee joint. An anatomical and histological study in the cat. J Anat 101:505–532
30. Galway HR, Beaupré A, MacIntosh DL (1972) Pivot shift: a clinical sign of symptomatic anterior cruciate insufficiency. J Bone Joint Surg Br 54:763–764
31. Galway HR, MacIntosh DL (1980) The lateral pivot shift. A symptom and sign of anterior cruciate ligament insufficiency. Clin Orthop 147:45–50
32. Gauffin H, Tropp H (1992) Altered movement and muscular-activation patterns during the one-legged jump in patients with an old anterior cruciate ligament rupture. Am J Sports Med 20:182–192
33. Gerber C, Matter P (1983) Biomechanical analysis of the knee after rupture of the anterior cruciate ligament and its primary repair. An instant-centre analysis of function. J Bone Joint Surg Br 65:391–399
34. irgis FG, Marshall JL, Monajem ARS (1975) The cruciate ligaments of the knee joint. Anatomical, functional and experimental analysis. Clin Orthop 106:216–231
35. Gomez-Barrena E, Munurea L, Martinez-Moreno E (1992) Neural pathways of anterior cruciate ligament traced to the spinal ganglia. Proc 38th Meet Orthop Res Soc:17–20
36. rabiner MD, Weiker GG (1993) Anterior cruciate ligament injury and hamstrings coactivation. Clin Biomech 8:215–219
37. Graf B (1987) Biomechanics of the anterior cruciate ligament. In: Jackson DW, Drez D Jr (eds) The anterior cruciate deficient knee. Mosby, St. Louis, pp 55–71
38. Haus J, Halata Z (1990) Innervation of the anterior cruciate ligament. Int Orthop 14:293–296
39. Houtz SJ, Fischer FJ (1959) An analysis of muscle action and joint excursion during exercise on a stationary bicycle. J Bone Joint Surg Am 41:123–131
40. Hughston JC, Barrett GR (1983) Acute anteromedial rotatory instability. J Bone Joint Surg Am 65:145–153
41. Johansson H, Lorentzon R, Sjölander P, Sojka P (1990) The anterior cruciate ligament. A sensor acting on the γ-muscle-spindle systems of muscles around the knee joint. Neuro-Orthopedics 9:1–23
42. Johansson H, Sjölander P, Sojka P (1991) Receptors in the knee joint ligaments and their role in the biomechanics of the joint. Crit Rev Biomed Eng 18:341–368
43. ohansson H, Sjölander P, Sojka P (1990) Activity in receptor afferents from the anterior cruciate ligament evokes reflex effects on fusimotor neurones. Neurosci Res 8:54–59
44. Johansson H, Sjölander P, Sojka P (1986) Actions on γ-motoneurones elicited by electrical stimulation of joint afferent fibres in the hind limb of the cat. J Physiol (Lond) 375:137–152

45. Jonsson H, Kärrholm J (1994) Three-dimensional knee joint movements during a step-up: evaluation after anterior cruciate ligament rupture. J Orthop Res 12:769–779
46. Jonsson H, Kärrholm J, Elmqvist L-G (1989) Kinematics of active knee extension after tear of the anterior cruciate ligament. Am J Sports Med 17:796–802
47. Kålund S, Sinkjaer T, Arendt-Nielsen L, Simonsen O (1990) Altered timing of hamstring muscle action in anterior cruciate ligament deficient patients. Am J Sports Med 18:245–248
48. Kärrholm J, Selvik G, Elmqvist L-G, Hansson LI (1988) Active knee motion after cruciate ligament rupture. Stereoradiography. Acta Orthop Scand 59:158–164
49. Kennedy JC, Alexander IJ, Hayes KC (1982) Nerve supply of the human knee and its functional importance. Am J Sports Med 10:329–335
50. Kennedy JC, Hawkins RJ, Willis RB, Danylchuk KD (1976) Tension studies of human knee ligaments. Yield point, ultimate failure, and disruption of the cruciate and tibial collateral ligaments. J Bone Joint Surg Am 58:350–355
51. Kennedy JC, Weinberg HW, Wilson AS (1974) The anatomy and function of the anterior cruciate ligament as determined by clinical and morphological studies. J Bone Joint Surg Am 56:223–235
52. Krauspe R, Schmidt M, Schaible H-G (1992) Sensory innervation of the anterior cruciate ligament. An electrophysiological study of the response properties of single identified mechanoreceptors in the cat. J Bone Joint Surg Am 74:390–397
53. Kuster M, Sakurai S, Wood GA (1995) The anterior cruciate ligament-deficient knee: compensatory mechanisms during downhill walking. Knee 2:105–111
54. Langrana NA, Bronfeld J (1985) Computer-assisted analysis of ligament constraints in the knee. Clin Orthop 196:42–50
55. Lass P, Kaalund S, leFevre S, Arendt-Nielsen L, Sinkjaer T, Simonsen O (1991) Muscle coordination following rupture of the anterior cruciate ligament. Acta Orthop Scand 62:9–14
56. Limbird TJ, Shiavi R, Frazer M, Borra H (1988) EMG profiles of knee joint musculature during walking: changes induced by anterior cruciate ligament deficiency. J Orthop Res 6:630–638
57. Madey SM, Cole KJ, Brand RA (1993) The sensory role of the anterior cruciate ligament. In: Jackson DW, Arnoczky SP, Frank CB, Woo SL-Y, Simon TM (eds) The anterior cruciate ligament: current and future concepts. Raven, New York, pp 23–33
58. Marans HJ, Jackson RW, Glossop ND, Young MC (1989) Anterior cruciate ligament insufficiency: a dynamic three-dimensional motion analysis. Am J Sports Med 17:325–332
59. Markolf KL, Bargar WL, Shoemaker SC, Amstutz HC (1981) The role of joint load in knee stability. J Bone Joint Surg Am 63:570–585
60. Markolf KL, Graff-Radford A, Amstutz HC (1978) In vivo knee stability. J Bone Joint Surg Am 60:664–674
61. Matsumoto H, Seedhom BB (1994) Treatment of the pivot-shift intraarticular versus extraarticular or combined reconstruction procedures. A biomechanical study. Clin Orthop 299:298–304
62. cNair PJ, Marshall RN (1994) Landing characteristics in subjects with normal and anterior cruciate ligament deficient knee joint. Arch Phys Med Rehabil 75:584–589
63. Morrison JB (1970) The mechanics of the knee joint in relation to normal walking. J Biomech 3(1):51–61
64. Müller W (1983) The knee: form, function, and ligament reconstruction. Springer, Berlin Heidelberg New York, p 314
65. Nisell R, Ericson MO, Németh G, Ekholm J (1989) Tibiofemoral joint forces during isokinetic knee extension. Am J Sports Med 17:49–54
66. Norwood LA, Cross MJ (1979) Anterior cruciate ligament: functional anatomy and its bundles in rotatory instabilities. Am J Sports Med 7:23–26
67. Noyes FR (1977) Functional properties of knee ligaments and alterations induced by immobilization: a correlative biomechanical and histological study in primates. Clin Orthop 123:210–242
68. Noyes FR, DeLucas JL, Torvik PJ (1974) Biomechanics of anterior cruciate ligament failure: an analysis of strain-rate sensitivity and mechanisms of failure in primates. J Bone Joint Surg Am 56:236–253

69. Noyes FR, Grood ES (1976) Strength of the anterior cruciate ligament in humans and rhesus. Age and species-related changes. J Bone Joint Surg Am 58:1074–1082
70. Noyes FR, Grood ES, Butler DL, Malek M (1980) Clinical laxity tests and functional stability of the knee: biomechanical concepts. Clin Orthop 146:84–89
71. O'Connor J, Biden E, Bradley J, Fitzpatrick D, Young S, Kershaw C, Daniel D, Goodfellow J (1990) The muscle-stabilized knee. In: Daniel DM, Akeson WH, O'Connor J (eds) Knee ligaments: structure, function, injury, and repair. Raven, New York, pp 239–277
72. Odensten M, Gillquist J (1985) Functional anatomy of the anterior cruciate ligament and a rationale for reconstruction. J Bone Joint Surg Am 67:257–261
73. Pope DF, Cole KJ, Brand RA (1990) Discharge properties of mechanoreceptive afferents in the anterior cruciate ligament. Transactions of the 36th annual meeting of the Orthopedic Research Society. 4–8 February 1990, New Orleans, p 31
74. Pope DF, Cole KJ, Brand RA (1990) Physiological loading of the anterior cruciate ligament does not activate quadriceps or hamstrings in the anesthetized cat. Am J Sports Med 18:595–599
75. Pope DF, Cole KJ, Brand RA (1990) Properties of ACL mechanoreceptive afferents. J Biomech 23:717
76. Pope DF, Cole KJ, Brand RA (1989) Properties of mechanoreceptive afferents. Proceedings of the 13th annual meeting of the American Society of Biomechanics. 23–25 August, Burlington, p 46
77. Pope DF, Cole KJ, Brand RA (1989) The ACL does not contribute to monosynaptic reflexes of the cat. Transactions of the 35th annual meeting of the Orthopedic Research Society. 6–9 February 1989, Las Vegas, p 22
78. Proske U, Schaible H-G, Schmidt RF (1988) Joint receptors and kinaesthesia. Exp Brain Res 72:219–224
79. Renström P, Arms SW, Stanwyke TS, Johnson RJ, Pope MH (1986) Strain within the anterior cruciate ligament during hamstring and quadriceps activity. Am J Sports Med 14:83–87
80. Rosenberg TD, Rasmussen GL (1984) The function of the anterior cruciate ligament during anterior drawer and Lachman's testing. An in vivo analysis in normal knees. Am J Sports Med 12:318–322
81. Schultz RA, Miller DC, Kerr CS, Micheli L (1984) Mechanoreceptors in human cruciate ligaments. J Bone Joint Surg Am 66:1072–1076
82. Schutte MJ, Dabezies EJ, Zimny ML, Happel LT (1987) Neural anatomy of the human anterior cruciate ligament. J Bone Joint Surg Am 69:243–249
83. Shiavi R, Limbird T, Borra H, Edmondstone MA (1991) Electromyography profiles of knee joint musculature during pivoting: changes induced by anterior cruciate ligament deficiency. J Electromyogr Kinesiol 1:49–57
84. Shiavi R, Limbird T, Frazer M, Stivers K, Strauss A, Abramovitz J (1987) Helical motion analysis of the knee. II. Kinematics of uninjured and injured knees during walking and pivoting. J Biomech 20:653–665
85. Shiavi R, Zhang L-Q, Limbird T, Edmondstone MA (1992) Pattern analysis of electromyographic linear envelopes exhibited by subjects with uninjured and injured knees during free and fast speed walking. J Orthop Res 10:226–236
86. Shoemaker SC, Markolf KL (1986) The role of the meniscus in the anterior-posterior stability of the loaded anterior cruciate-deficient knee. J Bone Joint Surg Am 68:71–78
87. Sinkjaer T, Arendt-Nielsen L (1991) Knee stability and muscle coordination in patients with anterior cruciate ligament injuries: an electromyographic approach. J Electromyogr Kin 1:209–217
88. Sjölander P, Djupsjöbacka M, Johansson H, Sojka P, Lorentzon R (1994) Can receptors in the collateral ligaments contribute to knee joint stability and proprioception via effects on the fusimotor-muscle-spindle system? An experimental study in the cat. Neuro Orthop 15:65–80
89. Sjölander P, Johansson H, Sojka P, Rehnholm A (1989) Sensory nerve endings in the cat cruciate ligaments: a morphological investigation. Neurosci Lett 102:33–38
90. Skinner HB, Barrack RL (1991) Joint position sense in the normal and pathological knee joint. J Electromyogr Kin 1:180–190

91. Skoglund S (1956) Anatomical and physiological studies of knee joint innervation in the cat. Acta Physiol Scand 36 [Suppl 124]:1–100
92. Sojka P, Johansson H, Sjölander P, Lorentzon R, Djupsjöbacka M (1989) Fusimotor neurones can be reflexly influenced by activity in receptor afferents from the posterior cruciate ligament. Brain Res 483:177–183
93. Sojka P, Sjölander P, Johansson H, Djupsjöbacka M (1991) Influence from stretch-sensitive receptors in the collateral ligaments of the knee joint on the γ-muscle-spindle systems of flexor and extensor muscles. Neurosci Res 11:55–62
94. Solomonow M, Baratta R, Zhou BH, Shoji H, Bose W, Beck C, D'Ambrosia R (1987) The synergistic action of the anterior cruciate ligament and thigh muscles in maintaining joint stability. Am J Sports Med 15:207–213
95. Stener B (1959) Experimental evaluation of the hypothesis of ligamento-muscular protective reflexes I. A method for adequate stimulation of tension receptors in the medial collateral ligament of the knee joint of the cat, and studies of the innervation of the ligament. Acta Physiol Scand 48 [Suppl 166]:5–26
96. Tamea CD, Henning CE (1981) Pathomechanics of the pivot shift maneuver: an instant center analysis. Am J Sports Med 9:31–37
97. Tibone JE, Antich TJ, Fanton GS, Moynes DR, Perry J (1986) Functional analysis of anterior cruciate ligament instability. Am J Sports Med 14:276–284
98. Tremblay GR, Laurin CA, Drouin G (1980) The challenge of prosthetic cruciate ligament replacement. Clin Orthop 147:88–92
99. Warren RF, Levy IM (1983) Meniscal lesions associated with anterior cruciate ligament injury. Clin Orthop 172:32–37
100. Welsh RP (1980) Knee joint structure and function. Clin Orthop 147:7–14
101. Yasuda K, Sasaki T (1987) Exercise after anterior cruciate ligament reconstruction. Biomechanics of the simultaneous isometric contraction method of the quadriceps and the hamstrings. Clin Orthop 220:266–274
102. Yasuda K, Sasaki T (1987) Exercise after anterior cruciate ligament reconstruction. The force exerted on the tibia by the separate isometric contractions of the quadriceps or the hamstrings. Clin Orthop 220:275–283
103. Zavatsky AB, O'Connor JJ (1995) Anteroposterior tibial translation during simulated isometric quadriceps contractions. Knee 2:85–91
104. Zimny ML (1988) Mechanoreceptors in articular tissues. Am J Anat 182:16–32
105. Zimny ML, Schutte M, Dabezies E (1986) Mechanoreceptors in the human anterior cruciate ligament. Anat Rec 214:204–209

I.2 Sensory Endings in Ligaments: Response Properties and Effects on Proprioception and Motor Control

P. Sjölander and H. Johansson

Introduction

As early as the 1950s joint ligaments were demonstrated to contain sensory endings that are sensitive to small changes in ligament tension and to passive joint movements. Since then the morphological characteristics and the functional properties of these sensory endings have received considerable interest. This chapter is to reviews the information available on the sensory properties of ligaments and their role in motor control and proprioception. Some clinical implications of these findings are also discussed.

Sensory Characteristics of Joint Receptors

Responses to Mechanical Stimuli

The knee joint of man as well as that of various animals (including the cat and the dog) contains four types of mechanosensitive nerve endings: Ruffini endings, Pacinian corpuscles, Golgi tendon organlike endings, and free nerve endings (for review see [149]). Each of these has been found in the cruciate ligaments and the medial meniscus. The joint capsule accommodates Ruffini endings, Pacinian corpuscles, and free nerve endings while the collateral ligaments shows Ruffini endings, Golgi tendon organlike endings, and free nerve endings.

These four types of receptors show different response characteristics to mechanical stimuli. Both the Ruffini endings and the Pacinian corpuscles have very low thresholds to mechanical stress, but while the Ruffini endings adapt slowly, the Pacinian corpuscles adapt rapidly [149]. Thus the Ruffini endings seem to function as static and dynamic mechanoreceptor which may signal static joint position, intraarticular pressure, and amplitude and velocity of joint rotations [75, 81, 90, 113, 114, 273, 251]. The Pacinian corpuscles, on the other hand, are silent during static conditions and when the joint is rotated at constant speed but are very sensitive to acceleration and deceleration [35, 90, 251]. Accordingly, these corpuscles behave as pure dynamic mechanoreceptors. The Golgi tendon organlike afferents adapt slowly, have high thresholds to mechanical stimuli, and are completely inactive in immobile joints [35, 90, 251]. Because of their high thresholds these receptors have been suggested to measure the extremes of the joint's normal movement range [90, 274, 275]. The majority of the free nerve endings remain silent during normal conditions but become active when the articular tissue is subjected to damaging mechanical deformations and to certain substances (see above and [9, 39, 90, 115, 230]).

Based on recordings from knee joint afferents of unknown origin (i.e., as regards receptor type) it has repeatedly been demonstrated that, although there are joint afferents (midrange units) that respond throughout the normal range of joint rotations, most of the receptors afferents (end-range units) become active only when the joint approaches the limits of its normal working range [9, 39, 47, 109, 112, 189, 251]. Therefore it is commonly believed that articular receptors operate as limit detectors, whose principal role is to initiate joint protective muscle reflexes and to contribute to the awareness of joint position and movements when the joint is close to the end of the normal working range. However, there are a number of experimental and theoretical arguments against this simple view. First, partly for methodological reasons (see [149, 153, 154]) it seems as the number of midrange units has been underestimated in some investigations. In fact, studies have demonstrated a considerable proportion of midrange units in several different joints, including the knee, hip, and finger joints [25, 41, 43, 80, 203]. It has also been shown that the midrange units can be sensitive both to movement directions and to velocity and acceleration [36, 185, 203, 251]. Secondly, "the limit sensor" hypothesis is based on response characteristics of joint receptors obtained during passive joint movements. When the joint is actively rotated, however, the response pattern from joint afferents is influenced by the muscles around the joint [109, 112], and as a result the active range appears to be considerably increased [70, 182]. Thirdly, since the total number of mechanosensitive receptors within a joint or a ligament is relatively low in comparison to the very much larger number of receptors in the surrounding muscles, it has been argued that these receptors are of less relevance for position and movement detection (e.g., [110]). However, the importance of the information transmitted in a population of afferents is by no means necessarily related to the number of afferents within the population. For instance, Macefield and coworkers [176] have demonstrated in human subjects that microstimulation of individual finger joint afferents often produce distinct sensations of alteration in joint movements, while stimulation of single muscle afferents only occasionally evoke movement related sensations (see "Segmental Effects Induced from Knee Ligament Receptors"). It has also been shown that physiological activation of small numbers of mechanosensitive sensory endings in the knee joint capsule or ligaments induce clearcut and potent reflex effects on lumbar γ-motoneurones (see "Effects on the γ-Muscle-Spindle System"). Fourthly, in the discussion of the function of joint receptors the main focus has been put on slowly adapting receptor afferents which are tonically active during joint rotations. However, phasic bursts of firing may be as relevant as tonic activity in providing the CNS with movement cues as well as in motor control and muscle coordination (see [183]). It is therefore interesting to noted that Pacinian corpuscles (phasically

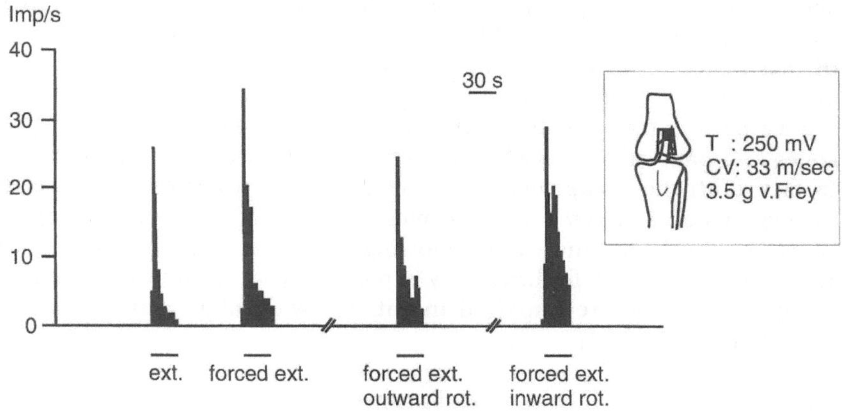

Fig. 1. A unit with a receptive field in the ACL. It had a conduction velocity (CV) of 33 m/s and a local mechanical threshold of 3.5 g. The graph shows the responses to the movements indicated in peristimulus time histograms, which show the number of action potentials in bins of 1 ms. *T*, Threshold for electrical stimulation. (From [161] with permission)

active) have been reported to dominate over Ruffini and Golgi tendon organlike endings (tonically active) in the human anterior cruciate ligament (ACL) [238]. It has also been reported that passive joint movements commonly evoke phasic responses in receptor afferents from both the medial collateral ligament (MCL) and the ACL in animals (see Fig. 1) [161, 162, 251]. Fifthly, a reduced activity from joint afferents during midrange movements per se may be important for movement and position detection, which in fact has been indicated for the proprioceptive ability of human finger joints [82].

In summary, the articular tissues are equipped with sensory endings with a wide range of different response characteristics; there are sensory endings that are sensitive to joint position, to movement directions, velocity, and acceleration of movement and to noxious mechanical stimuli and inflammation. Both direct and indirect evidence clearly suggests that at least in the knee joint there are a significant number of joint afferents active in the midrange of natural joint positions. Thus for this it seems likely that the joint receptor afferents have the potential to make significant contributions to the perception of movements and joint positions and to the normal motor control and muscle coordination, rather than functioning simply as limit detectors that serve protective functions.

Responses to Chemical Stimuli and Inflammation

Among the free nerve endings which have thin afferents with condition velocities in the group III and IV range there are a considerable number of chemosensitive units. These are activated by certain ions and a number of biochemical substances, such as the inflammatory mediators seretonin, histamine, bradykinin, and prostaglandin [34, 118, 204, 234–236]. It has also been shown that experimentally induced inflammation results in an increased activity and in reduced thresholds to mechanical probing and to joint movements of these chemosensitive joint receptors [34, 53, 98, 115, 231, 233, 279]. The altered afferent output from an inflamed joint would be expected not only to induce pain but also to influence the activity in spinal and supraspinal neurones and pathways both when the limb is stationary and during movements. In fact, both the background activity and the responsiveness to passive joint movements of α- and γ-motoneurones as well as of spinal ascending tract and supraspinal cells are increased following knee joint inflammation [65, 86, 120, 125, 234, 235]. These findings may explain why the proprioceptive acuity and the motor control are impaired in patients with inflammatory joint diseases [23, 83, 177, 178, 212].

Response Characteristics of Knee Ligament Receptors

Receptors in the Collateral Ligaments

The first reports on recordings from single knee collateral ligament receptor afferents were published in the early 1950s [8, 9]. In these studies the MCL in cats was stretched longitudinally while the activity in slowly adapting MCL receptor afferents were recorded from in the medial articular nerve. The stretch sensitive receptors were responsive to small loads applied to the ligament (see also [258]), and some of them showed a very slow adaptation behavior. Rapidly adapting stretch sensitive receptor afferents were later also demonstrated in the cat MCL [162, 251]. Krauspe et al. [162] reported that phasic responses are readily induced by passive movements in the working range of the knee joint. Sustained responses during passively altered positions are seen only when extension or flexion is combined with abduction, and the most powerful responses are obtained when extension or flexion is combined with both abduction and external rotation of the tibia (see Fig. 2).

Receptors in the Cruciate Ligaments

Direct recordings from cruciate ligament afferents have been made in a few animal studies. In these experiments single unit registrations were made from afferents running in the posterior articular nerve of the knee joint in the cat. Boyd [35] described slowly adapting afferents with relatively high thresholds to mechanical stress from afferents originating in the cruciate ligaments and suggested that these afferents probably originate from Golgi tendon organlike endings. Slowly adapting responses emanating from cruciate ligament receptors have also been reported by Skoglund [251], who found these afferents extremely sensitive to rotations and that their active range is larger than those of slowly adapting sensory endings in other knee joint structures. Ferrell [80] recorded from afferents both from the knee joint capsule and from the cruciate ligaments during passive extension-flexion movements. The relative frequency

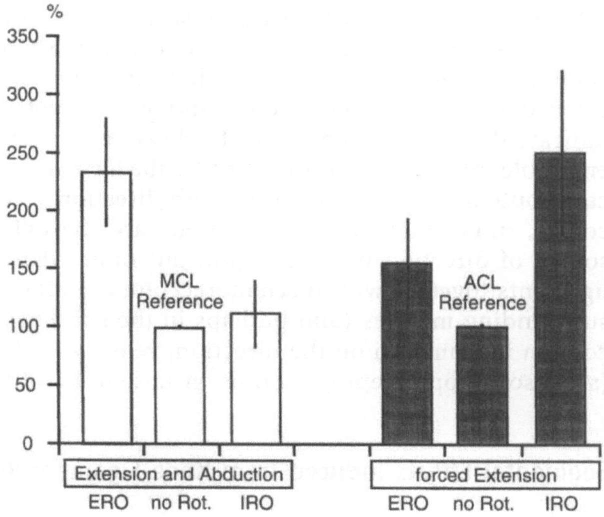

Fig. 2. Activity of afferent fibers of the MCL and ACL evoked by the joint positions indicated (means, standard deviations). The discharge rates of 11 MCL fibers during abduction and extension without rotation of the joint were set to 100%. The other discharge rates were determined as percentage of the reference value. For comparison of the discharge patterns of the responses evoked in seven ACL afferents by forced extension without rotation was set at 100%, and the other discharges were calculated as percentage of the reference value. *ERO*, Outward rotation; *IRO*, inward rotation. (From [162] with permission)

of midrange afferents was found to be larger in the cruciate ligaments than in the knee joint capsule, with a ratio of capsular units to ligament units of 1:1 for midrange units and 4:1 for end-range units. In a recent investigation, Krauspe et al. [161] analyzed responses from ACL receptor afferents during passive tibial rotation in different directions. The ACL afferents were activated when the knee joint was extended or flexed and externally or internally rotated (Fig. 1) but were silent when the knee joint was held in its resting position (30° of flexion). The thresholds to movements of the afferents were low and some of them showed strong dynamic responses to movements. The presence of both rapidly and slowly adapting mechanoreceptors in the ACL has also been demonstrated by selectively increasing the tension in the ACL [54]. Some of these receptor afferents are responsive to axial loading of the ligament and others to ligament probing or to both stimuli [54] (see also [197]).

It has been suggested that since ligament tension can be caused by a number of different types of rotations the mechanoreceptors in them are poor candidates to subserve proprioceptive sensations of movements in specific directions, i.e., they can be regarded as "nonspecific limit detectors" (e.g., [110]). This may be correct as long as only individual ligaments are considered into account. However, this view disregards the fact that individual ligament receptors are most sensitive to a particular direction of movement, and that there are general differences between the discharge pattern of receptor afferents in different ligaments [161, 162]. Thus, for instance, while the majority of the movement sensitive receptors in the cat ACL are most sensitive to extension combined with inward tibial rotation, those in the MCL are most sensitive to extension combined to abduction and outward rotation (Fig. 2). These findings indicate that, if all ligaments of the knee joint is taken into consideration, the ensemble information transmitted by the ligament afferents might significantly contribute in providing the CNS with direction specific cues. This does not, of course, imply that ligament afferents are the only (or the most important) source of direction relevant input, but rather that sensory ending within the ligaments together with mechanoreceptors in other articular structures and in surrounding muscles (and perhaps in the skin) generate sensory codes which contain information on the direction, velocity, and acceleration of movements (see also "Proprioceptive Coding in Ensembles of Receptor Afferents").

Segmental Effects Induced from Knee Ligament Receptors

Effects on the α-Motoneurones

The knowledge about reflex effects evoked on α-motoneurones by activation of sensory endings in ligaments is rather limited. Early studies using EMG or monosynaptic testing technique observed that the skeletomotor activity in the hamstrings and the quadriceps muscles is altered when the tension in the MCL is selectively increased [73, 214]. Similar findings were later reported for the ACL [119, 197, 256]. Although most of these investigations lack quantitative information on the ligament loading, the study by Solomonow et al. [119] indicates that the ligaments need to be heavily stretched to evoke reflex effects on skeletomotoneurones. This observation is in line with experiments showing that ligament loading, at last at small and moderate loads, does not cause any effects directly onto the skeletomotor system [6, 106, 216, 218, 258]. Based on results from recent animal studies Miyatsu et al. [197] drew the same general conclusion despite the fact that they demonstrated clearcut EMG alterations in the hamstrings and quadriceps muscles when modest loads (axial loading of 10–60 N) were applied to the cat ACL. However, since the EMG responses as well as the delay between the muscle responses and the ACL loading varied both between and within experiments, it was suggested that stretch sensitive receptor in ACL would modulate the EMG response via effects on the γ-muscle-spindle system rather than by direct effects on the α-motoneurones. Their suggestion seems also plausible due to the fact that the muscles were prestretched, resulting in an increased sensitivity of the muscle spindles to excitation of fusimotor neurons.

Additional evidence against direct effects on the skeletomotor system from knee joint receptors was recently obtained in a series of animal experiments by Marshall and coworkers (results to be published). In experiments on alert cats latency measurements were made from the start of an electrically induced activation of PAN afferents to the appearance of excitatory effects on single γ-efferents and to the onset of the EMG response in the gastrocnemius muscle. It was found that the start of the EMG response was always preceded by an increased activity in the γ-efferents, and the latencies were in good correspondence with the delay expected to occur if the EMG responses were mediated through the γ-muscle spindle loop. These results are in accordance with results from our laboratory, showing potent and frequent effects on the γ-muscle-spindle system, without accompanying direct effects on α-motoneurones, during activation of mechanoreceptors in different knee joint ligaments (e.g., [145, 156, 248]).

Effects on the γ-Muscle-Spindle System

From their classic experiments on the behavioral characteristics of various types of articular sensory endings and their effects on the muscle activity [88–90] Freeman and Wyke concluded that reflexes from joint mechanoreceptors contribute considerably to the normal coordination of the muscle activity in posture and movements. On indirect evidence they furthermore proposed that joint afferents influence muscle coordination through the γ-motoneurones (see also [73]). The existence of reflex pathways from knee joint afferents to γ-motoneurones has since been demonstrated in a number of studies (for references see [249]). By graded electrical activation of knee joint afferents (mostly in the PAN) it has been shown that both low- and high-threshold joint afferents evoke frequent and strong reflex effects on static and dynamic lumbar γ-motoneurones [24, 116, 150, 239, 266]. Recordings from muscle spindle afferents have revealed that the excitatory reflex effects evoked on γ-motoneurones, by electrical activation of the PAN, are powerful enough to significantly modulate the activity in both primary and secondary muscle spindle afferents from muscles around the knee joint [151, 190]. Although these findings clearly support the suggestion that articular sensory endings can influence motor control and muscle coordination via the γ-muscle-spindle system, the use of electrical activation of joint nerves have several distinct limitations, for example, the fact that the method allows only crude assessments of the specific tissue and the receptor types from which the joint afferents originate (see [249]).

A few years ago in our laboratory a method was developed for assessing fusimotor reflexes elicited by stretch sensitive sensory endings in cat knee joint ligaments [156, 247, 254]. In various series of experiments the intact ligaments (PCL, ACL, MCL, and LCL) were loaded transversely with traction forces of 5–40 N, while simultaneous recordings were made from single muscle spindle afferents from muscles around the knee [145, 152, 156, 248, 254, 255]. These muscles were subjected to sinusoidal stretching in order to determine, based on changes in the sinusoidal response of the spindle afferents, the size and character of the induced fusimotor reflex effects. The mean rate of discharge ("fitted mean") and the amplitude ("depth of modulation") of the sinusoidal response were taken as quantitative estimates of the spindle afferents sensitivity. The criteria used for the classification of the fusimotor effects (i.e., as static, dynamic, or mixed static and dynamic) were based on earlier investigations where effects on spindle afferents to controlled electrical activation of single γ-efferents were documented [13, 133].

One of the main results of these studies was that potent fusimotor effects regularly was evoked at low levels of ligament loadings. Figure 3 shows the results of a threshold test aimed at measuring the threshold load for eliciting significant fusimotor reflexes from stretch sensitive sensory endings in the MCL. Recordings were made from a primary muscle spindle afferent from the posterior biceps-semitendinosus muscles in the anesthetized cat. The threshold load for the appearance of fusimotor effects strong enough to influence this spindle afferent was found between 5 and 10 N. Similarly low thresholds were also demonstrated for the LCL, PCL, and ACL [145, 152, 156, 248]. In the example of Fig. 3 it can also be seen that small predominantly dynamic fusimotor effects were elicited at low loads (5–10 N), whereas static fusimotor reflexes dominated at higher loads (20 N). The changes in the fusimotor effects to increased ligament load are likely due to recruitment of stretch sensitive ligament receptors which activate previously silent fusimotor neurons and/or increase the activity in the already active ones [145, 248].

In the cat ligament loads of 5–10 N can most probably be regarded as low levels of mechanical stress, since loads of 200–300 N must be applied to the knee joint ligaments before they rupture [145, 156, 248, 256]. A load of 5–10 N is thus comparable to about 1.5%–5% of the ligament's ultimate strength, which provides a considerable "safety margin" indicating a none nociceptive origin of the effects found on the γ-muscle-spindle system. It is worth noting that during normal activities the human ACL is subjected to traction forces

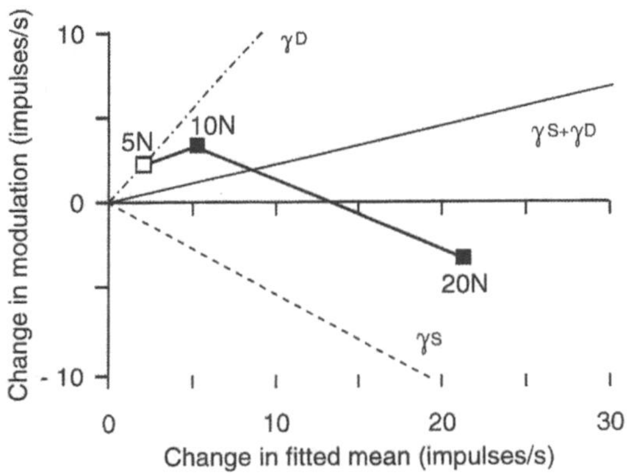

Fig. 3. Average changes in sinusoidal response for a primary spindle afferents from the posterior biceps and semitendinosus muscles during application of different loads to the MCL. For each load the mean change (i.e., mean values for five or six consecutive control-test responses; each control and test response being the result of the averaged sinusoidal stretching cycles) in modulation is plotted against the corresponding change in fitted mean. *Open symbol* (5 N), statistically nonsignificant ($p>0.05$; paired t test) alteration in the afferent response; *filled symbols* (10 and 20 N), significant changes ($p<0.05$). *Reference lines* (*gD*, *gS+gD*, *gS*), mean changes in soleus primary muscle spindle afferents elicited by selective and combined electrical stimulation of dynamic and static γ-efferents (133). (From [248] with permission)

amounting to about 25% of that necessary to tear the ligament [209]. Data are still largely missing for the cat but are available for another quadruped, the goat. In a recent study Holden et al. [128] made direct measurements of ACL forces during stance, walking, and trotting. They showed that the ACL was loaded in all three conditions, with the ligament tension positively correlated with the speed of gait. The lowest forces were observed in quite standing and during the stance phase of gait. In these conditions the maximum magnitude of the ACL force was comparable to about 6% of the ACLs ultimate strength. Thus these results suggest that the ligament loads needed for reflex activation of the γ-muscle-spindle system are low enough to be well comparable to loads occurring during normal activities, such as stance, walking, and trotting.

In addition to the low threshold of the fusimotor effects, a remarkably high responsiveness to stretch of different knee joints have been found among both primary and secondary muscle spindle afferents, from extensor as well as flexor muscles [145, 152, 156, 248]. Moreover, the changes in the sensitivity of the muscle spindles evoked by loading the knee joint ligaments are often of significant magnitudes. Examples of these features are illustrated in Fig. 4. The diagrams display statistically significant changes in the sinusoidal response characteristics (i.e., in fitted mean and in depth of modulation) induced on four simultaneously recorded muscle spindle afferents by application of a transversal load separately on the three knee joint ligaments (ACL, MCL, and LCL). The magnitude of the alterations was in the order of 5–35 impulses/s, which correspond to an increase in the both response parameters of 10%–130% in relation to control responses (i.e., sinusoidal responses in absence of ligament stimulation). The relatively modest absolute changes reflect a generally low background fusimotor drive, which is typical for the extensively denervated and anesthetized preparation. However, the large relative alterations in the sensitivity together with the high responsiveness of the muscle spindles indicate that the synaptic coupling between ligament mechanoreceptors and the γ-motor system is impressively strong.

Another principal observation that is illustrated in Fig. 4, is the considerable degree of complexity in the response patterns found to stimulation of the various knee joint ligaments [145, 152, 156, 248, 254, 255]. Thus the different ligaments often induce fusimotor effects which differ both qualitatively and quantitatively on the individual muscle spindle. Behind these heterogeneous response profiles, however, there was a trend towards a predominance of dynamic fusimotor effects to loading of the ACL, pure static, pure dynamic, and mixtures of static and dynamic effects (with an about even frequency) to loading of the PCL, and pure dynamic or mixed effects to loading of the collateral ligaments. In the light of the previously reported sensitivity differences between mechanoreceptors in different ligaments to various knee joint rotations (see "Response Characteristics of Knee Ligament Receptors"), these findings seem to suggest that the relative balance between static and dynamic sensitivity of the muscle spindles in muscles around the joint may vary depending on joint position and on different directions and qualities of the joint rotation (e.g., flexion-extension, adduction-abduction, and inward-outward rotation).

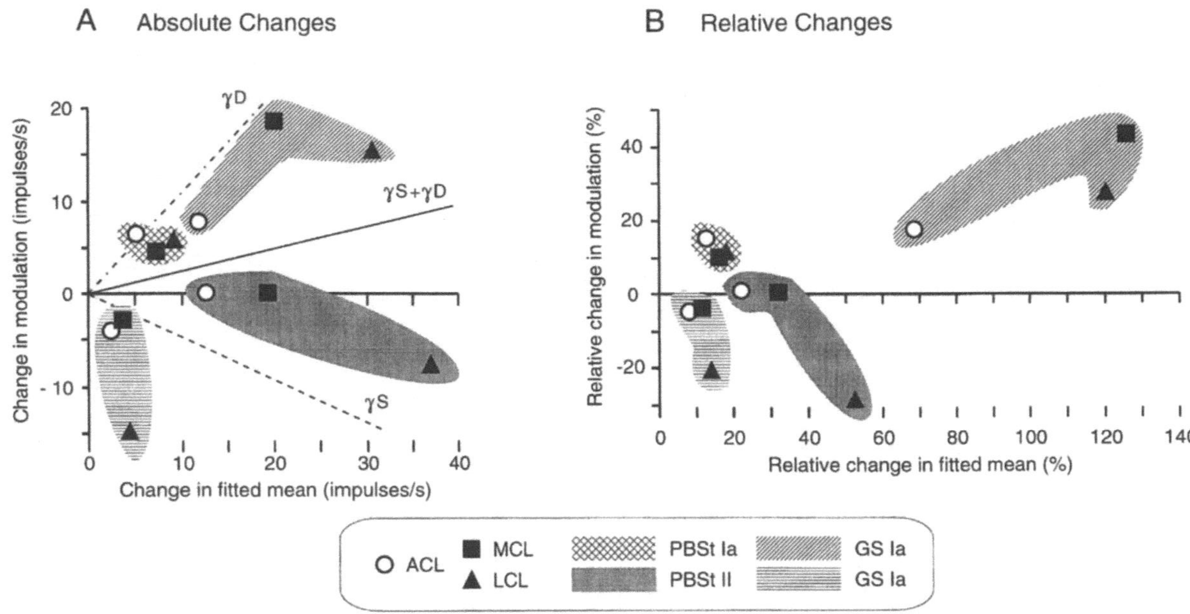

Fig. 4A,B. Simultaneously recorded fusimotor effects evoked on four muscle-spindle afferents by the ACL, MCL, and LCL. For each afferent and each ligament the mean change (i.e., mean values for 5–11 consecutive control-test responses; each control and test response being the result of the averaged sinusoidal stretching cycles) in modulation is plotted against the corresponding change in fitted mean. **A** Absolute changes. **B** Relative changes in comparison to the averaged control values. For explanation of the reference lines (*gD*, *gS*+*gD*, *gS*) in **A**, see Fig. 3. (Modified from [145] and [248] with permission)

Comparison of Reflexes on γ- and α-Motoneurones

A plethora of animal studies have shown that α- and γ-motoneurones to a large extent are differently and independently controlled from segmental and descending pathways (for reviews see [131, 140, 142, 153, 154]). One difference between α- and γ-motoneurones is that γ-motoneurones seem to be influenced more potently from low threshold joint receptors. There are three major lines of evidence in support for this conclusion. First, comparison of the responsiveness for α- and γ-motoneurones to low-threshold electrical stimulation of the PAN reveals that γ-motoneurones are much more frequently influenced at intensities below a level twice the PAN threshold [67, 150, 151, 174]. Secondly, a significantly higher responsiveness for γ-motoneurones than for α-motoneurones has been reported to mechanical stimulation of the knee joint and to passive knee joint movements [125]. Thirdly, several studies have shown that the activity in low threshold mechanoreceptors from the knee joint ligaments influences the γ-muscle-spindle system rather than causing reflex effects directly on the α-motoneurones (see "Effects on the α-Motoneurones").

The complexity of the reflex control of the γ-muscle-spindle system contrasts with the rather rigid segmental reflex control of the α-motoneurones. The γ-motoneurones have both wider receptive fields than the α-motoneurones (i.e., including input from various types of muscle, skin, and joint receptor afferents) and unique individual response profiles (for references see [140, 157]). Based on these features of the γ-muscle-spindle system it has been suggested that descending messages and peripheral receptor information are integrated in the γ-motoneurones and then transmitted to the muscle spindles, where the polymodal information is integrated with the muscle length input (i.e., "the final common input" hypothesis [16, 140]. Thereby the γ-muscle-spindle system may constitute a premotoneuronal integrative system which provides the CNS with polymodal feedback important for muscle coordination, muscle stiffness, and proprioception (see "Preprogramming of Muscle Stiffness and Coordination via the γ-Muscle Spindle System"). Furthermore, while it seems conceivable that the sophisticated receptor messages required for reflex adjustment of motor programs for finely tuned movements are received by, and integrated into, the γ-muscle-spindle system, it appears that the rather stereotypic reflexes directly to α-motoneurones are more suitable for induction of coarse automatic movements or protective motor responses.

In contrast to the early assumption that α- and γ-motoneurones are normally coactivated (e.g., [108, 195]), several studies have disclosed a high level of independence in the control, not only of α- and γ-motoneurones but also of static and dynamic γ-motoneurones. For instance, activation of a specific descending pathway from the mesencephalon in the cat is known to excite dynamic γ-motoneurones whereas static γ-motoneurones and α-motoneurones are left unaffected (for review, see [12]). In both animal studies and human subjects it has been demonstrated that the γ-muscle-spindle system can be activated without concomitant activation of skeletomotoneurones (for reviews see [131, 140]), and that the α-γ independence is task dependent [10, 224]. It is also known from both animal and human studies that the reflex gain in the γ-muscle-spindle system, as well as the relative activation of static and dynamic fusimotor neurons, is modified according to the motor task [11, 40, 132, 172, 202]. Together these findings indicate that there are independent motor programs for α- and γ-motoneurones (and perhaps for static and dynamic γ-motoneurones as well), and that the input to the CNS from muscle-spindle afferents is modulated in a task-dependent way.

Supraspinal Projections of Joint Receptors

Using electrical stimulation of various joint nerves or physiological activation of joint receptors, it has been shown in animals that joint afferents project to supraspinal centers through several ascending pathways, including the dorsal columns and the spinothalamic, spinoreticular, spinocervical and spinocerebellar pathways [29, 46, 51, 87, 95, 123, 163, 171, 184, 196, 234]. There is also direct evidence that information from joint afferent reaches the cerebral cortex, cerebellum, reticular formation, and thalamus [5, 51, 94, 96, 198–200]. Several of these studies have demonstrated clearcut projections from low threshold joint afferents (probably from Ruffini endings). Moreover, a significant number of cerebral neurons, at least in the thalamus, appear to be as sensitive to changes in the joint angle as the slowly adapting low threshold joint receptors themselves [199]. At the cortical level clear representations of the peripheral topography of joint receptors have been manifested [198, 200]. Thus it seems fair to conclude that the somatosensory system of the cortex has a good capability to adequately track the activity in joint afferents.

This conclusion has been bolstered by results elicited on human subjects with the use of a microstimulation technique which allows selective excitation of single identified peripheral afferents. Macefield et al. [176] stimulated 11 single afferents from the wrist or the finger joints and found 8 of them to evoke distinct and specific sensations; 4 induced innocuous deep pressure sensation while the other 4 induced sensations of joint displacement. The results suggest a potent and secure transmission of joint afferent information to perceptual levels and indicate that the brain may be able to obtain meaningful information on joint movements on the basis of the activation pattern in very few, or perhaps even single, joint afferents [176].

Cortical Projections of Ligament Receptors

The information on supraspinal projections of receptor afferents originating in specific joint structures is extremely sparse. To date there is only one conclusive study on this issue. Pitman and coworkers [217] recorded somatosensory evoked potentials (SEPs) elicited by electrical stimulation of the normal ACL during arthroscopic surgery. Clearcut SEPs were evoked from different parts of the ACL (i.e., from the tibial end, femoral end, and midsubstance of the ligament), but the largest SEPs were observed upon stimulation of the midsubstance of the ligament (Fig. 5). Control experiments in

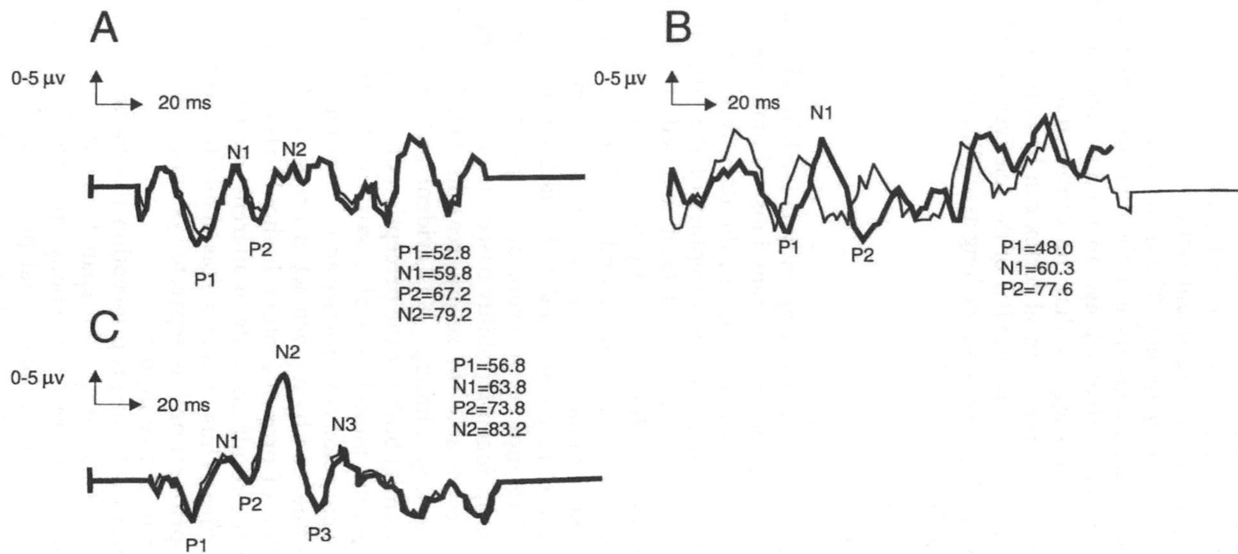

Fig. 5. Recording of somatosensory evoked potentials upon stimulation of the tibial third of the ACL (**A**), the middle third of the ACL (**B**), and the femoral third of the ACL (**C**). The time-locked onset of the SEP is at P1, and the latency is the time measured between the stimulus of the ACL and the cortical recording of the evoked potential. The amplitude is measured by the height of the spike between P1 and N1. (From [217] with permission)

which the intraarticular tissue fluid surrounding the ACL was stimulated showed that the SEPs induced by ACL stimulation are due to activation of sensory endings within the ligament rather than to excitation of receptors or afferents in surrounding tissues. These observations indicate not only that sensory endings in ligaments have projections to cortical levels, but also that these projections are potent enough to produce measurable cortical evoked potentials. The latter suggests that the central transmission of information from ACL afferents to the cerebral cortex can be impressively effective, which may be somewhat surprising in view of the relatively small number of sensory endings within the ligament and the long ascending pathway from the ligament afferents to the cortex.

Proprioception

The term proprioception was originally coined by Sherrington [242] to describe the sense (conscious or unconscious) of limb position and movements subserved by the proprioceptors, i.e., the deep receptors in the muscles, the tendons and the joints, and the receptors of the labyrinth. Since then proprioception has been widely used (and misused) to describe a number of different phenomena. Most frequently proprioception has been used to describe the conscious awareness of limb movement and/or positions. Others have investigated automatic motor responses evoked by proprioceptors (e.g., postural correction responses and EMG latencies), and claimed that such studies provide an indirect measure of the proprioceptive ability. However, assessment of proprioceptive ability based on automatic or unconscious motor responses is questionable for several reasons. One conceptual problem with this approach is that, although it is evident that proprioceptive afferent information is essential both for the perception of movements and positions and for the unconscious modulation of motor coordination, there is no obvious reason why the conscious experience and the unconscious motor effects should be similarly influenced by a given set of proprioceptive parameters or from a given population of proprioceptors. It should also be emphasized that motor responses are the result not only of afferent input but also of inherent motor control properties. Thus it seems reasonable to treat "conscious" and "unconscious proprioception" as separate systems, which not necessarily rely on the same proprioceptive inputs.

Below we use the term proprioception to describe the conscious ability to detect, without visual input, the spatial position and/or movements of the limbs in relation to the rest of the body. Admittedly, this definition is not fully satisfactory since it seems likely that proprioception should be divided into several different senses (e.g., static position sense and sensations of different qualities of motion), and that the afferent signals which give rise to them may well take origin in different sets of receptors. Based on proprioceptive tests on human joints, it has in fact been indicated that position sense and movement sense (e.g., sense of velocity) are separately coded and depend on different neuronal mechanisms [48, 82, 158, 186, 246]. However, since few reports distinguish between different proprioceptive qualities, it seems judicious the time being to retain a rather broad definition of proprioception. It should also be stressed that the lack of knowledge in this respect underlines the importance of developing novel methods by which distinctions can be made between different proprioceptive qualities.

Techniques Used to Evaluate Proprioceptive Acuity

Essentially two different approaches have been used in studies of the proprioceptive acuity: the ability (a) to detect passive movements imposed on a limb, i.e., measurements of movement threshold, and (b) to match the position of a limb to some reference position. By the former method the joint is rotated with a slow constant angular velocity, and the angular displacement that occurs before the subject is able to detect the movement is measured [19, 20, 48, 49, 52, 56, 122, 168, 169, 188]. This test is supposed to give an estimate mainly of the movement sense. However, effects on the position sense cannot be ruled out since the joint angle is actually altered.

The ability to reproduce a given joint position has been measured by various techniques. The most commonly used position-matching procedure is that in which a passively guided target position is kept constant while the subject actively reproduces the position either by the corresponding contralateral joint [20, 52, 56, 84, 85, 168, 169, 178] or by indicating the position on some kind of visual analog model [22, 82]. The accuracy that is measured by these position matching procedures certainly depends on position sense, but in the former it also depends unavoidably on the position sense of the contralateral limb. With the latter technique, on the other hand, the result is influenced by cognitive functions. A different matching procedure is applied in the so-called "assigned position test" [85], in which the joint is passively or actively rotated to a given position that is memorized. Thereafter the joint angle is returned to the control position and passively or actively positioned to the target angle. Although this method is probably a conceptually more correct method for assessing the position sense of a specific joint, the success of reproduction certainly depends on the memory of the subject.

This brief overview shows that proprioceptive testing techniques generally do not permit separate assessment of the various aspects of movement and position sense, and that the results of several of these testing procedures reflect not only the proprioceptive ability but also other factors such as inherent motor control properties and memory. An implication of this is that the results should be interpreted with care, especially regarding comparison of the outcome obtained by different methods and the interpretation of negative results. In this context it should be added that many reports provide no information on the sensitivity of the measuring system, which, for example, for optical movement analysis systems is a result of the marker size, the camera distance, and the resolution of the sampling system (see [122]). Other factors that should be taken into account when results from proprioception studies are evaluated are the possible effects of training of a given procedure, the angular range covered in the test, and the velocity of the movement imposed in the test (see [20, 82, 84, 85, 122, 168, 169, 178]).

Contribution from Muscle and Cutaneous Receptors

The relative importance of mechanoreceptor input from joints, muscles, and cutaneous tissues to proprioceptive sensations has been controversial for several decades. Views have changed over the years, sometimes favoring muscle receptors and sometimes articular and cutaneous receptors (for reviews, see [142, 149, 153, 154]).

Early in the twentieth century it was suggested that muscle receptors could take a major part in providing the CNS with proprioceptive cues. This belief was based on the complex morphology and high sensitivity to mechanical muscle stimuli of muscle receptors such as the muscle spindle and the Golgi tendon organ [59, 241, 242]. Over the past three decades this view has been substantially bolstered by a number of experimental results. In animals muscle afferents have been demonstrated to have projections to several supraspinal structures, including the somatosensory cortex and cerebellum [7, 146–148, 165, 166, 175, 211]. Activation of muscle receptors by longitudinal vibration of the muscle in a stationary limb was found to induce illusions of joint movements in man (e.g., [74, 101, 104, 121, 225, 226]). Since vibration excites mainly muscle spindle afferents, these findings indicate that information from these afferents might be especially important for movement sensations. It has also been observed that the frequency of muscle receptor activation (i.e., vibration frequency) is to some extent correlated to the perceived movement velocity [225, 226]. Similar movement illusions have been described when an exposed tendon is pulled (whereby tension sensitive muscle receptors are activated due to increased muscle length) in a stationary limb [187]. These observations have supported the view that the primary muscle-spindle afferents provide relatively more information on limb velocity whereas secondary muscle spindle afferents contribute mainly to limb position information [186, 246].

Studies on the ability to detect movements and positions of finger joints have provided evidence that anesthesia of the joint and the skin (i.e., leaving only the muscle afferents unaffected) results in only a partial reduction of the proprioceptive ability [48, 50, 82, 84, 85, 91, 92, 105, 188]. Other evidence of a major contribution from muscle receptors includes the demonstration of disturbed proprioception by muscle loading and the observation that the proprioceptive ability can be enhanced during contractions of muscles around the joint [85, 93, 186, 228]. The latter finding was suggested to be caused by increased sensitivity of muscle spindle afferents due to increased fusimotor activity [10, 11, 85, 93] but could also, at least partly, be a result of increased feedback from Golgi tendon organs [228, 270].

In contrast to the seemingly clearcut contribution to proprioception from muscle receptors, there is still controversy as to whether cutaneous afferents carry specific proprioceptive information. It has since long been commonly believed that the main role of afferents from the skin is to enhance the effects of other proprioceptive inputs [179, 180, 220]. The accuracy of the proprioception of finger joints has been shown to be reduced after skin anesthesia, and this has been suggested to be due to the loss of facilatory inputs from cutaneous receptors [48, 92] (see also [82, 84]). Since anesthesia of skin areas not directly overlying the joint induces the same blunting of the proprioception, it was suggested that the facilatory input from cutaneous afferents is relatively nonspecific. Lately, however, evidence has been reported for a more specific proprioceptive role of cutaneous receptors. Microneurographic recordings from cutaneous afferents emanating from the human wrist and finger joints have shown that receptors located in the dorsal skin have response properties which can provide information on movements of the hand and fingers [71, 72, 134]. However, this does not necessarily imply that CNS actually uses the cutaneous input as proprioceptive cues. The finding that microstimulation of individual skin afferents from the wrist and finger joints induce no perception of joint movements seems to indicate that cutaneous input to the CNS is used for other purposes [176].

In summary, there is enough evidence to safely conclude that mechanosensitive muscle afferents (primarily muscle spindle afferents) have a major role in movement and position sensations. However, optimal resolution of the movement and position sensations requires feedback also from joint and/or cutaneous afferents. This is born out by the facts that (a) the proprioceptive ability is reduced when muscle afferents are excited alone, as compared to when muscle, joint, and cutaneous afferents are activated together (see above), (b) when the finger muscles are paralyzed (by nerve block) or decoupled from the joint, whereby the muscle receptors cannot contribute to the proprioceptive feedback, the proprioceptive acuity is reduced but not abolished [48, 82, 91], and (c) tendon vibration can cause anatomically impossible movement sensations [58, 104].

Contribution from Joint Receptors

In addition to the above indirect support for a significant proprioceptive function of joint receptors, a number of studies have furnished more explicit evidence for this view. Removal of joint afferent input (e.g., by local intraarticular injection of anesthetics, nerve blockades, or surgical removal of the joint) has been found to reduce the proprioceptive acuity of different joints (e.g., [18, 21, 23, 82, 84, 91, 111, 220]). Another line of evidence is based on the findings that microstimulation of individual joint afferents from the human hand induces distinct sensations of joint movements [176]. Although these findings clearly demonstrate that joint afferents contribute to proprioceptive sensations, they are limited in that no conclusion can be drawn as to what extent sensory endings in specific joint structures are involved (i.e., capsule, ligaments, and menisci).

Proprioceptive Disturbances Following Knee Ligament Injuries

Several reports on the proprioceptive function of specific joint ligaments have recently been presented on the proprioceptive ability after ACL disruption. Barrack and coworkers [19] observed a statistically significant increase in the threshold to detection of passive motion in ACL deficient subjects. Essentially the same results were reported by Corrigan et al. [56], who in addition noted a reduction in static position sense in ACL deficient knee joints. Lephart et al. [169], using a similar experimental approach, demonstrated that the reduction in movement detection ability persists after patellar tendon autograft or allograft ACL reconstruction. A reduced proprioceptive acuity after ACL rupture was also reported by Barrett [22], who measured a significant difference in the ability to actively reproduce a passively guided knee joint position (Fig. 6). In this study the position sense was considerably improved, but not fully restored, after reconstruction surgery performed by a modified MacIntosh-Jones method (Fig. 6). Somewhat contradictory results were recently reported by Skinner and coworkers [52], who found no differences in the proprioceptive acuity between the ACL reconstructed and the intact contralateral knee joint when the patients were tested an average of 31 months after the surgery. The authors suggested that a well-performed ACL reconstruction with appropriate rehabilitation can result in knee joint proprioception that is essentially normal. However, to what extent the proprioceptive ability after an ACL rupture can be fully restored must be further investigated since it is likely that the proprioceptive ability also in the contralateral limb may be affected after an ipsilateral joint injury (see [151, 194, 277]).

Based on the above findings it may appear that disruption of the sensory feedback from the ACL reduces the proprioceptive ability of the knee joint, and consequently that sensory endings in the ACL make significant contributions to proprioceptive sensations. Although a proprioceptive function of the ACL seems very likely, it should be remembered that an ACL rupture results in disturbed sensory feedback not only from the ligamentous tissue but also from other structures within and around the joint. This is due partly to the fact that isolated ACL injuries are very uncommon, i.e., often the capsule, the menisci,

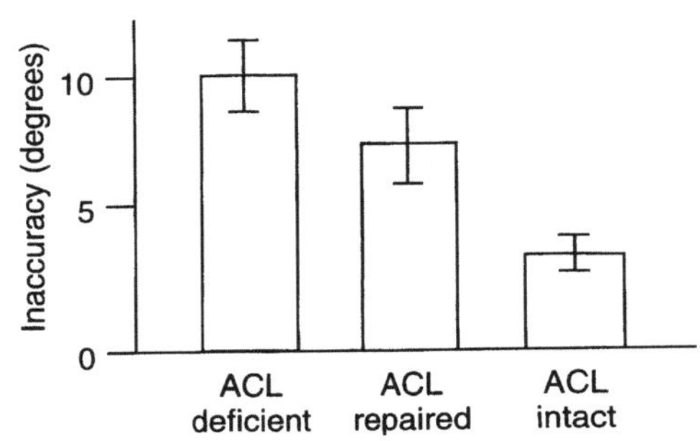

Fig. 6. Accuracy of joint sensation compared among patients with ACL-deficient knees, ACL-repaired knees and normal knees. (From [22] with permission)

and/or the collateral ligaments are also injured. More importantly, a ruptured ligament results in changes in the loads imposed on other articular structures and on surrounding muscles, causing alterations in the proprioceptive feedback from sensory endings also in other tissues.

The Role of Joint Afferent Projections to the γ-Muscle-Spindle System

It is quite likely that joint and ligament afferents contribute to perceptions of joint positions and movements also through their reflex effects on the γ-muscle-spindle system. Although explicit experimental evidence is still lacking, the logic behind this possibility has been strengthened by a number of observations. It has long been known that muscle spindle afferents have extensive supraspinal projections, and that these afferents contribute significantly to proprioceptive sensations (see "Contribution from Muscle and Cutaneous Receptors"). Furthermore, results from our and others experiments show that the sensitivity of muscle spindles can be strongly influenced by fusimotor reflexes evoked by activity in joint receptors afferents, both from the joint capsule and the ligaments (see "Effects on the γ-Muscle-Spindle System"). Thus, if it is accepted that the information transmitted by muscle spindle afferents are important for movement and position sensations, it would be expected that joint afferents should participate in the generation of these sensations via their influences on the fusimotor control of the muscle spindle sensitivity.

Proprioceptive Coding in Ensembles of Receptor Afferents

Research on the way in which peripheral stimuli are encoded in receptor afferent discharges has placed a strong emphasis on the response properties and maximal sensitivity of single receptors. This line of research must a large extent been based on the presumption that different stimuli or modalities are separately encoded in specific types of receptor afferents (i.e., the so-called "labeled line" theory). Consequently receptors in muscles, joints, and the skin have been separated into a number of modality specific categories, for instance, temperature, pressure, touch, muscle length, force, and chemosensitive receptors.

There are both theoretical and empirical arguments against the belief that, as a general rule, sensory information is encoded according to traditional modality principles, and that such information is transmitted to, and within, various structures in the CNS through private pathways or "labeled lines" [31, 32, 78, 143, 144, 222, 223]. The "labeled line" theory rely on the questionable hypothesis that there are decoding mechanisms or algorithms in the CNS which correlate to postulated modalities (see [257]). This is doubtful, partly, since most receptors and neurons seem to be sensitive to different types of stimuli and not only to a specific stimulus (i.e., the adequate stimulus). For instance, the majority of the cutaneous free nerve endings with C-fiber afferents is sensitive to mechanical, chemical, and thermal stimuli. It is also well known that the sensitivity of a receptor to the adequate stimulus can be significantly altered by other stimuli, for example, the response patterns of mechanosensitive free nerve endings in joints and muscles can be changed (sensitized or desensitized) by changes in the chemical environment and in the temperature (for references, see "Responses to Chemical Stimuli and Inflammation"). It should also be inferred theoretically that a sensory system which relies on separate information transmission in specific pathways ("labeled lines") may be very vulnerable to loss of even rather few receptors or neurons.

Another hypothesis on sensory coding is the "ensemble coding" or "population coding" theory, which places the emphasis on encoding in neural populations rather than by individual receptors and "labeled lines" [31, 32, 78, 143, 144, 222, 223]. From a teleological point of view the "ensemble coding" theory has several clearcut advantages in comparison to the "labeled line" hypothesis [78, 143, 144, 222, 223]. First, the theory implies that each receptor, afferent, and neuron participates in the coding of several different types of stimuli. This accords with the polymodal characteristics described for most peripheral sensory endings and has the advantage that the number of different messages that can be encoded in a given population of receptors would greatly exceed the number of receptors within the population. Secondly, since the sensory information is the result of the discharge pattern of many receptors, a loss of a few receptors or afferents would have only limited effects on the encoding. Finally, the signal-to-noise ratio is lower in ensemble responses than in single afferent responses, which implies that the variability in the coding of identical sensory inputs is reduced.

Although the concept of "ensemble coding" has been discussed for several decades, there are unfortunately still relatively few studies that have addressed the hypothesis in conclusive experiments. However, the concept has lately gained some experimental support due to the development of novel recording techniques that permit simultaneous recordings of large number of individual afferents and to the application of pattern recognition techniques in the analysis of coding properties in population responses, [63, 126, 143, 144, 222, 278]. In an early study using principal component analysis of simultaneously recorded slowly adapting knee joint afferent responses to tibial flexion-extension rotations and twists Heetderks [126] found that the ensemble responses from small number of individual afferents show unique response patterns over a wide range of different joint movements. His results also indicated that the information content in the ensemble response is larger than that in the individual afferents. This major observation has recently been substantiated in experiments conducted in our laboratory [31, 32, 143, 144]. These studies found that the ensemble responses of primary muscle spindle afferent have a considerably greater ability to discriminate between muscle length stimuli than the response patterns of single spindle afferents. Moreover, the information content in populations of primary muscle spindle afferents was greatly reduced when the fusimotor innervation of the spindles was eliminated, indicating that the γ-drive is important for the encoding properties in ensembles of muscle spindle afferents. Another important observation was that the ability to discriminate between passive muscle length inputs is significantly larger in populations containing both primary and secondary muscle spindle afferents and Golgi tendon organ afferents than in pure primary spindle afferent populations. The fact that Golgi tendon organ afferents contribute significantly to the information content may be somewhat surprising since these receptors have traditionally been regarded as detectors of contraction force (for review see [137]). However, it is well known that Golgi tendon organs can be responsive to passive forces imposed on the muscle, although in general the sensitivity of Golgi tendon organs are lower and different from that of muscle spindles to such inputs [137]. Thus the findings by Johansson and coworkers indicate that optimal resolution of the information content in ensemble responses is achieved by populations comprising individual receptor afferents with different but overlapping ranges of sensitivity (see [78]).

In summary, the classification of various sensory endings according to traditional modalities and submodalities appears to be of limited relevance for our understanding of how naturally occurring complex peripheral stimuli (e.g., proprioceptive stimuli) are encoded in sensory receptor afferents. Recent research indicates that sensory information may be encoded in ensembles of receptor afferents rather than in the discharge pattern of individual afferents and "labeled lines." In accordance with these ideas it seems possible that position- and movement-related information is transmitted to the CNS in population codes generated by those afferents (i.e., combination of different types of muscle, joint, and cutaneous afferents) that are activated in a given joint position or movement. This implies that rather than to look at the function of the individual components (i.e., the various receptor types), position- and movement-activated afferents should be viewed as functional units.

Motor Disturbances Following Knee Ligament Injuries

The idea that joint and ligament sensory endings are of importance for motor control and reflex protection of the joint was put forward early in this century [214, 215]. Since then a plethora of papers have been published on disturbances of the muscle activity and the muscular coordination patterns related to altered sensory feedback after joint injuries.

Effects on the Muscle Activity

Muscle Strength

Commonly reported long-term effects of knee ligament injuries include atrophy and weakness of knee flexor and particularly extensor muscles around the joint (e.g., [17, 42, 76, 77, 135, 159, 173, 192, 205, 213, 252, 253, 260, 261]). Muscle weakness has been demonstrated during both isometric and isokinetic conditions and during functional tests, and the reduction in muscle strength has been reported to be related to the functional ability of the knee (e.g., [17, 76, 97, 240]; see also [191]).

Lorentzon and coworkers reported a 6.3% reduction in the cross-sectional area of the quadriceps after ACL ruptures [77, 173]. The same study showed that along with the decreased isokinetic quadriceps strength there is a reduction in both the speed and the acceleration of contractions in the rectus femoris muscle. These findings could indicate that the atrophy is primarily the result a reduced number of type 2 muscle fibers [117, 193]. However, analysis of biopsy specimens from the quadriceps muscle reveals no differences in either the size or the proportions of different muscle fiber types in comparison to that found on the contralateral uninjured muscle or on muscles in subjects with healthy knee joints [173] (see also [99]). Interestingly, when the reduction in isokinetic muscle strength was corrected for the decreased cross-section area, the relative quadriceps muscle strength of the injured limb was still significantly lower than that in the uninjured limb [173]. The authors concluded that the strength deficit is caused by nonoptimal activation of the muscle during voluntary contraction rather than by muscle atrophy or structural changes per se and attributed this to changed afferent feedback from mechanoreceptors located around or within the injured knee joint, probably from the torn ACL.

Rehabilitation programs for improving muscle strength parameters such as peak torque and rate of torque development have been shown to improve the functional ability of the knee joint in ACL-deficient subjects [41, 102, 136, 260, 261]. However, it seems unlikely that the potential to generate large forces would be of much benefit for knee ligament injured patients since most activities in which the knee gives way involve submaximal levels of muscle activity [194]. However, strength training might contribute to an improvement in the functional properties of the joint by facilitating the development (or relearning) of an appropriate neuromuscular regulation of muscle stiffness and muscle coordination. In this context it should be noted that muscle exercise in general, regardless of whether endurance, strength, or coordination exercise, induces a muscle-specific increase in the iEMG (both magnitude and duration) which results in a closer interaction between preactivity and reflex-induced activation of the muscle (e.g., reduced drop in EMG between contraction cycles in rhythmic movements) [61, 237]. Such general effects of muscle exercise might explain why functional ability can be improved by very different rehabilitation programs, but it should not of course be taken to indicate that the choice of training program is unimportant. On the contrary, there is substantial evidence suggesting that the functional properties of the joint are regained most efficiently by applying rehabilitation regimes aimed at improving proprioception and muscle coordination (see "Rehabilitation After Ligament Injuries").

ACL-Hamstring Synergy

Based on the biomechanical fact that the ACL and the hamstrings muscles act as synergistic structures in preventing anterior tibial translation, it has been hypothesized that there is an ACL-hamstrings reflex loop that protects the ACL from excessive tensions [27, 66, 106, 107, 218, 256, 271]. It has recently been suggested that the latency in this reflex loop would provide an indirect measure of the "proprioceptive ability of the joint," and an "objective measure of the protective ability of the hamstrings muscles" [27] (see also [107, 271]). In an attempt to investigate to what extent the latency in this potentially protective reflex loop is affected after ACL ruptures, Beard and coworkers measured the delay from the application of a posteroanterior shear force at the knee joint to the appearance of on EMG response in the hamstrings muscles in uninjured and ACL-deficient knees [26–28]. They reported that the latency in the injured legs was an average nearly twice that in the unaffected limbs and suggested that the slowing of the reflex muscle activity could be a result of loss of sensory feedback from the ACL or other joint receptors. However, contradictory results have been presented by Jennings and Seedholm [139], who were unable to reproduce these results despite the use of a nearly identical method. These authors argued that the increased reaction time of the injured leg observed by Beard and coworkers [27, 28] may have been due to the lack of detectable short-latency autogenic hamstrings reflexes rather than to altered feedback from the ACL or to the fact that the experiments were carried out only 3 weeks after arthroscopy. The latter suggests that some residual effusion in the knee joint affected their results, either by mechanically reducing the degree of tibial-femoral movements [139] or by activating inhibitory muscle reflexes from pressure sensitive capsular receptor (for references see [149, 154]). Thus, as concluded by Jennings and Seedholm [139], the reflex hamstring latency does not seem to be a valid method for objective assessment of the sensory ability of the joint since there are too many unknown factors in such measurements.

In addition to the contradictory results described above, there are both experimental evidence and theoretical arguments against the concept of an ACL-hamstring protective reflex mechanism (for discussion of the functional relevance of protective reflexes, see "Ligamentomuscular Protective Reflexes"). In animal studies in which controlled and selective activation of tension sensitive ligament receptors were applied, no effects directly onto skeletomotor neurons have been demonstrated with small and moderate ligament loading (see "Effects on the α-Motoneurones"). Moreover, the clinical relevance of a potential prolongation of the reflex latency after ACL injury is also ambiguous in the light of recent EMG recordings during walking and running at different speeds and inclines. These studies show that ACL deficient patients have an earlier onset and a prolonged activity of the hamstrings muscles, which thus could compensate for an increase in reflex latency (see "Effects on the Muscular Activation Patterns").

Muscle Stiffness

The hypothesis that mechanoreceptors in ligaments are of importance for the regulation of the stiffness in muscles around the joint grew out the work by Johansson and coworkers in the late 1980s (for reviews, see [149, 153, 154, 247];

see also "Preprogramming of Muscle Stiffness and Coordination via the γ-Muscle Spindle System"). In a recent series of experiments McNair and coworkers [191, 193, 194] were the first to measured the active muscle stiffness in patients with ligament injury (i.e., complete ACL tear). The total angular muscle stiffness was calculated based on measurements of the oscillation induced by a brief downward force applied to the posterior aspect of foot which was preloaded (the limb extended in horizontal position). A model in which the measured damped frequency of oscillation, coefficient of damping, mass and center of mass of the lower leg and foot was used to calculate the total muscle stiffness. They reported no significant differences between ACL-deficient patients and normal subjects regarding either the total hamstring stiffness or regarding kinematics and EMG parameters during a landing task. The active hamstring stiffness was, however, positively correlated to the functional ability of the injured limb (i.e., Noyes knee rating questionnaire), and negatively correlated to the magnitude of the EMG activity. The authors suggested, in line with the ideas put forward by Johansson et al., that hamstring muscle stiffness may have an important role in the functional ability of ACL-deficient knee joints.

The lack of difference between the total hamstring stiffness of the uninjured and the injured limbs may indicate that the muscle stiffness is unaffected by an ACL rupture. However, the stiffness was measured in a limb position in which the ACL is unlikely to be under much tension, and the stiffness values were not corrected for differences in maximal muscle strength or in cross-section area of the hamstring muscles. It should also be noted that their method did not permit any differentiation of the total stiffness into intrinsic and reflex-mediated components (see "Preprogramming of Muscle Stiffness and Coordination via the γ-Muscle Spindle System"). A reduction in the reflex-mediated stiffness due to altered sensory feedback from the knee joint could therefore have passed undetected if a corresponding increase in the intrinsic stiffness occurred. An impaired reflex-mediated muscles stiffness could also have been compensated for by increased coactivation of the muscle around the joint [208, 244]. Thus it must still be shown experimentally whether muscle stiffness parameters are altered following ligament injuries.

Effects on the Muscular Activation Patterns

Dynamic Activities

A number of studies have shown more or less inconsistent differences in EMG patterns between ACL deficient limbs and normal limbs during movements (e.g., [37, 45, 164, 167, 170, 192, 244]). The reason for at least some of the inconsistencies is the fact that the muscular activation patterns have been investigated during very different activities, such as walking, running, cutting, and stair climbing. However, in studies assessing simple activities the number of contradictory results is larger than in studies in which more demanding tasks were performed. For instance, during slow horizontal walking Sinkjaer and coworkers reported that ACL deficient subjects demonstrated no differences in the EMG pattern in comparison to subjects with intact knee joints, but during uphill walking the ligament deficient subjects showed an earlier recruitment and prolonged activity in muscles around the knee joint [164, 167, 244]. The alterations in EMG onset and burst duration were most pronounced in the hamstring and the gastrocnemius muscles. Interestingly, when they compared the coordination patterns and relative levels of activity between a functionally excellent and a functionally poor-ACL patient group (according to the Lysholm score), significant differences were observed in both EMG onset and burst duration of the gastrocnemius medialis muscle (Fig. 7). The EMG profile of patients with poor functional knee joints was in general very similar to that of the control subjects, while the EMG profile of patients with an excellent knee joint function was clearly different (Fig. 7). These findings demonstrate that the coordination patterns of several muscles around the knee joint are affected by an ACL rupture, and that such effects are most clearly manifested in more demanding motor tasks. Sinkjaer and Arendt-Nielsen [244] suggested that the altered coordination patterns could be due to a compensatory mechanism which might be essential in the ACL-deficient patient to secure a functional knee joint stability (see also [37, 192, 262]).

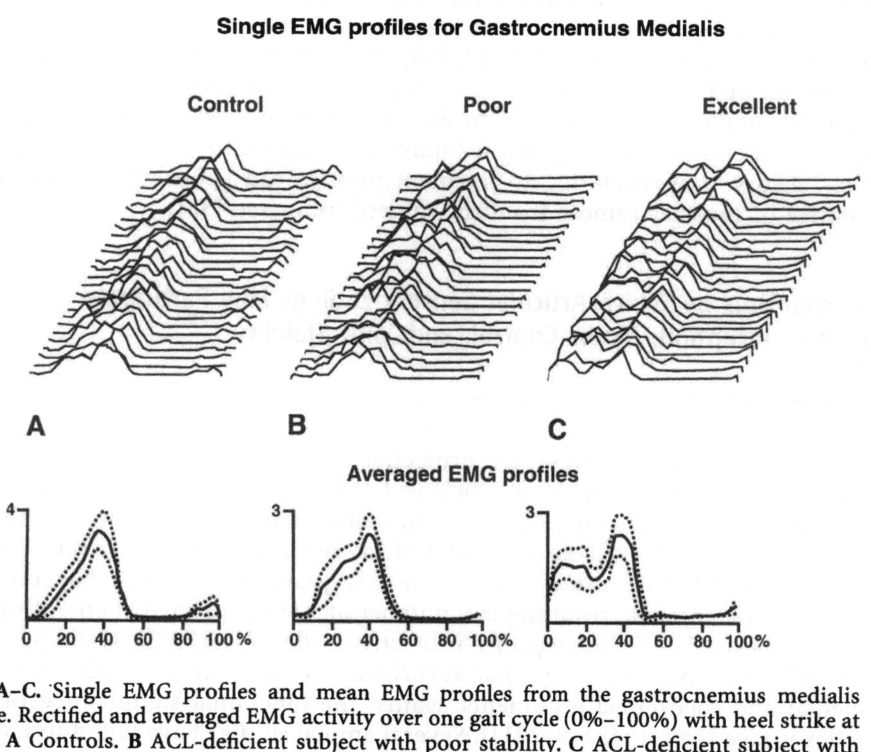

Fig. 7A–C. Single EMG profiles and mean EMG profiles from the gastrocnemius medialis muscle. Rectified and averaged EMG activity over one gait cycle (0%–100%) with heel strike at 100%. **A** Controls. **B** ACL-deficient subject with poor stability. **C** ACL-deficient subject with excellent stability. Mean profiles shown with ±1 SD of the mean. (From [244] with permission)

Kinematic data have also indicated that changes in movement strategies occur as a result of ACL deficiency. Based on angular information from ACL-deficient and uninjured limbs during running at different velocities, McNair and coworkers [192] concluded that the ACL-deficient subjects seem to adopt one of two solutions to the problem introduced by the ligament injury. The overall knee angle in the ACL deficient limb was increased in one group of subjects whereas the flexion angle of the injured knee was decreased at foot strike in the other group. Moreover, studies on dogs with unilateral ACL deficient knee joints have shown that changes in the kinematics are not confined to the unstable joint but also affect the other joints in that limb, the joints in the contralateral limb, and the vertical movements of the rump [160, 265]. Again, these results may indicate that modification of the movement patterns is functionally important to protect the knee joint from giving way and/or from rapid detoriation in the new sensory and biomechanical situation induced by the ligament rupture.

Postural Control

Effects on the postural control related to disturbed sensory input from ligaments have been reported in several recent studies. Both ACL-deficient patients and subjects with ankle ligament sprains have altered postural control compared to uninjured subjects [38, 97, 264, 277]. Zätterström and coworkers [277] demonstrated that the disturbed standing balance in patients with chronic ACL insufficiency was not restricted to the injured leg but was also present in the uninjured contralateral leg. These findings strengthen the view that changes in the sensory properties due to ligament injuries can cause alterations in the control of muscles remote from the injured joint (see [38]).

Mechanisms by Which Articular Sensory Endings May Participate in Proprioception, Motor Control, and Joint Stability

Ligamentomuscular Protective Reflexes

The idea of ligamentomuscular protective reflexes arises from early clinical observations [214, 215]. It was believed that when the joint is subjected to threatening displacements or hyperrotations, sensory endings within the ligaments (or other joint structures) induce reflex activation of muscles that counteract the displacement. Over the last years this concept has regained considerable interest, resulting in a number of attempts aimed at corroborating a hypothesized ACL-hamstrings protective reflex (see "ACL-Hamstring Synergy"). This may be somewhat surprising, however, in the light of experimental and theoretical arguments against the functional significance of such reflexes (see [149, 154, 219, 221]). Several animal studies have failed to confirm any direct reflex effects from the ACL to the skeletomotor neurons of the muscles around the knee joint (see "Effects on the α-Motoneurones"), and even if such reflexes did exist, they would be too slow to protect the joint from injuries except during very slow movements. Pope and coworkers [219] carried out a theoretical analysis of the latency between the initiation of load to the MCL and the occurrence of ligament rupture in a typical ski injury and found a delay of 34.6 ms. This should be compared with the results reported by Stener and Petersén [259] who during surgery demonstrated that a tap applied to the MCL induced an EMG response in the sartorius muscle with a minimal latency of 76 ms (see also [219]).

The fact that the process by which reflex activation of muscles can increase the joint stability is slow has been used to place the mechanical properties of the ligaments in the focus among the factors that subserve joint stability and prevent joint injuries. However, recent findings showing very low forces in the ACL under normal loading conditions indicate a neurophysiological rather than a purely mechanical role for the ligament [138]. It should also be pointed out that the emphasis on the mechanical properties of the ligaments disregards the importance of the muscles in the control of the functional joint stability (for review, see [154]), and the possibility that sensory endings in the ligaments contribute to the regulation of the muscle stiffness and coordination through preprogramming mechanisms.

Feedback Regulation of Movements via Servoregulation and Preprogramming

There are a number of reports suggesting that feedback information from muscles and joints is required for an adequate execution, as well as for learning, of various motor tasks (for review, see e.g., [61]). The peripheral feedback can influence the motor pattern through essentially two different, but not mutually exclusive, control mechanisms, i.e., through servoregulation and through preprogramming. Servoregulation is an analogous process by which movement parameters (e.g., the amplitude and duration of joint rotation) are continuously adjusted in both a spatial and a temporal sense (e.g., [1, 129, 180]). Feedback control of movements through servoregulation implies a certain latency between the activation of the peripheral receptors and the adjustment of the movement and is therefore involved mainly in the control of slow movements [246]. For fast movements, on the other hand, it seems reasonable to assume that the motor programs require information about the condition of the limbs prior to or early in the movement [124, 229, 246, 268]. Since fast movement at least to a certain degree results from preparatory feedback adjustments of the motor programs, feedforward control mechanisms (preprogramming) bypass problems of feedback delay. Evidence have been presented that afferent position information is dominant in the servoregulation of slow movements whereas mainly afferent velocity information is used in the preprogramming of fast movements [246].

One type of preparatory feedback mechanism that has recently gained some experimental support is that of "proprioceptive triggering," which is a process by which proprioceptive input about the position and velocity of a moving joint can be used by that CNS to trigger other motor responses during the execution of a movement [33, 55]. In contrast to servoregulation, proprioceptive triggering of movements is a digital process which arbitrates between two states (movement or no movement) at a discrete moment in time. The results reported by Cordo and coworkers [55] indicate that the coordination of a wide range of natural movements in normal behavior can be achieved by proprioceptive triggering, and that there are very few normally occurring movement sequences that are too fast to be coordinated through this feedback mechanism. It seems likely that phasic and tonic inputs from mechanosensitive sensory endings in joints and muscles should have an important role in this preprogramming mechanism.

Preprogramming of Muscle Stiffness and Coordination via the γ-Muscle Spindle System

In the light of the result reviewed here it seems safe to conclude that the information emanating from sensory endings in ligaments and other joint structures is of relevance for the maintenance of a normal muscular activation pattern in movements and posture. On indirect evidence Freeman and Wyke [89] suggested that the normal, smooth coordination of muscle tone was accomplished via polysynaptic reflexes on γ-motoneurones evoked by discharges in articular receptor afferents (see also [73]). Recent studies in our laboratory has provided this concept with substantial experimental support (see "Effects on the γ-Muscle-Spindle System"). Sensory endings (in ligaments and in joint capsules) with low threshold to mechanical stimuli exert potent reflex effects simultaneously on the γ-muscle-spindle systems of flexor and extensor muscles, and activation of knee ligament receptors can evoke both excitatory and inhibitory effects. Moreover, the relative balance between static and dynamic fusimotor activation, and thereby the alteration of the static and dynamic sensitivity of the muscle spindles, depends both on which of the knee ligaments is stretched and on the amount (and possibly the speed) of stretch. These features of the reflex control of the γ-muscle-spindle system from joint/ligament receptors may permit a sophisticated regulation of the muscular activation pattern during various movements and posture.

The total muscle stiffness (change in force/change in length) consists of a reflex-mediated and an intrinsic component [3, 127]. While the intrinsic component depends on the viscoelastic properties of the muscle and on the number of actomyosin bonds, the reflex-mediated stiffness is determined by the excitability of the α-motoneuron pool. Due to modulation of the reflex-mediated component, the total muscle stiffness has been shown to vary during cocontractions, voluntary dynamic contractions and rhythmic movements [3, 30, 206, 208]. Experiments on humans have demonstrated that the reflex-mediated component can increase the overall stiffness in muscles around the wrist and ankle joints of up to 100% [208, 243, 245, 263]. Since it is well known that the excitability of the skeletomotoneurons and hence the reflex-mediated stiffness depends largely on autogenic and heterogenic input from muscle spindle afferents [4, 130, 206, 207], it would be expected that the fusimotor control of the muscle spindle sensitivity should have an impact on the overall muscle stiffness [3]. The importance of the input from muscle spindle afferents is evident in the fact that half of the total muscle stiffness during contraction levels comparable to standing or phases of walking is caused by the stretch reflex [263]. Considering the capability of ligament receptors to signal joint positions and movements and the potent fusimotor effects evoked by these receptors, it is quite likely that sensory endings in ligaments participate in the regulation of muscle stiffness through their reflex actions on the γ-muscle-spindle system [149, 153, 154, 247]. However, it should be remembered that both γ-motoneurones and interneurons influenced by joint receptor discharges also receive input from muscle and cutaneous receptors (e.g., [57, 69, 141, 155, 234, 235, 250, 267]), which implies that information from more than one receptor category is involved in the regulation of the reflex-mediated muscle stiffness.

The intrinsic muscle stiffness is largely dependent on the existing acto-myosin bonds, which in turn is the results of the level of muscle contraction at a given moment. This stiffness component is thereby caused by the preceding reflex and descending influences on the α-motoneuron pool. As the intrinsic stiffness is always present (but not constant during a movement), it has been regarded as the body's first line of defense against perturbations ([2, 3]; see also [243]). Thus, if ligament receptors significantly contribute to the control of the reflex-mediated muscle stiffness, they are also important for the amount of intrinsic muscle stiffness and thereby for the protection of the joint against potentially harmful joint rotations. It has been suggested that a major function of the sensory system of the ligaments in the control of joint stability is to continuously preprogram (both in a spatial and a temporal sense) the intrinsic muscle stiffness through reflex modulation of the reflex-mediated stiffness [142, 149, 157].

The effects on secondary muscle spindle afferents from stretch sensitive ligament receptors may have especially interesting implications for the regulation of muscle stiffness [145, 152, 156, 248, 254, 255, 267]. This is due to the fact that the secondary muscle spindle afferents, in contrast to the primaries [14], project back to the γ-muscle-spindle system, i.e., they cause strong effects on static and dynamic γ-motoneurones of both homonymous and heteronymous muscles [15, 250]. The secondary spindle afferents may thereby be part of a positive feedback loop (secondary spindle afferents–fusimotor neurons–muscle spindles), which may maintain a high sensitivity of the spindle and a high level of muscle stiffness. It seems also possible that certain "threshold" exists in the activation of the γ-muscle-spindle system at which the activity in this positive feedback loop begins to function in a way resembling an amplifier that increases the effect of other inputs. It has been suggested that this potential "vicious circle" may be involved in the pathophysiological mechanisms behind the onset and spread of muscle tension and pain in chronic musculoskeletal pain disorders [62, 64, 157, 267]. Thus, it is conceivable that, for instance, a whiplash trauma causes a massive afferent input from mechanosensitive and nociceptive sensory endings in ligament and capsules in the cervical region that triggers the positive feedback circuit, which in turn results in increased muscle stiffness. In this context it should be recalled that the amount of fusimotor excitation seems to be related to the ligament tension (see Fig. 3). When stiffness (or tension) in the muscles is elevated, the production and liberation of ions and metabolites are also increased (e.g., K ions, lactic acid, arachidonic acid, histamine, and seretonin). These substances are known to activate chemosensitive muscle afferents that, via their reflex effects on fusimotor neurons, further increase the excitatory load on the γ-muscle-spindle system (e.g., [62, 64]). This could tentatively result in a chronic state of high muscle stiffness and pain which is maintained by the positive feedback loop (secondary spindle afferents–fusimotor neurons–muscle spindles) and by the increased activity in the chemosensitive muscle afferents.

Figure 8 summarizes the mechanisms by which ligaments can contribute to proprioception and to regulation of joint stability. There is substantial evidence for the existence of ligament sensory endings with sufficiently low mechanical thresholds to be active during normal movements. These mechanoreceptors do not seem to cause any direct reflex effects on the motoneurones but rather on the γ-muscle-spindle system. Therefore it seems likely that the ligaments, in addition to their well-known mechanical function, work as sensors that contribute to the regulation of the joint stability by preadjusting the stiffness of the muscles around the joint. The proprioceptive input from ligaments is probably mediated both through direct effects from the receptor afferents on ascending pathways and indirectly via influences on the γ-muscle-spindle system and the supraspinal projections of the spindle afferents.

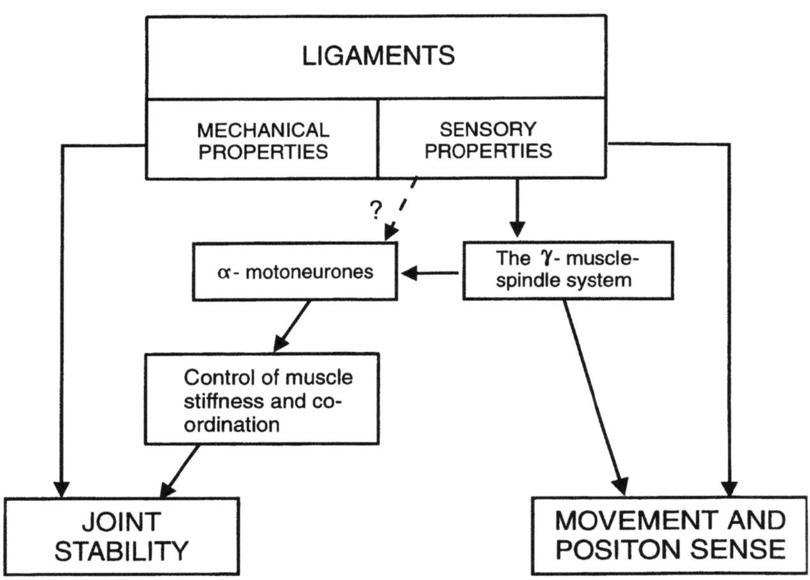

Fig. 8. Basic organizational plan of the mechanisms by which ligaments may contribute to regulation of joint stability and to proprioception. (Modified from [153] with permission)

Clinical Implications

A ligament injury would be expected to result both in altered sensory feedback and in changed biomechanical properties of the joint. These effects are interconnected since the sensory disturbances are due not only to denervation of the ligament but also to altered feedback from the surrounding articular tissue and the muscles as a result of the changed biomechanical functions. If it is accepted that different motor programs are adapted to receive specific sensory input for the accurate execution of various motor tasks, it would be expected that a ligament injury causes an ensemble sensory feedback which does not fit the existing motor programs. The disturbed ensemble sensory input thereby induces errors in the normal motor pattern (e.g., coordination and stiffness control) and functional joint stability, as well as in the perception of limb movements and position.

Surgical Repair After Ligament Rupture

Considering the rich innervation of various joint structures, it seems unavoidable that surgical treatment of an injured joint would cause further sensory damage. Therefore conservative treatment should be advocated as the first choice. Nevertheless, whenever surgical treatment is necessary, it should be performed with methods (e.g., arthroscopic techniques) that minimize any further sensory damages. For the same reason it seems questionable to use substitution techniques to restore ruptured cruciate ligaments to remove a part of an intact tendon to create an ACL graft that will certainly generate a further reduction in the afferent feedback from the limb. From a sensory preserving point of view, primary repair of teared ligaments is the most reasonable approach. In this context it should be noted that 3 months after an ACL rupture the number of mechanoreceptors in the ligament stumps gradually decreases [60]. Thus, if the primary repair is performed within 3 months after the lesion, there is good chance to restore, at least partly, the normal sensory capacity of the ligament. In addition, since the ligamentous sensory endings have their sensitivity range within a limited range of ligament tensions, and since it is important to restore the geometry of the joint, it appears crucial to reestablish the normal tension profile in the repaired ligament.

A well-known problem with primary repair of cruciate ligaments is the relatively low mechanical strength of the reconstructed ligament. To satisfy the biomechanical demands it might be necessary to combine the repair with some kind of augmentation. Excellent results have been reported with recently developed surgical techniques in which primary ACL repairs are augmented with grafts (for references, see [154]). Interestingly, ligamentoplastics on animals have shown that ACL reconstruction performed with bone-patellar tendon-bone graft (autologous patellar tendon) and with deep-frozen bone-ACL-bone allograft create an ACL substitution which contains both mechanoreceptors and free nerve endings [60, 103]. It is thereby possible that an augmentation with allografts and autologous grafts, in contrast to artificial grafts in which no mechanoreceptors are developed [60], could be beneficial in the sensory reconstruction of the ligament. However, this remains conjectural until it is demonstrated that the mechanoreceptors in such grafts have sensory properties and segmental and central connections that are functionally relevant.

Rehabilitation After Ligament Injuries

As an injured ligament causes a persistently disturbed sensory feedback from the joint and sometimes also from surrounding structures the existing motor programs and the proprioceptive decoding mechanisms must be modified in accordance with the new sensory situation. The role of sensory input in "retraining" the CNS has repeatedly been confirmed in both animal and human experiments [68, 181, 201, 227, 269, 270], and a special need of joint afferent input for regaining an appropriate timing and positioning of the limbs has been suggested [1]. A number of reports have indicated that rehabilitation programs designed to improve the neuromuscular control in a wide variety of tasks and situations are beneficial in the relearning process (for references, see [154, 210]). Thus, this process seems to be positively influenced by coordination training, aimed at fitting the motor programs and proprioceptive decoding mechanisms to a diversified sensory input in the new sensory situation.

The adaptation of the motor programs to persistent alteration in the sensory input appears to include a muscle specific increase in magnitude and duration of the muscle activity and of changes of the timing between programmed and reflex induced muscle activity (see [61, 100, 210]). This was recently demonstrated for ACL deficient subjects [244]. Subjects with a poor Lysholm score (poor knee joint functional) had an EMG activation pattern of the gastrocnemius muscle during gait which was very similar to that of subjects with normal knee joint (i.e., intact ACL) but very different to that of ACL-deficient persons with excellent Lysholm score (see Fig. 7). With the aid of biofeedback training, patients with poor functional knee joints can learn to adopt a gastrocnemius activation pattern characteristic for the ACL-deficient knee with excellent functional properties. The results suggest that if ACL-deficient persons who do not change their muscular activation pattern after the injury will suffer from a functionally instable knee joint.

Biofeedback training has also been shown to be efficient in modifying other neuromuscular properties. For instance, Basmajian demonstrated that biofeedback training can alter the muscle recruitment order [23a], and others have shown that biofeedback training can induce persistent changes in the magnitude of the short latency reflex response to muscle stretch [79, 273]. These findings seem to indicate that biofeedback training is a potent tool by which a variety of specific neuromuscular properties can be efficiently modified. It is therefore likely that biofeedback training aimed at improving the proprioceptive ability and motor coordination can offer an efficient way to modify the CNS to the new sensory situation.

Acknowledgements. This research was supported by grants from the Swedish Work Environment Fund, Swedish Medical Research Council, Swedish Sports Research Council, Inga Britt and Arne Lundbergs Forskningsstiftelse, and King Gustaf V's Research Fund.

References

1. Adams JA (1977) Feedback theory of how joint receptors regulate the timing and positioning of a limb. Psychol Rev 84:504–523
2. Akazawa K, Aldridge JW, Steeves JD, Stein RB (1982) Modulation of stretch reflexes during locomotion in the mesencephalic cat. J Physiol (Lond) 329:553–567
3. Akazawa K, Milner TE, Stein RB (1983) Modulation of reflex EMG and stiffness in response to stretch of human finger muscle. J Neurophysiol 49:16–27
4. Allum JHJ, Mauritz K-H (1984) Compensation for intrinsic muscle stiffness by short-latency reflexes in human triceps surae muscles. J Neurophysiol 52:797–818
5. Andersen P, Andersson SA, Lömo T (1967) Nature of thalamo-cortical relations during spontaneous barbiturate spindle activity. J Physiol (Paris) 192:283–307
6. Andersson S, Stener B (1959) Exp evaluation of the hypothesis of ligamento-muscular protective reflexes. II. A study in cat using the medial collateral ligament of the knee joint. Acta Physiol Scand 48:27–49
7. Andersson SA, Landgren S, Wolsk D (1966) The thalamic relay and cortical projection of group I muscle afferents from the forelimb of the cat. J Physiol (Paris) 183:576–591
8. Andrew BL (1954) The sensory innervation of the medial ligament of the knee joint. J Physiol (Paris) 123:241–250
9. Andrew BL, Dodt E (1953) The deployment of sensory nerve endings at the knee joint of the cat. Acta Physiol Scand 28:287–296
10. Aniss AM, Diener H-C, Hore J, Burke D, Gandevia SC (1990) Reflex activation of muscle spindles in human pretibial muscles during standing. J Neurophysiol 64:671–679
11. Aniss AM, Diener H-C, Hore J, Gandevia SC, Burke D (1990) Behavior of human muscle receptors when reliant on proprioceptive feedback during standing. J Neurophysiol 64:661–670
12. Appelberg B (1981) Selective central control of dynamic gamma motoneurons utilised for the functional classification of gamma cells. In: Taylor A, Prochazka A (eds) Muscle receptors and movement. Macmillan, London, pp 97–108
13. Appelberg B, Hulliger M, Johansson H, Sojka P (1982) Fusimotor reflexes in triceps surae elicited by natural stimulation of muscle afferents from the cat ipsilateral hind limb. J Physiol (Lond) 329:211–229
14. Appelberg B, Hulliger M, Johansson H, Sojka P (1983) Actions on γ-motoneurones elicited by electrical stimulation of group I muscle afferent fibres in the hind limb of the cat. J Physiol (Lond) 335:237–253
15. Appelberg B, Hulliger M, Johansson H, Sojka P (1983) Actions on γ-motoneurones elicited by electrical stimulation of group II muscle afferent fibres in the hind limb of the cat. J Physiol (Lond) 335:255–273
16. Appelberg B, Hulliger M, Johansson H, Sojka P (1983) Actions on γ-motoneurones elicited by electrical stimulation of group III muscle afferent fibres in the hind limb of the cat. J Physiol (Lond) 335:275–292
17. Arvidsson I, Eriksson E, Haggmark T, Johnson RJ (1981) Isokinetic thigh muscle strength after ligament reconstruction in the knee joint: results from a 5–10 year follow-up after reconstructions of the anterior cruciate ligament in the knee joint. Int J Sports Med 2:7–11
18. Barrack RL, Skinner HB, Brunet ME, Haddad RJ (1983) Functional performance of the knee after intraarticular anesthesia. Am J Sports Med 11:258–261
19. Barrack RL, Skinner HB, Buckley SL (1989) Proprioception in the anterior cruciate deficient knee. Am J Sports Med 17:1–6
20. Barrack RL, Skinner HB, Cook SD (1984) Proprioception of the knee joint. Paradoxical effect of training. Am J Med 63:175–181

21. Barrack RL, Skinner HB, Cook SD, Haddad RJ (1983) Effect of articular disease and total knee arthroplasty on knee joint-position sense. J Neurophysiol 50:684–687
22. Barrett DS (1991) Proprioception and function after anterior cruciate reconstruction. J Bone Joint Surg Br 73:833–837
23. Barrett DS, Cobb AG, Bentley G (1991) Joint proprioception in normal, osteoarthritic and replaced knees. J Bone Joint Surg Br 73:53–56
23a. Basmajian JV (1963) Control and training of individual motor units. Science 141:440–441
24. Baxendale RH, Davey NJ, Ellaway PH, Ferrell WR (1992) The interaction between joint and cutaneous afferent input in the regulation of fusimotor neurone discharge. In: Jami L, Pierrot-Deseilligny E, Zytnicki D (eds) Muscle afferents and spinal control of movement. Pergamon, Oxford, pp 95–104
25. Baxendale RH, Ferrell WR (1983) Discharge characteristics of the elbow joint nerve of the cat. Brain Res 261:195–203
26. Beard DJ, Dodd CAF, Trundle HR, Simpson AHRW (1994) Proprioception enhancement for anterior cruciate ligament deficiency: a prospective randomised trial of two physiotherapy regimes. J Bone Joint Surg Br 76:654–659
27. Beard DJ, Kyberd PJ, Fergusson CM, Dodd CAF (1993) Proprioception after rupture of the anterior cruciate ligament. An objective indication of the need for surgery. J Bone Joint Surg Br 75:311–315
28. Beard DJ, Kyberd PJ, O'Connor JJ, Fergusson CM, Dodd CAF (1994) Reflex hamstring contraction latency. J Orthop Res 12:219–228
29. Belcari P, Carli G, Strata P (1974) The projection of the posterior knee joint nerve to the cerebellear cortex. J Physiol (Paris) 237:371–384
30. Bennett DJ (1994) Stretch reflex responses in the human elbow joint during a voluntary movement. J Physiol (Lond) 474:339–351
31. Bergenheim M (1995) Encoding of stimulus discrimination in ensembles of muscle afferents. Its relation to proprioceptive coding and to the genesis and spread of muscle tension and pain. Umeå University Medical Dissertations, no 446, pp 1–48
32. Bergenheim M, Johansson H, Pedersen J (1995) The role of the γ-system for improving information transmission in populations of Ia afferents. Neurosci Res 23:207–215
33. Bevan L, Cordo P, Carlton L, Carlton M (1994) Proprioceptive coordination of movement sequences: discrimination of joint angle versus angular distance. J Neurophysiol 71:1862–1872
34. Birrell GJ, McQueen DS, Iggo A, Grubb BD (1990) The effects of 5-HT on articular sensory receptors in normal and arthritic rats. Br J Pharmacol 101:715–721
35. Boyd IA (1954) The histological structure of the receptors in the knee-joint of the cat correlated with their physiological response. J Physiol (Paris) 124:476–488
36. Boyd IA, Roberts TDM (1953) Proprioceptive discharges from stretch-receptors in the knee-joint of the cat. J Physiol (Paris) 122:38–58
37. Branch TP, Hunter R, Donath M (1989) Dynamic EMG analysis of anterior cruciate deficient legs with and without bracing during cutting. Am J Sports Med 17:35–41
38. Bullock-Saxton JE (1994) Local sensation changes and altered hip muscle function following severe ankle sprain. Phys Ther 74:17–31
39. Burgess PR, Clark FJ (1969) Characteristics of knee joint receptors in the cat. J Physiol (Paris) 203:317–335
40. Burke D, Gandevia SC (1992) Selective activation of fusimotor neurones innervating human tibialis anterior. In: Jami L, Pierrot-Deseilligny E, Zytnicki D (eds) Muscle afferents and spinal control of movement. Pergamon, Oxford, pp 151–156
41. Burke D, Gandevia SC, Macefield G (1988) Responses to passive movement of receptors in joint, skin and muscle of the human hand. J Physiol (Lond) 402:347–361
42. Campbell DE, Glenn W (1982) Rehabilitation of knee flexor and knee extensor muscle strength in patients with meniscectomies, ligamentous repairs, and chondromalacia. Phys Ther 62:10–16
43. Carli G, Farabollini F, Fontani G, Meucci M (1979) Slowly adapting receptors in cat hip joint. J Neurophysiol 42:767–778

44. Carlsöö S, Nordstrand A (1968) The coordination of the knee-muscles in some voluntary movements and in the gait in cases with and without knee joint injuries. Acta Chir Scand 134:423–426
45. Ciccotti MG, Kerlan RK, Perry J, Pink M (1994) An electromyographic analysis of the knee during functional activities. II. The anterior cruciate ligament-deficient and -reconstructed profiles. Am J Sports Med 22:651–658
46. Clark FJ (1972) Central projection of sensory fibers from the cat knee joint. J Neurobiol 3:101–110
47. Clark FJ, Burgess PR (1975) Slowly adapting receptors in cat knee joint: can they signal joint angle. J Neurophysiol 38:1448–1463
48. Clark FJ, Burgess RC, Chapin JW, Lipscomb WT (1985) Role of intramuscular receptors in the awareness of limb position. J Neurophysiol 54:1529–1540
49. Clark FJ, Grigg P, Chapin JW (1989) The contribution of articular receptors to proprioception with the fingers in humans. J Neurophysiol 61:186–193
50. Clark FJ, Horch KW, Bach SM, Larson GF (1979) Contributions of cutaneous and joint receptors to static knee-position sense in man. J Neurophysiol 42:877–888
51. Clark FJ, Landgren S, Silfvenius H (1973) Projections to the cat's cerebral cortex from low threshold joint afferents. Acta Physiol Scand 89:504–521
52. Co FH, Skinner HB, Cannon WD (1993) Effect of reconstruction of the anterior cruciate ligament on proprioception of the knee and the heel strike transient. J Orthop Res 11:696–704
53. Coggeshall RE, Hong KAP, Langford LA, Schaible H-G, Schmidt RF (1983) Discharge characteristics of fine medial articular afferents at rest and during passive movements of inflamed knee joints. Brain Res 272:185–188
54. Cole KJ, Pope D, Brand RA (1989) Discharge properties of mechanoreceptive afferents in the cat anterior cruciate ligament. Soc Neurosc Abstr 15:441
55. Cordo P, Carlton L, Bevan L, Carlton M, Kerr GK (1994) Proprioceptive coordination of movement sequences: role of velocity and position information. J Neurophysiol 71:1848–1861
56. Corrigan JP, Cashman WF, Brady MP (1992) Proprioception in the cruciate deficient knee. J Bone Joint Surg Br 74:247–250
57. Craig AD, Heppelmann B, Schaible H-G (1988) The projection of the medial and posterior articular nerves of the cat's knee to the spinal cord. J Comp Neurol 276:279–288
58. Craske B (1977) Perception of impossible limb positions induced by tendon vibration. Science 169:71–73
59. Creed RS, Denny-Brown D, Eccles JC, Liddell EGT, Sherrington CS (1932) Reflex activity of the spinal cord. Clarendon, Oxford
60. Denti M, Monteleone M, Berardi A, Panni SA (1994) Anterior cruciate ligament mechanoreceptors. Histological studies on lesions and reconstruction. Clin Orthop 308:29–32
61. Dietz V (1992) Human neuronal control of automatic functional movements: interaction between central programs and afferent input. Physiol Rev 72:33–69
62. Djupsjöbacka M, Johansson H, Bergenheim M (1994) Influences on the γ-muscle-spindle system from muscle afferents stimulated by increased intramuscular concentrations of arachidonic acid. Brain Res 663:293-302
63. Djupsjöbacka M, Johansson H, Bergenheim M, Sandström U (1994) A multichannel hook electrode for simultaneous recording of up to 12 nerve filaments. J Neurosci Methods 52:69–72
64. Djupsjöbacka M, Johansson H, Bergenheim M, Wenngren BI (1995) Influences on the γ-muscle-spindle system from muscle afferents stimulated by increased intramuscular concentrations of bradykinin and 5-HT. Neurosci Res 22:325–333
65. Dougherty PM, Sluka KA, Sorkin LS, Westlund KN, Willis WD (1992) Neural changes in acute arthritis in monkeys. I. Parallel enhancement of responses of spinothalamic tract neurons to mechanical stimulation and excitatory amino acids. Brain Res Rev 17:1–13
66. Draganich LF, Jaeger RJ, Kralj AR (1989) Coactivation of the hamstrings and quadriceps during extension of the knee. J Bone Joint Surg Am 71:1075–1081

67. Eccles RM, Lundberg A (1959) Synaptic actions in motoneurones by afferents which may evoke the flexion reflex. Arch Ital Biol 97:199–221
68. Edgerton VR, de Guzman CP, Gregor RJ (1991) Trainability of the spinal cord to generate hindlimb stepping patterns in adult spinalized cats. In: Shimamura M, Grillner S, Edgerton VR (eds) Neurophysiological bases of human locomotion. Japan Scientific Societies, Tokyo, pp 411–423
69. Edgley SA, Jankowska E (1987) Field potentials generated by group II muscle afferents in the middle lumbar segments of the cat spinal cord. J Physiol (Lond) 385:393–413
70. Edin BB (1990) Finger joint movement sensitivity of non-cutaneous mechanoreceptor afferents in the human radial nerve. Exp Brain Res 82:417–422
71. Edin BB (1992) Quantitative analysis of static strain sensitivity in human mechanoreceptors from hairy skin. J Neurophysiol 67:1105–1113
72. Edin BB, Abbs JH (1991) Finger movement responses of cutaneous mechanoreceptors in the dorsal skin of the human hand. J Neurophysiol 65:657–670
73. Ekholm J, Eklund G, Skoglund S (1960) On the reflex effects from the knee joint of the cat. Acta Physiol Scand 50:167–174
74. Eklund G (1972) Position sense and the state of contraction: the effect of vibration. J Neurol Neurosurg Psychiatry 35:606–611
75. Eklund G, Skoglund S (1960) On the specificity of the Ruffini like joint receptors. Acta Physiol Scand 49:184–191
76. Elmqvist L-G, Lorentzon R, Johansson C, Fugl-Meyer AR (1988) Does a torn cruciate ligament lead to change in the central nervous drive of the knee extensors? Eur J Appl Physiol 58:203–207
77. Elmqvist L-G, Lorentzon R, Långström M, Fugl-Meyer AR (1988) Reconstruction of the anterior cruciate ligament. Long-term effects of different knee angles at primary immobilization and different modes of early training. Am J Sports Med 16:455–462
78. Erickson RP (1968) Stimulus coding in topographic and nontopographic afferent modalities: on the significance of the activity of individual sensory neurons. Psychol Rev 75:447–465
79. Evatt ML, Wolf SL, Segal RL (1989) Modification of human spinal stretch reflex: preliminary studies. Neurosci Lett 105:350–355
80. Ferrell WR (1980) The adequacy of stretch receptors in the cat knee joint for signaling joint angle throughout a full range of movement. J Physiol (Paris) 299:85–99
81. Ferrell WR (1987) The effect of acute joint distension on mechanoreceptor discharge in the knee of the cat. Q J Exp Physiol 72:493–499
82. Ferrell WR, Craske B (1992) Contribution of joint and muscle afferents to position sense at the human proximal interphalangeal joint. Exp Physiol 77:331–342
83. Ferrell WR, Crighton A, Sturrock RD (1992) Position sense at the proximal interphalangeal joint is distorted in patients with rheumatoid arthritis of finger joints. Exp Physiol 77:675–680
84. Ferrell WR, Milne SE (1989) Factors affecting the accuracy of position matching at the proximal interphalangeal joint in human subject. J Physiol (Lond) 411:575–583
85. Ferrell WR, Smith A (1989) The effect of loading on position sense at the proximal interphalangeal joint of the human index finger. J Physiol (Lond) 418:145–161
86. Ferrell WR, Wood L, Baxendale RH (1988) The effect of acute joint inflammation on flexion reflex excitability in the decerebrate, low-spinal cat. Q J Exp Physiol 73:95–102
87. Fields HL, Clanton CH, Anderson SD (1977) Somatosensory properties of spinoreticular neurones in the cat. Brain Res 120:49–66
88. Freeman MAR, Wyke B (1966) Articular contributions to limb muscle reflexes. The effects of partial neurectomy of the knee-joint on postural reflexes. Br J Surg 53:61–69
89. Freeman MAR, Wyke B (1967) Articular reflexes at the ankle joint: an electromyographic study of normal and abnormal influences of ankle-joint mechanoreceptors upon reflex activity in the leg muscles. Br J Surg 54:990–1001
90. Freeman MAR, Wyke B (1967) The innervation of the knee joint. An anatomical and histological study in the cat. J Anat 101:505–532
91. Gandevia SC, Hall LA, McCloskey DI, Potter EK (1983) Proprioceptive sensation at the terminal joint of the middle finger. J Physiol (Lond) 335:507–517

92. Gandevia SC, McCloskey DI (1976) Joint sense, muscle sense, and their combination as position sense, measured at the distal interphalangeal joint of the middle finger. J Physiol (Paris) 260:387–407
93. Gandevia SC, McCloskey DI, Burke D (1992) Kinaesthetic signals and muscle contraction. Trends Neurosci 15:62–65
94. Gardner E, Haddad B (1953) Pathways to the cerebral cortex for afferent fibres from the hindleg of the cat. Am J Physiol 172:475–482
95. Gardner E, Latimer F, Stilwell D (1949) Central connections for afferent fibers from the knee joint of the cat. Am J Physiol 159:195–198
96. Gardner E, Noer R (1952) Projection of afferent fibers from muscles and joints to the cerebral cortex of the cat. J Physiol (Paris) 168:437–441
97. Gauffin H, Pettersson G, Tegner Y, Tropp H (1990) Function testing in patients with old rupture of the anterior cruciate ligament. Int J Sports Med 11:73–77
98. Gentle MJ, Thorp BH (1994) Sensory properties of ankle joint capsule mechanoreceptors in acute monoarthritic chickens. Pain 57:361–374
99. Gerber C, Hoppeler H, Claassen H, Robotti G, Zehnder R, Jakob RP (1985) The lower-extremity musculature in chronic symptomatic instability of the anterior cruciate ligament. J Bone Joint Surg Am 67:1034–1043
100. Ghez C, Hening W, Favilla M (1990) Parallel interacting channels in the initiation and specification of motor response features. In: Jeennerod M (ed) Attention and performance, vol 12. Erlbaum, Hillsdale, pp 265–293
101. Gilhodes JC, Roll JP, Tardy-Gervet MF (1986) Perceptual and motor effects of agonist-antagonist muscle vibration in man. Exp Brain Res 61:395–402
102. Giove TP, Miller SJI, Kent BE, Sanford TL, Garrick JG (1983) Nonoperative treatment of the torn anterior cruciate ligament. J Bone Joint Surg Am 65:184–192
103. Goertzen M, Gruber J, Dellmann A, Clahsen H, Schulitz K-P (1992) Neurohistological findings after experimental anterior cruciate ligament allograft transplantation. Arch Orthop Trauma Surg 111:126–129
104. Goodwin GM, McCloskey DI, Matthews PBC (1972) The contribution of muscle afferents to kinaesthesia shown by vibration induced illusions of movement and by the effects of paralysing joint afferents. Brain 95:705–748
105. Goodwin GM, McCloskey DI, Matthews PBC (1972) The persistence of appreciable kinesthesia after paralysing joint afferents but preserving muscle afferents. Brain Res 37:326–329
106. Grabiner MD, Campbell KR, Hawthorne DL, Hawkins DA (1989) Electromyographic study of the anterior cruciate ligament-hamstrings synergy during isometric knee extension. J Orthop Res 7:152–155
107. Grabiner MD, Koh TJ, Miller GF (1992) Further evidence against a direct automatic neuromotor link between the ACL and hamstrings. Med Sci Sports Exerc 24:1075–1079
108. Granit R (1979) Interpretation of supraspinal effects on the gamma system. In: Granit R, Pompeiano O (eds) Progress in brain research. Reflex control of posture and movement. Elsevier, Amsterdam, pp 147–154
109. Grigg P (1975) Mechanical factors influencing response of joint afferent neurons from cat knee. J Neurophysiol 38:1473–1483
110. Grigg P (1994) Peripheral neural mechanisms in proprioception. J Sport Rehabil 3:2–17
111. Grigg P, Finerman GA, Riley LH (1973) Joint-position sense after total hip replacement. J Bone Joint Surg Am 55:1016–1025
112. Grigg P, Greenspan BJ (1977) Response of primate joint afferent neurons to mechanical stimulation of knee joint. J Neurophysiol 40:1–8
113. Grigg P, Hoffman AH (1982) Properties of Ruffini afferents revealed by stress analysis of isolated sections of cat knee capsule. J Neurophysiol 47:41–54
114. Grigg P, Hoffman AH, Fogarty KE (1982) Properties of Golgi-Mazzoni afferents in cat knee joint capsule, as revealed by mechanical studies of isolated joint capsule. J Neurophysiol 47:31–40
115. Grigg P, Schaible H-G, Schmidt RF (1986) Mechanical sensitivity of group III and IV afferents from posterior articular nerve in normal and inflamed cat knee. J Neurophysiol 55:635–643

116. Grillner S, Hongo T, Lund S (1969) Descending monosynaptic and reflex control of γ-motoneurones. Acta Physiol Scand 75:592–613
117. Grimby G, Gustafsson E, Peterson L, Renström P (1980) Quadriceps function and training after knee ligament surgery. Med Sci Sports Exerc 12:70–75
118. Grubb BD, Birrell GJ, McQueen DS, Iggo A (1991) The role of PGE2 in the sensitization of mechanoreceptors in normal and inflamed ankle joints of the rat. Exp Brain Res 84:383–392
119. Grüber J, Wolter D, Lierse W (1988) In vivo study on the proprioceptive function of knee ligaments. Abstracts of the Third Congress of the European Society of Knee Surgery and Arthroscopy, Amsterdam, p 6
120. Guilbaud G (1991) Central neurophysiological processing of joint pain on the basis of studies performed in normal animals and in models of experimental arthritis. Can J Physiol Pharmacol 69:637–646
121. Hagbarth KE, Eklund G (1966) Motor effects of vibratory muscle stimuli in man. In: Granit R (ed) Muscular afferents and motor control. Almqvist, Stockholm, pp 177–186
122. Hall MG, Ferrell WR, Baxendale RH, Hamblen DL (1994) Knee joint proprioception: threshold detection levels in healthy young subjects. Neuro Orthop 15:81–90
123. Harrison PJ, Hultborn H, Jankowska E, Katz R, Storai B, Zytnicki D (1984) Labelling of interneurones by retrograde transsynaptic transport of horseradish peroxidase from motoneurones in rats and cats. Neurosci Lett 45:15–19
124. Hasan Z, Stuart DG (1988) Animal solutions to problems of movement control: the role of proprioceptors. Annu Rev Neurosci 11:199–199
125. He X, Proske U, Schaible H-G, Schmidt RF (1988) Acute inflammation of the knee joint in the cat alters responses of flexor motoneurons to leg movements. J Neurophysiol 59:326–340
126. Heetderks W (1978) Principal component analysis of neural population responses of knee joint proprioceptors in cat. Brain Res 156:51–65
127. Hoffer JA, Andreassen S (1981) Limitations in the servo-regulation of soleus muscle stiffness in premammillary cats. In: Taylor A, Prochazka A (eds) Muscle receptors and movement. Macmillan, London, pp 311–324
128. Holden JP, Grood ES, Korvick DL, Cummings JF, Butler DL, Bylski-Austrow DI (1994) In vivo forces in the anterior cruciate ligament: direct measurements during walking and trotting in a quadruped. J Biomech 27:517–526
129. Houk JC (1979) Regulation of stiffness by skeletomotor reflexes. Annu Rev Physiol 41:99–114
130. Houk JC, Crago PE, Rymer WZ (1981) Function of the dynamic response in stiffness: a predictive mechanism provided by non-linear feedback. In: Taylor A, Prochazka A (eds) Muscle receptors and movement. Macmillan, London, pp 299–309
131. Hulliger M (1984) The mammalian muscle spindle and its central control. Rev Physiol Biochem Pharmacol 101:1–110
132. Hulliger M, Dyrmyller N, Prochazka A, Trend P (1989) Flexible fusimotor control of muscle spindle feedback during a variety of natural movements. Prog Brain Res 80:87–101
133. Hulliger M, Matthews PBC, Noth J (1977) Effects of combining static and dynamic fusimotor stimulation on the response of the muscle spindle primary ending to sinusoidal stretching. J Physiol (Paris) 267:839–856
134. Hulliger M, Nordh E, Thelin A-E, Vallbo ÅB (1989) The responses of afferent fibres from the glabrous skin of the hand during voluntary finger movements in man. J Physiol (Paris) 291:233–249
135. Hurley MV, Jones DW, Wilson D, Newham DJ (1992) Rehabilitation of quadriceps inhibited due to isolated rupture of the anterior cruciate ligament. J Orthop Rheumatol 5:145–154
136. Ihara H, Nakayama A (1986) Dynamic joint control training for knee ligament injuries. Am J Sports Med 14:309–315
137. Jami L (1992) Golgi tendon organs in mammalian skeletal muscle: functional properties and central actions. Physiol Rev 72:623–666

138. Jennings AG (1994) A proprioceptive role for the anterior cruciate ligament: a review of the literature. J Orthop Rheumatol 7:3–13
139. Jennings AG, Seedholm BB (1994) Proprioception in the knee and reflex hamstring contraction latency. J Bone Joint Surg Br 76:491–494
140. Johansson H (1985) Reflex integration in the γ-motor system. In: Boyd IA, Gladden MH (eds) The muscle spindle. Macmillan, London, pp 297–301
141. Johansson H (1988) Rubrospinal and rubrobulbospinal influences on dynamic and static γ-motoneurones. Behav Brain Res 28:97–107
142. Johansson H (1991) Role of knee ligaments in proprioception and regulation of muscle stiffness. J Electromyogr Kinesiol 1:158–179
143. Johansson H, Bergenheim M, Djupsjöbacka M, Sjölander P (1995) Analysis of stimulus separation in ensembles of muscle afferents. In: Taylor A, Gladden M, Durbaba R (eds) Alpha and gamma motor systems. Plenum, New York, pp 287–293
144. Johansson H, Bergenheim M, Djupsjöbacka M, Sjölander P (1995) A method for analysis of stimulus separation in ensembles of afferents. J Neurosci Methods 63:67–74
145. Johansson H, Lorentzon R, Sjölander P, Sojka P (1990) The anterior cruciate ligament. A sensor acting on the γ-muscle-spindle systems of muscles around the knee joint. Neuro Orthop 9:1–23
146. Johansson H, Silfvenius H (1977) Axon-collateral activation by dorsal spinocerebellar tract fibres of group I relay cells of nucleus Z in the cat medulla oblongata. J Physiol (Paris) 265:341–369
147. Johansson H, Silfvenius H (1977) Input from ipsilateral proprio- and exteroceptive hind limb afferents to nucleus Z of the cat medulla oblongata. J Physiol (Paris) 265:371–393
148. Johansson H, Silfvenius H (1977) Connexions from large, ipsilateral hind limb muscle and skin afferents to the rostral main cuneate nucleus and to the nucleus X region in the cat. J Physiol (Paris) 265:395–428
149. Johansson H, Sjölander P (1993) The neurophysiology of joints. In: Wright V, Radin EL (eds) Mechanics of human joints: physiology, pathophysiology, and treatment. Dekker, New York, pp 243–290
150. Johansson H, Sjölander P, Sojka P (1986) Actions on γ-motoneurones elicited by electrical stimulation of joint afferent fibres in the hind limb of the cat. J Physiol (Lond) 375:137–152
151. Johansson H, Sjölander P, Sojka P (1988) Fusimotor reflexes in triceps surae muscle elicited by natural and electrical stimulation of joint afferents. Neuro Orthop 6:67–80
152. Johansson H, Sjölander P, Sojka P (1990) Activity in receptor afferents from the anterior cruciate ligament evokes reflex effects on fusimotor neurones. Neurosci Res 8:54–59
153. Johansson H, Sjölander P, Sojka P (1991) A sensory role for the cruciate ligaments. Clin Orthop 268:161–178
154. Johansson H, Sjölander P, Sojka P (1991) Receptors in the knee joint ligaments and their role in the biomechanics of the joint. Crit Rev Biomed Eng 18:341–368
155. Johansson H, Sjölander P, Sojka P (1991) Fusimotor reflex profiles of individual triceps surae primary muscle spindle afferents assessed with multi-afferent recording technique. J Physiol (Paris) 85:6–19
156. Johansson H, Sjölander P, Sojka P, Wadell I (1989) Reflex actions on the γ-muscle-spindle systems of muscles acting at the knee joint elicited by stretch of the posterior cruciate ligament. Neuro Orthop 8:9–21
157. Johansson H, Sojka P (1991) Pathophysiological mechanisms involved in genesis and spread of muscle tension in occupational muscle pain and in chronic musculoskeletal pain syndromes: a hypothesis. Med Hypotheses 35:196–203
158. Jones LA (1988) Motor illusions: what do they reveal about proprioception? Psychol Bull 103:72–86
159. Kannus P (1988) Knee flexor and extensor strength ratios with deficiency of the lateral collateral ligament. Arch Phys Med Rehabil 69:928–931
160. Korvick DL, Pijanowski GJ, Schaffer DJ (1994) Three-dimensional kinematics of the intact and cranial cruciate ligament-deficient stifle of dogs. J Biomech 27:77–87

161. Krauspe R, Schmidt M, Schaible H-G (1992) Sensory innervation of the anterior cruciate ligament. An electrophysiological study of the response properties of single identified mechanoreceptors in the cat. J Bone Joint Surg Am 74:390–397
162. Krauspe R, Schmidt M, Schaible H-G (1993) Different patterns of activation of sensory nerve endings with thick myelinated afferent axons supplying the medial collateral and anterior cruciate ligament of the cat's knee joint. Neuro Orthop 14:67–80
163. Kuno M, Munoz-Martinez EJ, Randic M (1973) Sensory inputs to neurones in Clarke's column from muscle, cutaneous and joint receptors. J Physiol (Paris) 228:327–342
164. Kålund S, Sinkjaer T, Arendt-Nielsen L, Simonsen O (1990) Altered timing of hamstring muscle action in anterior cruciate ligament deficient patients. Am J Sports Med 18:245–248
165. Landgren S, Silfvenius H (1969) Projection to cerebral cortex of group I muscle afferents from the cat's hind limb. J Physiol (Paris) 200:353–372
166. Landgren S, Silfvenius H (1971) Nucleus z, the medullary relay in the projection path to the cerebral cortex of group I muscle afferents from the cat's hind limb. J Physiol (Paris) 218:551–571
167. Lass P, Kålund S, LeFevre S, Arendt-Nielsen L, Sinkjaer T, Simonsen O (1991) Muscle coordination following rupture of the anterior cruciate ligament. Electromyographic studies of 14 patients. Acta Orthop Scand 62:9–14
168. Lephart SM, Fu FH (1995) The role of proprioception in the treatment of sports injuries. Sports Exerc Inj 1:96–102
169. Lephart SM, Kocher MS, Fu FH (1992) Proprioception following ACL reconstruction. J Sport Rehabil 1:186–196
170. Limbird TJ, Shiavi R, Frazer M, Borra H (1988) EMG profiles of knee joint musculature during walking: changes induced by anterior cruciate ligament deficiency. J Orthop Res 6:630–638
171. Lindström S, Takata M (1972) Monosynaptic excitation of dorsal spinocerebellar tract neurones from low threshold joint afferents. Acta Physiol Scand 84:430–432
172. Loeb GE (1984) The control and responses of mammalian muscle spindles during normally executed motor tasks. In: Terjung RL (ed) Exercise and sport sciences reviews. Collamore, Toronto, pp 157–204
173. Lorentzon R, Elmqvist L-G, Sjöström M, Fagerlund M, Fugl-Meyer AR (1989) Thigh musculature in relation to chronic anterior cruciate ligament tear: muscle size, morphology, and mechanical output before reconstruction. Am J Sports Med 17:423–429
174. Lundberg A, Malmgren K, Schomburg ED (1978) Role of joint afferents in motor control exemplified by effects on reflex pathways from Ib afferents. J Physiol (Paris) 284:327–343
175. Lundberg A, Oscarsson O (1960) Functional organization of the dorsal spino-cerebellar tract in the cat. VII. Identification of units by antidromic activation from the cerebellar cortex with recognition of five functional subdivisions. Acta Physiol Scand 50:356–374
176. Macefield G, Gandevia SC, Burke D (1990) Perceptual responses to microstimulation of single afferents innervating joints, muscles and skin of the human hand. J Physiol (Lond) 429:113–129
177. Marks R, Percy JS, Semple J, Kumar S (1994) Comparison between the surface electromyogram of the quadriceps surrounding the knees of healthy women and the knees of women with osteoarthrosis. Clin Exp Rheumatol 12:11–15
178. Marks R, Quinney HA, Wessel J (1993) Proprioceptive sensibility in women with normal and osteoarthritic knee joints. Clin Rheumatol 12:170–175
179. Marsden CD, Merton PA, Morton HB (1972) Servo action in human voluntary movement. Nature 238:140–143
180. Marsden CD, Merton PA, Morton HB (1977) The sensory mechanism of servo action in human muscle. J Physiol (Paris) 265:521–535
181. Marsden CD, Rothwell JC, Day BL (1984) The use of peripheral feedback in the control of movement Trends Neurosci 7:253–257
182. Marshall KW, Tatton WG (1990) Joint receptors modulate short and long latency muscle responses in the awake cat. Exp Brain Res 83:137–150
183. Matthews PBC (1982) Where does Sherrington's "muscular sense" originate? Muscles, joints, corollary discharges? Annu Rev Neurosci 5:189–218

184. Maunz RA, Pitts NG, Peterson BW (1978) Cat spinoreticular neurones: locations, responses and changes in responses during repetitive stimulation. Brain Res 148:365–379
185. McCall WDJ, Farias MC, Williams WJ, BeMent SL (1974) Static and dynamic responses of slowly adapting joint receptors. Brain Res 70:221–243
186. McCloskey DI (1973) Differences between the senses of movement and position shown by the effects of loading and vibration of muscles in man. Brain Res 63:119–131
187. McCloskey DI, Cross MJ, Honner R, Potter E (1983) Sensory effects of pulling or vibrating exposed tendons in man. Brain 106:21–37
188. McCloskey DI, Macefield G, Gandevia SC, Burke D (1987) Sensing position and movements of the fingers. News Physiol Sci 2:226–230
189. McIntyre AK, Proske U, Tracey DJ (1978) Afferent fibres from muscle receptors in the posterior nerve of the cat's knee joint. Exp Brain Res 33:415–424
190. McIntyre AK, Proske U, Tracey DJ (1978) Fusimotor responses to volleys in joint and interosseous afferents in the cat's hindlimb. Neurosci Lett 10:287–292
191. McNair PJ, Marshall RN (1994) Landing characteristics in subjects with normal and anterior cruciate ligament deficient knee joints. Arch Phys Med Rehabil 75:584–589
192. McNair PJ, Marshall RN, Matheson JA (1989) Gait of subjects with anterior cruciate ligament deficiency. Clin Biomech 4:243–248
193. McNair PJ, Marshall RN, Maguire K (1994) Knee effusion and quadriceps muscle strength. Clin Biomech 9:331–334
194. McNair PJ, Wood GA, Marshall RN (1992) Stiffness of the hamstring muscles and its relationship to function in anterior cruciate ligament deficient individuals. Clin Biomech 7:131–137
195. Merton P A (1953) Speculations on the servo-control of movement. In: Wolotenholme GEW (ed) The spinal cord. Churchill, London, pp 247–255
196. Meyers DER, Snow PJ (1982) The responses to somatic stimuli of deep spinothalamic tract cells in the lumbar spinal cord of the cat. J Physiol (Lond) 329:355–371
197. Miyatsu M, Atsuta Y, Watakabe M (1993) The physiology of mechanoreceptors in the anterior cruciate ligament. An experimental study in decerebrate-spinalised animals. J Bone Joint Surg Br 75:653–657
198. Mountcastle VB (1957) Modality and topographic properties of single neurons of cat's somatic sensory cortex. J Neurophysiol 20:408–434
199. Mountcastle VB, Poggio GF, Werner G (1963) The relation of thalamic cell response to peripheral stimuli varied over an intensive continuum. J Neurophysiol 26:807–834
200. Mountcastle VB, Powell TPS (1959) Central nervous mechanisms subserving position sense and kinesthesia. Bull Johns Hopkins Hosp 105:173–200
201. Mulder T, Hulstyn W (1984) Sensory feedback therapy and theoretical knowledge of motor control and learning. Am J Phys Med 63:226–244
202. Murphy PR, Stein RB, Taylor J (1984) Phasic and tonic modulation of impulse rates in γ-motoneurons during locomotion in premammillary cats. J Neurophysiol 52:228–243
203. Nade S, Newbold PJ, Straface SF (1987) The effects of direction and acceleration of movement of the knee joint of the dog on medial articular nerve discharge. J Physiol (Lond) 388:505–519
204. Neugebauer V, Schaible H-G, Schmidt RF (1989) Sensitization of articular afferents to mechanical stimuli by bradykinin. Pflugers Arch 415:330–335
205. Newham DJ, Hurley MV, Jones DJ (1989) Ligamentous knee injury and muscle inhibition. J Orthop Rheumatol 2:163–173
206. Nichols TR (1987) The regulation of muscle stiffness. Implications for the control of limb stiffness. Med Sci Sport Exerc 26:36–47
207. Nichols TR (1989) The organization of heterogenic reflexes among muscles crossing the ankle joint in the decerebrate cat. J Physiol (Lond) 410:463–477
208. Nielsen J, Sinkjaer T, Toft E, Kagamihara Y (1994) Segmental reflexes and ankle joint stiffness during co-contraction of antagonistic ankle muscles in man. Exp Brain Res 102:350–358
209. Noyes FR, Keller CS, Grood ES, Butler DL (1984) Advances in the understanding of knee ligament injury, repair, and rehabilitation. Med Sci Sports Exerc 16:427–443

210. Nyland J, Brosky T, Currier D, Nitz A, Caborn D (1994) Review of the afferent neural system of the knee and its contribution to motor learning. J Orthop Sports Phys Ther 19:2–11
211. Oscarsson O, Rosén I (1963) Projection to cerebral cortex of large muscle spindle afferents in forelimb nerves of the cat. J Physiol (Paris) 169:924–945
212. Pai Y-C, Chang HJ, Chang RW, Sinacore JM, Lewis JL (1994) Alteration in multijoint dynamics in patients with bilateral knee osteoarthritis. Arthritis Rheum 37:1297–1304
213. Palmer I (1938) In the injuries to the ligament of the knee joint, actual study. Acta Chir Scand 81 [Suppl 53]:1–282
214. Palmer I (1958) Pathophysiology of the medial ligament of the knee joint. Acta Chir Scand 115:312–318
215. Partridge EJ (1924) Joints – the limitation of their range of movement and an explanation of certain surgical conditions. J Anat 58:346–354
216. Petersén I, Stener B (1959) Experimental evaluation of the hypothesis of ligamento-muscular protective reflexes. III. A study in man using the medial collateral ligament of the knee joint. Acta Physiol Scand 48:51–61
217. Pitman MI, Nainzadeh N, Menche D, Gasalberti R, Song EK (1992) The intraoperative evaluation of the neurosensory function of the anterior cruciate ligament in human using somatosensory evoked potentials. Arthroscopy 8:442–447
218. Pope DF, Cole KJ, Brand RA (1990) Physiological loading of the anterior cruciate ligament does not activate quadriceps or hamstrings in the anesthetized cat. Am J Sports Med 18:595–599
219. Pope MH, Johnson RJ, Brown DW, Tighe C (1979) The role of the musculature in injuries to the medial collateral ligament. J Bone Joint Surg Am 61:398–402
220. Provins KA (1958) The effect of training and handedness on the performance of two simple motor tasks. Q J Exp Psychol 10:29–39
221. Rack PMH (1981) Limitations of somatosensory feedback in control of posture and movement. In: Brooks VB (ed) Handbook of physiology. The nervous system II. American Physiological Society, Bethesda, pp 229–256
222. Ray RH, Doetsch GS (1990) Coding of stimulus location and intensity in populations of mechanosensitive nerve fibers of the raccoon. I. single fiber response properties. Brain Res Bull 25:517–532
223. Ray RH, Doetsch GS (1990) Coding of stimulus location and intensity in populations of mechanosensitive nerve fibers of the raccoon. II. Across-fiber response patterns. Brain Res Bull 25:533–550
224. Ribot E, Roll J-P, Vedel J-P (1986) Efferent discharges recorded from single skeletomotor and fusimotor fibres in man. J Physiol (Lond) 375:251–268
225. Roll JP, Vedel JP (1982) Kinaesthetic role of muscle afferents in man, studied by tendon vibration and microneurography. Exp Brain Res 47:177–190
226. Roll JP, Vedel JP, Ribot E (1989) Alteration of proprioceptive messages induced by tendon vibration in man: a microneurographic study. Exp Brain Res 76:213–222
227. Rothwell JC, Traub MM, Day BL, Obeso JA, Thomas PK, Marsden CD (1982) Manual motor performance in a deafferented man. Brain 105:515–542
228. Rymer WZ, D'Almeida A (1980) Joint position sense. The effects of muscle contraction. Brain 103:1–22
229. Sanes JN, Jennings VA (1984) Centrally programmed patterns of muscle activity in voluntary motor behavior of humans. Exp Brain Res 54:23–32
230. Schaible H-G, Schmidt RF (1983) Activation of groups III and IV sensory units in medial articular nerve by local mechanical stimulation of knee joint. J Neurophysiol 49:35–44
231. Schaible H-G, Schmidt RF (1985) Effects of an experimental arthritis on the sensory properties of fine articular afferent units. J Neurophysiol 54:1109–1122
232. Schaible H-G, Schmidt RF (1988) Time course of mechanosensitivity changes in articular afferents during a developing experimental arthritis. J Neurophysiol 60:2180–2195
233. Schaible H-G, Schmidt RF (1988) Excitation and sensitization of fine articular afferents from cat's knee joint by prostaglandin E2. J Physiol (Lond) 403:91–104

234. Schaible H-G, Schmidt RF, Willis WD (1987) Convergent inputs from articular, cutaneous and muscle receptors onto ascending tract cells in the spinal cord. Exp Brain Res 66:479–488
235. Schaible H-G, Schmidt RF, Willis WD (1987) Enhancement of the responses of ascending tract cells in the cat spinal cord by acute inflammation of the knee joint. Exp Brain Res 66:489–499
236. Schepelmann K, Messlinger K, Schaible HG, Schmidt RF (1992) Inflammatory mediators and nociception in the joint: excitation and sensitization of slowly conducting afferent fibers of cats knee by prostaglandin I2. Neuroscience 50:237–247
237. Schmidtbleicher D, Gollhofer A, Frick U (1988) Effects of a stretch-shortening typed training on the performance capability and innervation characteristics of leg extensor muscles. In: de Groot G, Hollander AP, Huijing PA, van Ingen Schenau GJ (eds) International series on biomechanics. Biomechanics XI-A. Free University Press, Amsterdam, pp 185–189
238. Schutte MJ, Dabezies EJ, Zimny ML, Happel LT (1987) Neural anatomy of the human anterior cruciate ligament. J Bone Joint Surg Am 69:243–247
239. Scott DT, Ferrell WR, Baxendale RH (1994) Excitation of soleus/gastrocnemius γ-motoneurones by group II knee joint afferents is suppressed by group IV joint afferents in the decerebrate, spinalized cat. Exp Physiol 79:357–364
240. Seto JL, Orofino AS, Morrissey MC, Medeiros JM, Mason WJ (1988) Assessment of quadriceps/hamstring strength, knee ligament stability, functional and sports activity levels five years after anterior cruciate ligament reconstruction. Am J Sports Med 16:170–180
241. Sherrington CS (1894) On the anatomical constitution of nerves of skeletal muscles: with remarks on recurrent fibres in the ventral spinal nerve root. J Physiol (Paris) 17:211–258
242. Sherrington CS (1906) The integrative action of the nervous system. Constable, London
243. Sinkjaer T, Toft E, Andreassen S, Hornemann BC (1988) Muscle stiffness in human ankle dorsiflexors: intrinsic and reflex components. J Neurophysiol 60:1110–1121
244. Sinkjaer T, Arendt-Nielsen L (1991) Knee stability and muscle coordination in patients with anterior cruciate ligament injuries: an electromyographic approach. J Electromyogr Kinesiol 1:209–217
245. Sinkjaer T, Hayashi R (1989) Regulation of wrist stiffness by the stretch reflex. J Biomech 22:1133–1140
246. Sittig AC, Denier van der Gon JJ, Gielen CCAM (1987) The contribution of afferent information on position and velocity to the control of slow and fast human forearm movements. Exp Brain Res 67:33–40
247. Sjölander P (1989) A sensory role for the cruciate ligaments. Regulation of joint stability via reflexes onto the γ-muscle-spindle system. Umeå Univerity Medical Dissertaions, no 245, pp 1–48.
248. Sjölaner P, Djupsjöbacka M, Johansson H, Sojka P, Lorentzon R (1994) Can receptors in the collateral ligaments contribute to knee joint stability and proprioception via effects on the fusimotor-muscle-spindle system? An experimental study in the cat. Neuro Orthop 15:65–80
249. Sjölander P, Johansson H (1995) Influences on the γ-muscle spindle system from joint mechanoreceptors. In: Taylor A, Gladden M, Durbada R (eds) Alpha and gamma motor systems. Plenum, New York, pp 137–144
250. Sjölander P, Johansson H (1995) Ipsi- and contralateral reflexes on individual static and dynamic γ-motoneurones of the cat. Neurosci Lett (submitted)
251. Skoglund S (1956) Anatomical and physiological studies of knee joint innervation in the cat. Acta Physiol Scand 36:1–101
252. Snyder-Mackler L, De Luca PF, Williams PR, Eastlack ME, Bartolozzi AR (1994) Reflex inhibition of the quadriceps femoris muscle after injury or reconstruction of the anterior cruciate ligament. J Bone Joint Surg Am 76:555–560
253. Snyder-Mackler L, Ladin Z, Schepsis AA, Young JC (1991) Electrical stimulation of the thigh muscles after reconstruction of the anterior cruciate ligament. Effects of electrically elicited contraction of the quadriceps femoris and hamstring muscles on gait and on strength of the thigh muscles. J Bone Joint Surg Am 73A:1025–1036

254. Sojka P, Johansson H, Sjölander P, Lorentzon R, Djupsjöbacka M (1989) Fusimotor neurones can be reflexly influenced by activity in receptor afferents from the posterior cruciate ligament. Brain Res 483:177–183
255. Sojka P, Sjölander P, Johansson H, Djupsjöbacka M (1991) Influence from stretch-sensitive receptors in the collateral ligaments of the knee joint on the γ-muscle-spindle systems of flexor and extensor muscles. Neurosci Res 11:55–62
256. Solomonow M, Baratta R, Zhou BH, Shoji H, Bose W, Beck C, D'Ambrosia R (1987) The synergistic action of the anterior cruciate ligament and thigh muscles in maintaining joint stability. Am J Sports Med 15:207–213
257. Stein RB (1982) What muscle variable(s) does the nervous system control in limb movements? Behav Brain Sci 5:537–577
258. Stener B (1959) Experimental evaluation of the hypothesis of ligamento-muscular protective reflexes. Acta Physiol Scand 48:5–26
259. Stener B, Petersén I (1962) Electromyographic investigation of reflex effects upon stretching the partially ruptured medial collateral ligament of the knee joint. Acta Chir Scand 124:396–415
260. Tegner Y, Lysholm J, Gillquist J, Öberg B (1984) Two-year follow-up of conservative treatment of knee ligament injuries. Acta Orthop Scand 55:176–180
261. Tegner Y, Lysholm J, Lysholm M, Gillquist J (1986) Strengthening exercises for old cruciate ligament tears. Acta Orthop Scand 57:130–134
262. Tibone JE, Antich TJ, Fanton GS, Moynes DR, Perry J (1986) Functional analysis of anterior cruciate ligament instability. Am J Sports Med 14:276–284
263. Toft E, Sinkjaer T, Andreassen S, Larsen K (1991) Mechanical and electromyographic responses to stretch of the human ankle extensors. J Neurophysiol 65:1402–1410
264. Tropp H (1985) Functional instability of the ankle joint. Linköping University, disseration no 202
265. Vilensky JA, O'Connor BL, Brandt KD, Dunn EA, Rogers PI (1994) Serial kinematic analysis of the trunk and limb joints after anterior cruciate ligament transection: temporal, spatial, and angular changes in a canine model of osteoarthritis. J Electromyogr Kinesiol 4:181–192
266. Voorhoeve PE, van Kanten RW (1962) Reflex behaviour of fusimotor neurones of the cat upon electrical stimulation of various afferent fibers. Acta Physiol Pharmacol Neerlandica 10:391–407
267. Wadell I, Johansson H, Sjölander P, Sojka P, Djupsjöbacka M, Niechaj A (1991) Fusimotor reflexes influencing secondary muscle spindle afferents from flexor and extensor muscles in the hind limb of the cat. J Physiol (Paris) 85:223–234
268. Wadman WJ (1979) Control of fast goal-directed arm movements. J Hum Mov Stud 5:3–17
269. Wall PD (1975) Signs of plasticity and reconnection in spinal cord damage. Ciba Found Symp 34:35–63
270. Wall PD, Egger MD (1971) Formations of new connexions in adult rat brains after partial deafferentation. Nature 232:542–545
271. Watson JDG, Colebath JG, McCloskey DI (1984) Effects of externally imposed elastic loads on the ability to estimate position and force. Behav Brain Res 13:267–271
272. Wojtys EM, Huston LJ (1994) Neuromuscular performance in normal and anterior cruciate ligament-deficient lower extremities. Am J Sports Med 22:89–104
273. Wolpaw JR (1985) Adaptive plasticity in the spinal stretch reflex: an accessible substrate of memory? Cell Mol Neurobiol 5:147–165
274. Wood L, Ferrell WR (1984) Response of slowly adapting articular mechanoreceptors in the cat knee joint to alterations in intra-articular volume. Ann Rheum Dis 43:327–332
275. Wyke B (1981) The neurology of joints: a review of general principles. Clin Rheum Dis 7:223–239
276. Zimny ML (1988) Mechanoreceptors in articular tissues. Am J Anat 182:16–32
277. Zätterström R, Fridén T, Lindstrand A, Moritz U (1994) The effect of physiotherapy on standing balance in chronic anterior cruciate ligament insufficiency. Am J Sports Med 22:531–536

278. Öhberg F, Johansson H, Bergenheim M, Pedersen J, Djupsjöbacka M (1996) A neural network approach to real-time spike discrimination during simultaneous recording from several multiunit nerve filaments. J Neurosci Methods 64:181–187
279. Özaktay AC, Cavanaugh JM, Blagoev DC, Getchell TV, King AI (1994) Effects of a carrageenan-induced inflammation in rabbit lumbar facet joint capsule and adjacent tissues. Neurosci Res 20:355–364

I.3 Morphological Evaluation of Bone-Ligament Interaction in Biological and Artificial Reconstruction

A. Schiavone Panni, S. Franzese, M. Denti, and C. Fabbriciani

Introduction

The junction between ligament and bone is one of the most complex of all biological structures. It is well known that there is a transition through the interposed cartilage from soft ligament to hard bone within a space of 1 mm [5, 9, 10]. This gradual change from soft to hard tissue serves to reduce the concentration of stresses and makes the area particularly resistant to damage. Nevertheless this region is a critical area during loading [21, 22, 33] because the bone-ligament junction has a very important function in the distribution of mechanical loads applied on the ligament to diminish stress concentration and shearing at the interface [2, 24, 30]. Butler [6] and Noyes [22] observed that the strain at the zones of tibial and femoral insertion is three or four times greater than that at the center of the ligament. Classically the bone-ligament junction has been divided into four distinct zones: ligament, fibrocartilage, calcified fibrocartilage, and bone [5, 9]. Frequently the microstructures of these insertion sites or, sometimes, the two ends of the same ligament, differ from each other. The insertion of ligaments into bone is usually direct, whereas the tibial insertion of the medial collateral ligament, such as that of the teres minor tendon, is indirect. There are marked microscopic differences between these two types of insertions.

Insertion

Both tendon and ligament fibers forming the insertion site can be divided into superficial and deep fibers. Superficial fibers are generally a continuation of the periosteum, whereas the deep fibers, which are usually more numerous, join the bone at right angles to form four zones [5, 9, 10] (Fig. 1).

Zone 1 consists of tendon and ligament containing type I collagen surrounded by an extracellular matrix made up of proteoglycans, particularly dermatan sulfate [15]. Collagen fibers (25–300 nm) intermingled with elastin fibers bear the greatest tensile forces and are accompanied in parallel by capillaries, arterioles, and venules. Fibroblasts are virtually the only cells in this zone.

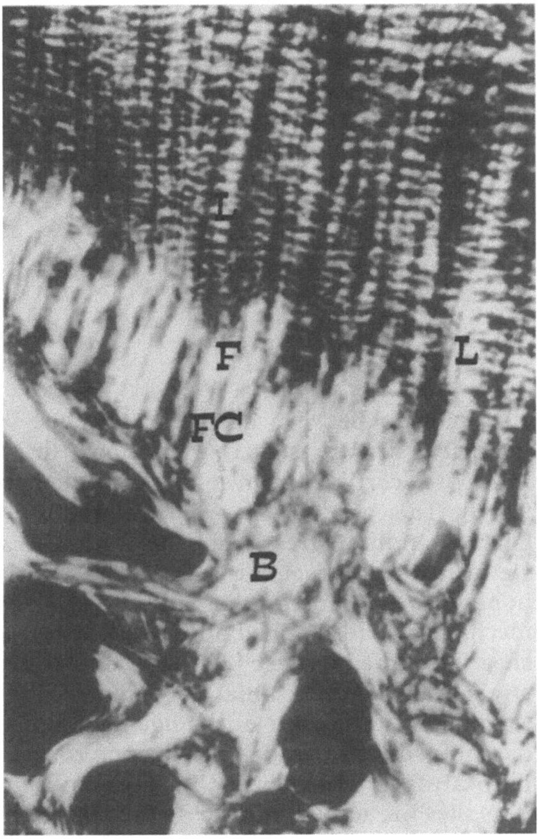

Fig. 1. Physiological junction; polarized light. The junction is composed of four zones: ligament (*L*), fibrocartilage (*F*), mineralized fibrocartilage (*FC*), and bone (*B*). The Sharpey fibers link ligament to bone

Zone II consists of fibrocartilage. Collagen fibrils resemble those in zone 1. The main difference between zone I and zone II is that the cells change from thin, flat fibroblasts to wider oval and round cells resembling chondrocytes that are often paired or in the form of a crown. Although these characteristics define the morphology of chondroid cells, the fibrous matrix in which these cells are embedded characterizes them as fibrochondrocytes. The transition from the ovoid fibrocytes of the ligament to the fibrochondrocytes of the insertion is gradual, and the demarcation between these two zones is indistinct. The fibrochondrocytes are arranged in columns containing different numbers of cells, the fibrochondrocytes being more numerous on the side of the line of mineralization. The collagen fiber bundles gradually blend to meet the mineralization front of the insertion. It has been hypothesized that the pericellular matrix of this zone supports a smooth bending of the fibers, so they penetrate into the mineralized fibrocartilage at angles almost perpendicular to the mineralization front [5, 34], probably reducing the mechanical stress of the collagen fibers as they pass from nonmineralized to mineralized tissue. No vascular or neural elements have been observed in this zone [1].

Zone III is composed of calcified fibrocartilage. A distinct basophilic "blue line" or "tide-mark" separates it from zone II and marks the mineralization (calcification) boundary. This is usually smooth, but may be irregular (Fig. 2). The transition from zone II to zone III tissue is very rapid. There are very few mineral crystals at the beginning of zone III, but these increase in number and size lower down, forming clumps that settle between and within the collagen fibers and make them hard to distinguish. Fibrocartilage cells are still active in this zone despite their inclusion in lacunae surrounded by a mineralized matrix (Fig. 3).

Zone IV consists of bone, its matrix proteoglycan being mainly chondroitin sulfate. Here the collagen of the inserting tissue blends imperceptibly with the collagen of the bone matrix fibers. There is no separating layer between the two. Using polarized light it is possible to observe the Sharpey fibers that firmly connect bone and ligament.

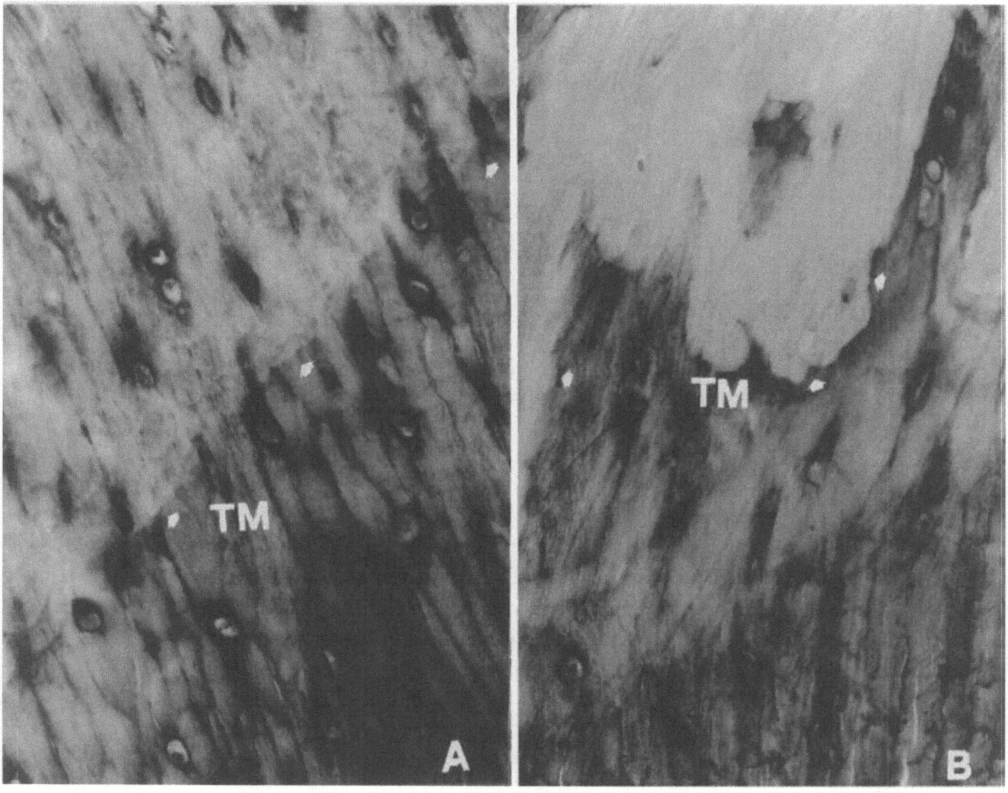

Fig. 2A,B. Physiological junction; light microscopy. Between mineralized and nonmineralized fibrocartilage there is a tide-mark (*TM*). This "blue line" can be smooth in contour (**A**) or very irregular (**B**)

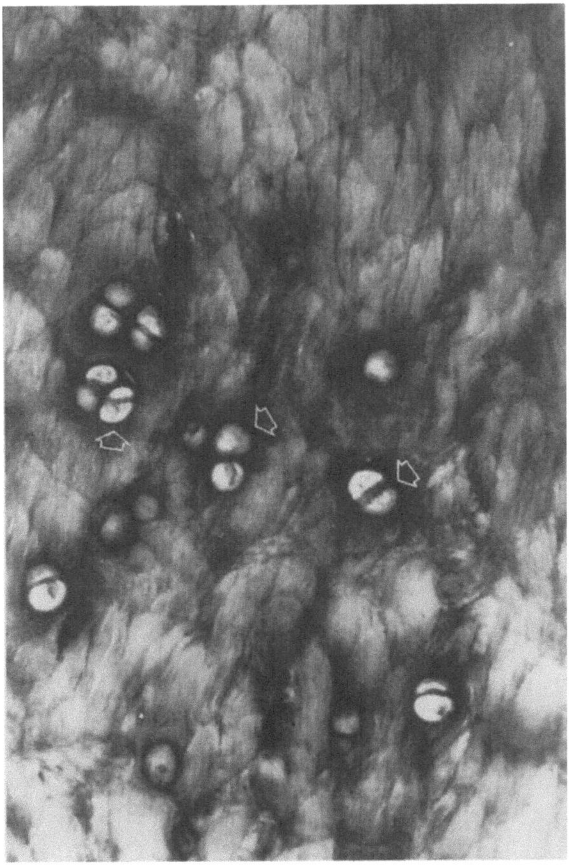

Fig. 3. Physiological junction; light microscopy. High magnification of zone III (calcified fibrocartilage); the fibrocondrocytes are paired and are surrounded by mineralized matrix (*arrows*)

Direct Insertions

Direct insertions are made up of superficial and deep fibers [5, 34]. The superficial fibers predominate and blend with the periosteum to insert into the bone. The Sharpey fibers (deep), which perforate the underlying bone, may also play an important role in periosteal attachment to bone and anchor the periosteum to the lamellar bone (Fig. 1).

Indirect Insertions

In indirect insertions the deep fibers attach to bone with little or no transitional zone of fibrocartilage [5, 34]. A tide-mark still exists between the mineralized and nonmineralized tissue.

This tide-mark separates the soft tissue fibers from bone without fibrocartilage.

Biomechanics of Insertion Sites

Forces are transmitted between bone and soft tissue by the insertion site and as is well known this function is very important in joint kinetics and kinematics. The insertions serve to diminish and dissipate stresses, and therefore lesion at the bone-ligament junction is uncommon. The insertion site is wider than the ligament (about two to four times wider than the ligament substance), and the forces therefore are distributed over a larger area [14]. Moreover, the composite material of the insertion provides a gradual increase in stiffness from ligament to bone, thus causing a diminution of stress concentration and tearing or shearing at the interface [21, 22, 24, 33]. However, it is very difficult to evaluate the tensile deformation of the insertion site with any accuracy because of the difficulty encountered in measuring irregular shapes and geometry and cross section. Another problem is the tensile deformation analysis method; generally the mechanical properties are resumed by analyzing the surface behavior, but the insertion is not an isotropic body so the surface behavior may be different from that of the deeper layer. Hence insertion site deformation can only be estimated and not measured. Using markers applied to the insertion sites and midsubstance, Woo and coworkers [33] demonstrated that the strain at failure of the medial collateral ligament is 14% in dogs, 12% in swine, and 7% in rabbits, whereas the strain at failure measured for the complex "bone-insertion-medial collateral ligament-insertion-bone" is 21%, 30%, and 16%, respectively. Therefore we can hypothesize that the strain at the insertion site is twofold or more the strain at midsubstance [24, 35]. Data have been obtained by analyzing human anterior cruciate ligament (ACL) and posterior cruciate ligament (PCL) elongation behavior [6, 7, 29]. The strain at the insertion site in human ACL complex (bone-ACL-bone) is three or four times higher than at midsubstance [33]. Thus we can conclude that the bone-ligament junction bears stresses better than the ligament substance. The reason for this peculiar tensile response of the insertion is unknown, but we can posit the following possibilities:

- Hypothesis I: there is a different collagen-fiber crimp period and angle; probably the periods are smaller and the angles larger at the insertion site; therefore, when a load is applied, the strain is larger at insertion sites [7].
- Hypothesis II: the proteoglycans and the number of elastic fibers may explain the tensile behavior of the bone-ligament junction.

Further investigation is needed to confirm these hypotheses. The data currently available, however, suggest that the insertions are very important to joint function, and thus, if possible, in knee surgery the bone-ligament junction must either be preserved or its reformation be favored.

Fig. 4

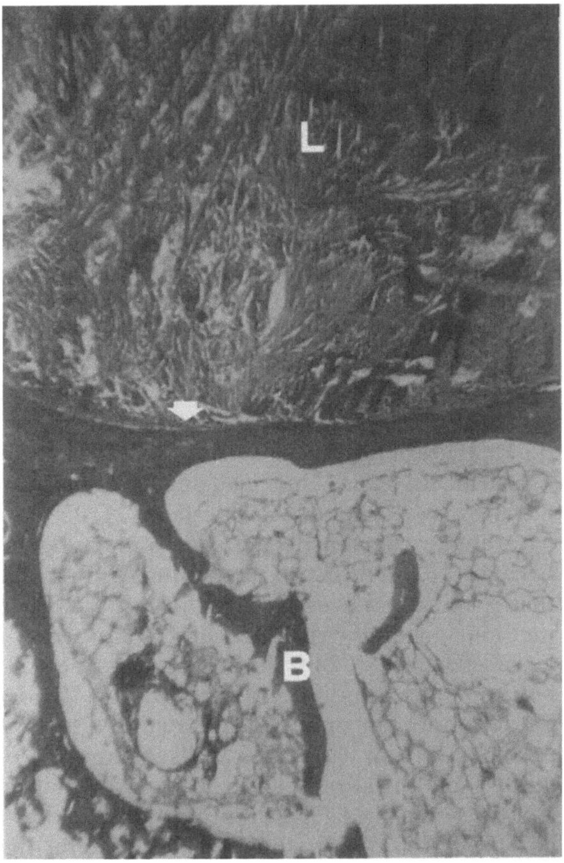

Fig. 5 ▼

Effect of Mobilization on Insertion Sites

It is well known that the postoperative formation of the junction is facilitated by a stable fixation that allows early joint mobilization, whereas biomechanical studies on animals [19, 23] have demonstrated a rapid reduction of bone-ligament-bone strength after immobilization. In fact, as demonstrated by Woo, after immobilization there is a weakening of the ligament, especially in indirect (periosteal) insertions due to bone reabsorption and destruction of the periosteal fibers [34]. Other authors [17, 30] also emphasize the beneficial effects on the junction of exercise and mobilization.

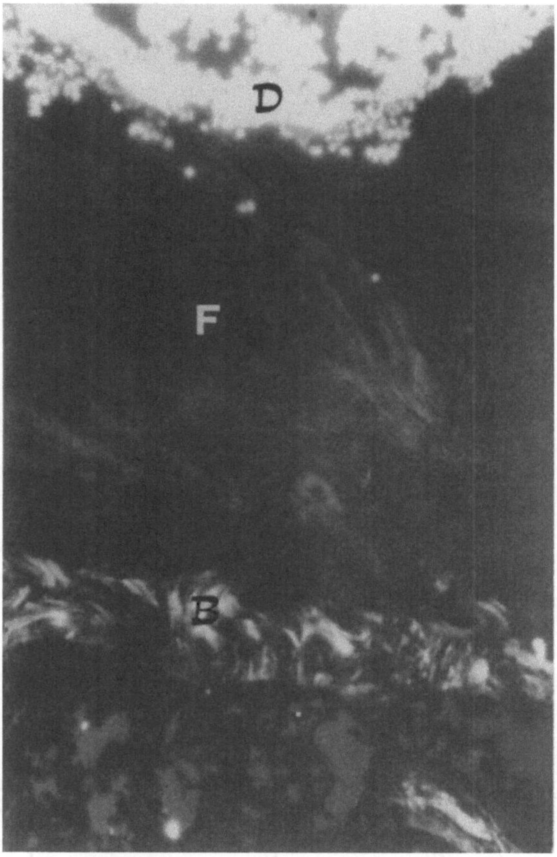

Fig. 6. Dacron prosthesis; polarized light. At 9 months there is only unoriented fibrous tissue (*F*) between bone (*B*) and prosthesis (*D*)

Fig. 4. Patellar tendon; light microscopy. At 2 months there is a thin layer of fibrocartilage between PT (*L*) and bone (*B*)

Fig. 5. Patellar tendon; light microscopy. At 6 months the junction is similar to the physiological insertion

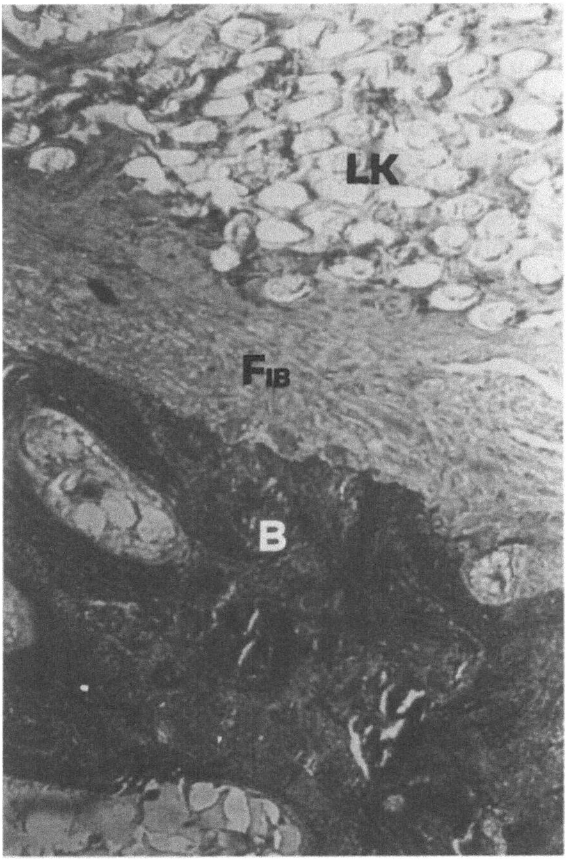

Fig. 7. Leeds-Keio scaffold; Light microscopy. At 2 months the scaffold (*L-K*), is surrounded by a layer of fibrous tissue (*Fib*) that penetrates the Leeds-Keio and separates the single bundles. *B*, Bone

Experimental Studies

There are few investigations in the literature concerning the bone-ligament junction and the results are sometimes discordant. A recent study in rabbits showed that after ACL reconstruction with patellar tendon (PT), regarded by many surgeons as "the gold standard," it is possible to form a junction that is morphologically similar to a physiological junction [26] and that 4 months are necessary for its complete reformation (Figs. 4, 5). Moreover, the study showed that in the physiological insertion of the ACL and the tibial collateral ligament four zones can be clearly identified. Whether regular (Fig. 2) or very irregular (Fig. 2), the tide-mark always borders the passage from zone II to zone III. Furthermore, the above study demonstrates in a cadaver evaluation that in the indirect insertion, for instance the tibial insertion of the tibial collateral ligament, there is only a thin layer of fibrocartilage between bone and ligament, while most of the ligament fibers are continuous with the periosteum.

In another experimental study in sheep [27] a prosthesis (Dacron), a scaffold (Ledds-Keio), a patella tendon augmented with Ligament Augmentation Device (LAD) and a PT were implanted as ACL substitutes. This study demonstrated the following.

- When a Dacron prosthesis is used, only amorphous fibrous tissue surrounds the neoligament and completely separates it from the bone tunnels (Fig. 6). (The unoriented fibrous tissue distribution, that resembles "distance osteogenesis" around a cemented hip prosthesis, is caused by an absence of mechanical stress due to the stress shielding effect of the prosthesis [4]).
- When the Leeds-Keio is used, the neoligament is surrounded only by fibrous tissue which penetrates the ligament without any signs of inflammation (Fig. 7).

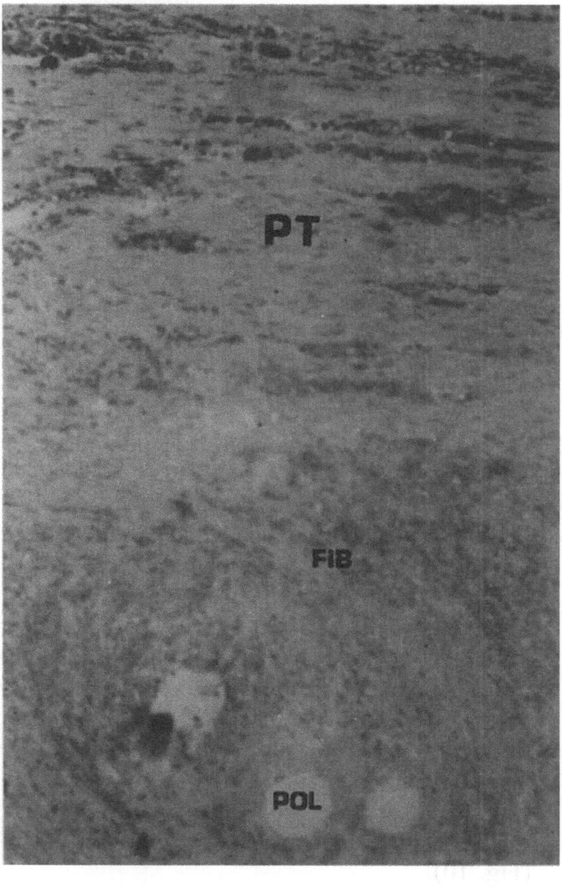

Fig. 8. Patellar tendon augmented with LAD; light microscopy. At 6 months an amorphous fibrous tissue (*Fib*) separates the LAD (*Pol*) from the PT (PT different orientation). Note the ingrowth of fibrous tissue into the LAD

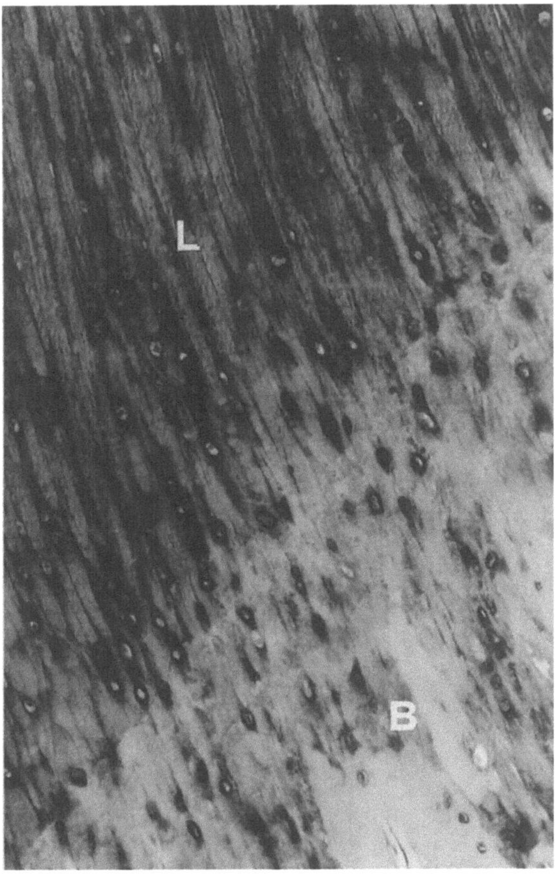

Fig. 9. Patellar tendon, light microscopy. At 9 months the new junction is indistinguishable from the physiological insertion. *L*, patellar tendon; *B*, bone

- With PT augmented with LAD the neoligament is inserted into the bone tunnel by way of a layer of fibrocartilage resembling a physiological bone-ligament insertion, whereas a conspicuous ingrowth of fibrous tissue is noted between the LAD and PT and around the polypropylene bundles [11, 20] (Fig. 8). This underscores the importance of the tubulation, not only of the intraarticular but also of the extraarticular portion of the LAD.
- When the PT is used in ACL reconstruction, the junction is almost indistinguishable from a physiological insertion and all layers have been formed: ligament, fibrocartilage with tide-mark, calcified fibrocartilage and bone (Fig. 9). Moreover, the study showed that free ending and Pacinian corpuscles can be observed only in PT and PT+LAD ACL reconstructions (Fig. 10).

Other authors studies [3, 31] also report that only fibrous tissue forms around artificial ligaments, with very little bone ingrowth in the more per-

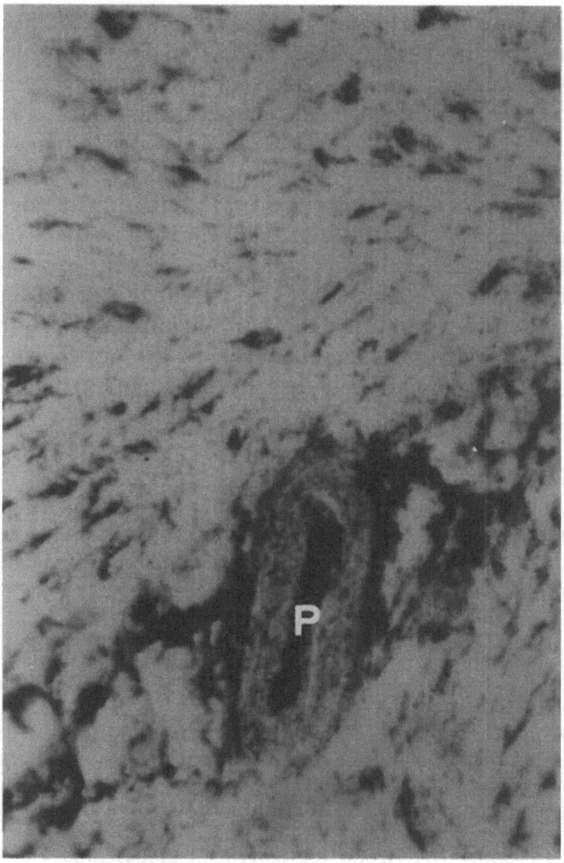

Fig. 10. Pacinian corpuscles, TEM. In the patellar tendon after 9 months Pacinian corpuscles (*PC*) can be observed

ipheral parts. In rabbit a thin layer of fibrocartilage between tendon and ligament was found by Jones et al. [16] 2 months after tendon implantation into bone. Some authors [1, 2, 25, 32] report the formation of an almost physiological bone-ligament junction 52 weeks after surgery, whereas most authors [8, 12, 13, 18, 22, 28] report the same result after about 20 weeks. These findings can explain the high percentage of artificial ligament reconstruction failures that can be attributed to the fact that nonelastic fibrous tissue cannot "cushion" the stresses to which a ligament is exposed and therefore its rupture is easier. Moreover, they provide further evidence that the PT forms a "physiological" junction in tibial and femoral tunnels after 4 months. This would confirm the superiority of the PT and biological tissues over synthetic ligaments in ACL reconstruction. In fact after several months artificial structures are still separated from bone and tendon by fibrous tissue [3, 31] that lacks the tensile characteristics of physiological insertion and is thus unable to allow optimum distribution of forces to which the bone and ligament are subjected. Furthermore, fixation of an artificial ligament is achieved only with staples and screws, since the fibrous tissue surrounding the ligament and attaching it to the bone cannot absorb and distribute the forces acting on it.

The PT reestablishes an almost physiological insertion into the tunnel that ensures optimal distribution of forces and their discharge over a larger surface, thus protecting both the bone and the ligament from excessive strain. Furthermore, it should be remembered that the strength of both the new insertion and the new ligament increases with time and that their partial damage may be followed by spontaneous repair. In our opinion, the fact that the PT forms a structure resembling the bone ligament insertion capable of further improving tensile response again confirms the superiority of biological reconstruction over that obtained with artificial ligaments.

References

1. Arnoczky SP, Rubin RM, Marshall JL (1979) Microvascolature of cruciate ligaments and its response to injury: an experimental study in dogs. J Bone Joint Surg Am 61:1221–1229
2. Arnoczky SP, Tarvin GB, Marshall JL (1982) Anterior cruciate ligament replacement using patellar tendon; an evaluation of graft revascularization in the dog. J Bone Joint Surg Am 64:217–224
3. Arnoczky SP, Warrem F, Minei JP (1986) Replacement of anterior cruciate ligament using a syntetic prosthesis. An evaluation of graft biology in the dog. Am J Sports Med 14(1):1–6
4. Arnoczky SP, Torzilli P, Warren R, Allen A (1988) Biological fixation of ligament prosthesis and augmentation: an evaluation of bone ingrowth in the dog. Am J Sports Med 16:106–112
5. Benjamin M, Evans EJ, Cooper L (1986) The histology of tendon attachments to bone in man. J Anat 149:89–100
6. Butler DL, Grood ES, Noyes FR (1984) Effects of structure and strain measurements technique on the material properties of young human tendon and fascia. J Biomech 17:579–596
7. Butler DL, Stouffer DC, Wukusick PM (1983) Analysis on nonhomogeneus strain response of human patellar tendon. Presented at the American Society of Mechanical Engeneers Joint Biomechanics Symposium, Houston, Texas
8. Clancy WG, Narechania RG, Rosenberg TD (1981) Anterior and posterior cruciate ligament reconstruction in rhesus monkeys: a histological microangiogrphic, and biomechanical analysis. J Bone Joint Surg Am 63:1270–1284
9. Cooper RR, Misol S (1960) Tendon and ligament insertion: a light and electron microscopic study. J Bone Joint Surg Am 52:1–21
10. Dolgo-Saburoff B (1929) ber Ursprung und Insertion der Skelettmuskeln: Mikroskopische Untersuchung. Anat Anz 68:30–87
11. Engebresten L, Low W, Hunter L, Hunter R (1989) Knee mechanics aftert repair of the anterior cruciate ligament – a cadaver study of ligament augmentation. Acta Orthop Scand 60:703–709
12. Forward AD, Cowan RJ (1963) Tendon suture to bone: an experimental investigation in rabbits. J Bone Joint Surg Am 45:807–823
13. Fukubayashy T, Kondo H, Nakajima M (1980) Experimental repair of medial collateral ligament of knee in rabbits. J Jpn Orthop Assoc 54:59–69
14. Grood ES, Butler DL, Noyes FR (1985) Models of ligament repairs and graft. In: Funk FJ (ed) American Academy of Orthopaedic Surgeons Symposium on Sport Medicine: the knee. Mosby, St Louis, pp 169–181

15. Heinegaard D, Paulsson M (1984) Structure and metabolism of proteoglycans. In: Pier KA, Reddi AH (eds) Extracellular matrix biochemistry. Elsevier, New York, pp 278–328
16. Jones JR, Smibert JG, Mccullough CJ, Price AB, Wutton WC (1987) Tendon implantation into bone: an experimental study. J Hand Surg [Br] 12:306–316
17. Kernwein GA (1942) A study of tendon implantations into bone. Surg Gynecol Obstet 75:794–796
18. McPherson GK, Mendenhall HV, Gibbons DF (1985) Experimental mechanical and histological evaluation of the Kennedy ligament augmentation. Clin Orthop 196:186–195
19. Laros GS, Tripton CM, Cooper RR (1971) Influence of physical activity on ligament insertions in the knee of dogs. J Bone Joint Surg Am 53:275–286
20. Mendenhall HV, Roth JH, Kennedy JC, Winter GD (1987) Evaluation of the polypropilene braid as prosthetic anterior cruciate ligament replacement in the dog. Am J Sports Med 15:543–546
21. Noyes FR, DeLucas F, Torvik PJ (1974) Biomechanics of anterior cruciate ligament failure: an analysis of strain-rate sensitivity and mechanism of failure in primates. J Bone Joint Surg Am 56:236–253
22. Noyes FR, Grood ES, Butler DL (1980) Clinical biomechanics of the knee-ligament restrains and functional stability. In: Funk FJ (ed) American Academy of Orthopaedic Surgeons Symposium on the Athlete's Knee: surgical repair and reconstruction. Mosby, St Louis, pp 1–55
23. Noyes FR, Grood ES (1976) The strength of anterior cruciate ligament in human and rhesus monkeys: age-related and species-related changes. J Bone Joint Surg Am 58:1074–1082
24. Noyes FR, Butler DL, Grood ES (1984) Biomechanical analysis of human ligament graft used in knee-ligament repairs and reconstructions. J Bone Joint Surg Am 66:344–352
25. O'Donoghue DH, Rockwood CA, Frank GR (1966) Repair of the anterior cruciate ligament in dogs. J Bone Joint Surg Am 48:503–519
26. Schiavone Panni A, Fabbriciani C, Delcogliano A, Franzese S (1993) The bone ligament interaction in ACL reconstruction with patellar tendon, knee surgery, sport traumatology, arthroscopy, vol 1. Springer, Berlin Heidelberg New York, pp 4–8
27. Schiavone Panni A, Denti M, Monteleone M, Franzese S (1993) The bone ligament junction comparison between biological and artificial ACL reconstruction with patellar tendon, knee surgery, sport traumatology, arthroscopy, vol 1. Springer, Berlin Heidelberg New York, pp 9–12
28. Shino K, Kawasaki T, Hirose H (1984) Replacement of anterior cruciate ligament by an allogenic tendon graft: an experimental study in the dog. J Bone Joint Surg Br 66:672–681
29. Stouffer DC, Butler DL, Hosny D (1985) Relationship between crimp pattern and mechanical response of human patellar tendon-bone units. J Biomech Eng 107:158–165
30. Tipton CM, Matheus RD (1974) In situ measurement of junction strength and ligament elongation in rats. J Appl Physiol 7:758–761
31. Thomas NP, Turner IG, Jones CB (1987) Prosthetic anterior cruciate ligaments in the rabbit: a comparison of four types of replacement. J Bone Joint Surg Br 69:312–316
32. Whiston TB, Walmsley R (1960) Some observation on reaction of bone and tendon after tunnelling of bone and insertion of tendon. J Bone Joint Surg Br 42:377–386
33. Woo SL-Y, Gomez MA (1983) Measurement of mechanical properties of ligament substance from a bone-ligament-bone preparation. J Orthop Res 1:22–29
34. Woo SL-Y, Gomez MA, Sites TJ (1987) The biomechanical and morphological changes in the medial collateral ligament of rabbit after immobilization and rimobilization. J Bone Joint Surg Am 69:1200–1211
35. Zernicke RF, Butler D, Grood ES (1984) Strain topography of human tendon and fascia. J Biomech Eng 106:177–180

I.4 Mesomechanical Modeling of Living Soft Tissues

V. Kafka and J. Jírová

Introduction

Mathematical modeling of the mechanical properties of living tissues is usually attacked either on the macroscale or on the microscale. The macroscopic phenomenological approach involves looking for suitable mathematical formulae that describe the observed macroscopic properties. However, when inelastic properties are considered, this is far from easy, and even in the event of some success it still has the drawback that the model provides no information as to that which occurs inside the tissue. On the other hand, the drawback of the micromechanical approach is that the microstructure of living tissues is so complicated that it is very difficult to describe the microscopic geometry and its changes and to correlate them with the macroscopic properties.

Therefore, following the viewpoint of the proverb *via media via optima* we have come to the conclusion that for living tissues the mesoscopic approach is very suitable (on the general concept see [1]). This concept does not go into detail in describing the microstructure; it describes the microstructure in a specific, averaged way by structural parameters. By definition these parameters are integral forms in the distribution functions descriptive of the microstructure; they are not to be determined by microscopic experimental methods, however, but by a mathematical analysis of the macroscopic stress-strain curves or flow-curves that can easily be measured on samples of the tissues in question. The structural parameters are nonnegative and combinations of their finite, vanishing or infinite values describing inclusions, matrix, or continuous infrastructures, or homogeneous stress fields (Reuss' model) or homogeneous strain fields (Kelvin's model). Kafka [1] applied the concept only to the processes and materials where the structural parameters remained unchanged, i.e., to processes with statistically stable structure. However, modeling living tissues requires in some cases taking into account changes in the structural parameters, such as straightening of fibers means substantial changes of the structure. These changes do not only have a local character but substantially influence the overall statistical description of the structure.

In our concept the deformation processes are described not only by the apparent macroscopic variables, which are not sufficient for inelastic deformation, but also by tensorial latent variables that have clear physical meaning: these represent statistical averages of microstresses in the material constituents.

To complete the general characterization of our approach it is important to emphasize that in our model the stresses and strains in individual material constituents are described not only by their averages in these constituents but also by other tensorial quantities that describe the influence of their fluctuations. It is shown in [1] that the influence of fluctuations is by no means negligible.

Symbolic Notation

$\overline{\sigma_{ij}}(\overline{\varepsilon_{ij}})$	macroscopic stress (strain)
$\sigma_{ijf}/\sigma_{ijm}(\varepsilon_{ijf}/\varepsilon_{ijm})$	mesoscopic stress (strain) in the fibers/matrix - averaged values in the respective material constituent
$\overline{s_{ij}}(s_{ijf}, s_{ijm})$	deviatoric stress - macroscopic (mesoscopic)
$\overline{e_{ij}}(e_{ijf}, e_{ijm})$	deviatoric strain - macroscopic (mesoscopic)
$\delta_{ij}\overline{\sigma}(\delta_{ij}\sigma_f, \delta_{ij}\sigma_m)$	isotropic part of the stress tensor, macroscopic (mesoscopic)
δ_{ij}	Kronecker's delta
ε'_{ijf}, ε'_{ijm}	defined by Eqs. 3, 4
e'_{ijf}, e'_{ijm}	deviatoric parts of ε'_{ijf}, ε'_{ijm}
$\delta_{ij}\varepsilon'_f$, $\delta_{ij}\varepsilon'_m$	isotropic parts of ε'_{ijf}, ε'_{ijm}
σ'_{ijf}, σ'_{ijm}	stresses descriptive of the influence of fluctuations (for details see [1]), implicitly defined by Eqs. 5, 8, 9
s'_{ijf}, s'_{ijm}	deviatoric parts of s'_{ijf}, s'_{ijm}
$\delta_{ij}\sigma'_f$, $\delta_{ij}\sigma'_m$	isotropic parts of σ'_{ijf}, σ'_{ijm}
v_f, v_m $(=1- v_f)$	volume fractions of the respective material constituents
η_f, η_m	nonnegative "structural parameters" (for details see Eq. 10 and Ref. [1])
μ_f, μ_m, ρ_f, ρ_m	elastic compliances corresponding to the deviatoric and isotropic parts, defined by Eq. 20
E_f, E_m, ν_f, ν_m	Young's moduli and Poisson's ratios of the respective material constituents
β_f, β_m	defined by Eqs. 21, 22
H_f, H_m	coefficients of viscosity of the respective material constituents

c_f , c_m	elastic limits defined by Eqs. 21, 22
$\bar{\mu}$	macroscopic deviatoric complience (see Eq. 28)
M, M′, N, A, B, C, D, R	expressions defined by Eq. 28
t	time
k_1, k_2, k_3, k	some formal constants the quantitative values of which are not and need not to be specified
a, b	constants that can be determined from the experimental stress-strain diagram

General Mathematical Model

It is clear that the mathematical models differ for different living tissues. However, it is possible to formulate a very general form and the special models can be derived from it as special cases.

Our general model is based on the assumption that it is possible to approximately describe many tissues as transversely isotropic two-phase materials. The two-material constituents are assumed to have different properties, but the general type of their constitutive equations is assumed to be the same: elastic properties in some limits and viscous deformation if these limits are exceeded. One of the material constituents is labeled by the index "m" (matrix), the other by the index "f" (fibers). The f constituent is always less compliant, but it does not describe fibers in all applications. To be concrete we assume the possibility of applying this general model to trabecular bone, tendons, ligaments, articular cartilage, and similar tissues.

On the basis of these assumptions the set of fundamental "user's equations" (see [1]) can be formulated as follows:

$$v_f \sigma_{ijf} + v_m \sigma_{ijm} = \overline{\sigma}_{ij} \tag{1}$$

$$v_f \varepsilon_{ijf} + v_m \varepsilon_{ijm} = \overline{\varepsilon}_{ij} \tag{2}$$

$$\varepsilon'_{ijf} = \varepsilon_{ijf} - \overline{\varepsilon}_{ij} \tag{3}$$

$$\varepsilon'_{ijm} = \varepsilon_{ijm} - \overline{\varepsilon}_{ij} \tag{4}$$

$$\sigma_{ijm} - \sigma_{ijf} + \sigma'_{ijm}/[\eta_m] - \sigma'_{ijf}/[\eta_f] = 0 \tag{5}$$

$$\dot{e}_{ijf} = \dot{e}_{ijf}(\dot{s}_{ijf}, s_{ijf}) \ , \ \varepsilon_f = \varepsilon_f(\sigma_f) \tag{6}$$

$$\dot{e}_{ijm} = \dot{e}_{ijm}(\dot{s}_{ijm}, s_{ijm}) \ , \ \dot{\varepsilon}_m = \dot{\varepsilon}_m(\sigma_m, \dot{\sigma}_m) \tag{7}$$

$$\dot{e}'_{ijf} = \dot{e}'_{ijf}(\dot{s}'_{ijf}, s'_{ijf}) \ , \ \varepsilon'_f = \varepsilon'_f(\sigma'_f) \tag{8}$$

$$\dot{e}'_{ijm} = \dot{e}'_{ijm}(\dot{s}'_{ijm}, s'_{ijm}) \ , \ \dot{\varepsilon}'_{m} = \dot{\varepsilon}'_{m}(\sigma'_{m}, \dot{\sigma}'_{m}) \tag{9}$$

The elementary symbols are explained above. Stresses and strains are measured simply as in classical elasticity assuming small strains.

Equations 1 and 2 are well known from the mechanics of composite or heterogeneous materials and follow from the conditions of equilibrium and continuity of displacements respectively. Equations 3 and 4 are to be understood simply as definitions of the symbols with primes.

Equation 5, deduced in [1], relates the average stresses in the material constituents with the tensorial quantities with primes that describe the influence of fluctuations. This relation depends on the structure of the tissue and therefore on the structural parameters $[\eta_f]$ and $[\eta_m]$ that differ for different tensorial directions as follows:

$$\begin{array}{llll} [\eta_f]=\eta_f & \text{and} & [\eta_m]=\eta_m & \text{for } ij=11 \\ [\eta_f]=\eta^+{}_f & \text{and} & [\eta_m]=\eta^+{}_m & \text{for } ij=22,23,33 \\ [\eta_f]=\eta^*{}_f & \text{and} & [\eta_m]=\eta^*{}_m & \text{for } ij=12,13 \end{array} \tag{10}$$

Equations 6 and 7 are constitutive equations of the two material constituents, written for stresses and strains averaged in the subvolumes of the constituents. Equations 8 and 9 are deduced in the quoted monograph [1] and together with Eq. 5 describe the influence of fluctuations of stresses and strains.

The system of nine tensorial Eqs. 1–9 corresponds to nine unknown rates of tensorial quantities:

$$\dot{\sigma}_{ijf}, \dot{\sigma}_{ijm}, \dot{\varepsilon}_{ijf}, \dot{\varepsilon}_{ijm}, \dot{\sigma}'_{ijf}, \dot{\sigma}'_{ijm}, \dot{\varepsilon}'_{ijf}, \dot{\varepsilon}'_{ijm}, \dot{\bar{\varepsilon}}_{ij}$$

supposing that $\dot{\bar{\sigma}}_{ij}$ is given.

The model of trabecular bone is presented in [2] and [3], and the related problem of hydraulic strengthening is discussed in [4]. Below we present the model of tendon; this discussion is based largely on our previous report [5].

Mesomechanical Model of Tendon

The mechanical function of tendons, observed on the scale of bodies, is clear enough, but it seems that when the scale shifts to that of tendons, and the structure is described by mathematical models, it is possible to find some finer features of the function resulting from biological optimization. Below we apply our general model to this problem.

Tendons, similarly as other biological soft tissues, are formed by a complex three-dimensional structure of wavy collagen fibers, elastin, proteoglycans, glycoproteins, and glycosaminoglycans. In connective tissues the content of elastin is usually very low, about 2% (see [6]), and therefore it is usual in micromechanical studies to consider tendon as a two-phase tissue composed of collagen fibers and viscoelastic matrix. A specific feature of tendons is that the bundles of collagen fibers are aligned in one direction, the fibers themselves being crimped.

A general macroscopic description of the mechanical behavior of tendon is presented, for example, by Fung [7]. A microscopic description of the tendon-structure and its changes in the course of deformation was given, for example, by Ault and Hoffman [8]. The characteristic features are:

- A nonlinear course of the stress-strain diagram (increase of tangent modulus) at the beginning of loading (stretching) followed by a linear part.
- Hysteresis loop in the cycle of loading and unloading.
- Significant creep under constant load and significant relaxation of stress under constant elongation.
- Dependence of the stress-strain diagram on the rate of deformation.
- Dependence of the instantaneous properties on the preceding deformation; after sufficient "preconditioning" they tend to become constant.
- Straightening of the crimped collagen fibers in the course of stretching.

Our approach takes into consideration all these specific features. The last one of these is described by a change of the structural parameter descriptive of the structure of the fibers.

The aim of our study was to correlate the real heterogeneous structure of tendon with its real macroscopic mechanical behavior, using our general concept, and to determine why tendons grow and form their microstructure as they do- Why are they optimized in this way? Tendon is modeled as a two-phase material consisting of fibers with changing infrastructure (straightening), and matrix. Both these material constituents can be looked upon as viscoelastic materials of the thixotropic type, i.e., behaving as elastic in some limits and viscoelastic if these limits are exceeded. The properties and hence the parameters of these two material constituents differ substantially.

The only loading process that we have in mind is uniaxial tension, the only nonvanishing macroscopic stress tensor component being $\overline{\sigma}_{11}$. Furthermore, we accept in such simplification that not only the macroscopic stress components in the orthogonal transverse directions but also the respective orthogonal microscopic and mesoscopic components are vanishing, i.e., they are neglected.

On the basis of these assumptions and the general scheme given in [1] the set of fundamental Eqs. 1–9 is transformed as follows:

$$v_f \sigma_{ijf} + v_m \sigma_{ijm} = \overline{\sigma}_{ij} \tag{11}$$

where $\sigma_{ijf} = \sigma_{ijm} = \overline{\sigma}_{ij} = 0$ for $ij \neq 11$

$$v_f \varepsilon_{ijf} + v_m \varepsilon_{ijm} = \overline{\varepsilon}_{ij} \tag{12}$$

$$\varepsilon'_{ijf} = \varepsilon_{ijf} - \overline{\varepsilon}_{ij} \tag{13}$$

$$\varepsilon'_{ijm} = \varepsilon_{ijm} - \overline{\varepsilon}_{ij} \tag{14}$$

$$\sigma_{11m} - \sigma_{11f} + \sigma'_{11m}/\eta_m - \sigma'_{11f}/\eta_f = 0 \tag{15}$$

$$\begin{aligned} \dot{e}_{11f} &= \mu_f \dot{s}_{11f} + \beta_f (s_{11f} - c_f) \\ \varepsilon_f &= \rho_f \sigma_f \end{aligned} \tag{16}$$

$$\begin{aligned} \dot{e}_{11m} &= \mu_m \dot{s}_{11m} + \beta_m (s_{11m} - c_m) \\ \varepsilon_m &= \rho_m \sigma_m \end{aligned} \tag{17}$$

$$\begin{aligned} \dot{e}'_{11f} &= \mu_f \dot{s}'_{11f} + \beta_f s'_{11f} \\ \varepsilon'_f &= \rho_f \sigma'_f \end{aligned} \tag{18}$$

$$\begin{aligned} \dot{e}'_{11m} &= \mu_m \dot{s}'_{11m} + \beta_m s'_{11m} \\ \varepsilon'_m &= \rho_m \sigma'_m \end{aligned} \tag{19}$$

The microscopic changes of structure (straightening of the crimped fibers) are not described by some displacements that would be large in relation to the microscale, but all the analysis remains on the macroscale and the changes of geometry of the fibers are described by the changes of the structural parameters that characterize the structure in a specific averaged, statistical way.

The elastic compliances in Eqs. 16–19 are defined by the well-known relations:

$$\begin{aligned} \mu_m &= (1 + \nu_m)/E_m \;, \quad \rho_m = (1 - 2\nu_m)/E_m \\ \mu_f &= (1 + \nu_f)/E_f \;, \quad \rho_f = (1 - 2\nu_f)/E_f \end{aligned} \tag{20}$$

E and ν being Young's moduli and Poisson's ratios of the respective constituents.

The thixotropic character of the material is characterized by coefficients β with the following definitions:

$$\begin{aligned} \beta_m &= 0 && \text{for } |s_{11m}| < c_m \\ \beta_m &= 1/(2H_m) && \text{for } |s_{11m}| \geq c_m \end{aligned} \tag{21}$$

$$\begin{aligned} \beta_f &= 0 && \text{for } |s_{11f}| < c_f \\ \beta_f &= 1/(2H_f) && \text{for } |s_{11f}| \geq c_f \end{aligned} \tag{22}$$

H_m and H_f being coefficients of viscosity of the respective material constituents.

For a further simplification of the analysis let us assume that the volumetric changes can be considered identical:

$$\varepsilon_m = \varepsilon_f = \bar{\varepsilon} \tag{23}$$

This approximate assumption is justified by the fact that in soft tissues the predominant differences in deformations lie in their deviatoric parts; the volumetric deformations do not differ substantially. Then from Eqs. 13, 14, 18, and 19:

$$\sigma'_m = \sigma'_f = 0,$$
$$\sigma'_{11m} = s'_{11m} + \sigma'_m = s'_{11m}$$
$$\sigma'_{11f} = s'_{11f} + \sigma'_f = s'_{11f}$$
$$\sigma_{11m} = 3s_{11m}/2$$
$$\sigma_{11f} = 3s_{11f}/2$$

and the use of these relations in Eq. 15 leads to:

$$3(s_{11m} - s_{11f})/2 + s'_{11m}/\eta_m - s'_{11f}/\eta_f = 0 \tag{24}$$

Equations 11–24 represent our mesomechanical model of tendon. For practical calculations it seems reasonable to accept a further simplification: neglecting volumetric changes, as they are of second-order meaning in relation to the deviatoric parts, i.e.:

$$\nu_m = \nu_f = 0.5\ ,\ \text{giving}\ \rho_m = \rho_f = 0 \tag{25}$$

Comparisons of the resulting theoretical curves with experimental data showed that this simplification does not cause observable discrepancies.

After a straightforward rearrangement we arrive at the sought after macroscopic constitutive equation with tensorial deviatoric internal variables and their evolution equations:

$$\begin{aligned}
\dot{\bar{\varepsilon}}_{11} &= \dot{\bar{e}}_{11} = \mu_f \dot{\bar{s}}_{11} + v_m(\mu_m - \mu_f)\dot{s}_{11m} + \\
&+ v_m \beta_m (s_{11m} - c_m) + v_f \beta_f (s_{11f} - c_f) = \\
&= \bar{\mu}\dot{\bar{s}}_{11} + \beta_m v_m [M(s_{11m} - c_m) + M' s'_{11m}] + \\
&+ \beta_f \{ v_f N(s_{11f} - c_f) + \\
&+ v_m(\mu_m - \mu_f)[-C s'_{11m} + D(\bar{s}_{11} - s_{11m})]\}
\end{aligned} \tag{26}$$

$$\begin{aligned}
\dot{s}_{11m} &= M\dot{\bar{s}}_{11} + \beta_m[-A(s_{11m} - c_m) + B s'_{11m}] + \\
&+ \beta_f [A(s_{11f} - c_f) - C s'_{11m} + D(\bar{s}_{11} - s_{11m})]
\end{aligned} \tag{27}$$

with the following meanings of the new symbols:

$$\bar{\mu} = \mu_m \mu_f [3\eta_m \eta_f (v_f \mu_f + v_m \mu_m) + 2(v_m \mu_m \eta_m + v_f \mu_f \eta_f)]/R$$
$$M = [3\mu_m \mu_f \eta_m \eta_f + 2\mu_f (v_m \mu_m \eta_m + v_f \mu_f \eta_f)]/R$$
$$N = [3\mu_m \mu_f \eta_m \eta_f + 2\mu_m (v_m \mu_m \eta_m + v_f \mu_f \eta_f)]/R$$
$$M' = 2 v_f \mu_f \eta_f (\mu_m - \mu_f)/R$$
$$A = 2 v_f (v_m \mu_m \eta_m + v_f \mu_f \eta_f)/R$$
$$B = 2 v_f \mu_f \eta_f / R$$
$$C = 2 v_f \mu_m \eta_f / R$$
$$D = 3 \mu_m \eta_m \eta_f / R$$
$$R = 3\mu_m \mu_f \eta_m \eta_f + 2(v_m \mu_f + v_f \mu_m)(v_m \mu_m \eta_m + v_f \mu_f \eta_f)$$

$$\dot{s}'_{11m} = \{-\mu_f \dot{\bar{s}}_{11} + (v_m\mu_f + v_f\mu_m)\dot{s}_{11m} + \beta_m[v_f(s_{11m} - c_m) - s'_{11m}] - v_f\beta_f(s_{11f} - c_f)\}/\mu_m \tag{29}$$

The value of s_{11f} can in any state be calculated from Eq. 11, i.e.:

$$s_{11f} = (\bar{s}_{11} - v_m s_{11m})/v_f \tag{30}$$

and therefore it is not a proper internal variable.

The general form of our constitutive equation given by Eqs. 26–30 is to be modified in different ways for different segments of the loading process:

The order of magnitude of β_f is lower than that of β_m and for short-time loading and unloading processes β_f can be considered vanishing, i.e., fibers are assumed elastic.

At the beginning of a loading process the stress-strain diagram is nonlinear, which is connected on the microscale with straightening of fibers and in our model with the decrease of the structural parameter η_f. The nonlinear part of the diagram is followed by a linear part, which is characterized on the microscale by a straight unidirectional configuration of the fibers and in our model by the zero value of the parameter η_f. This zero value means, in the general understanding of the model, that the strain-rate field is homogeneous.

In the linear part of the loading process the value of s_{11m} is substantially lower than that of s_{11f} and if loading stops and is followed by unloading, the first segment is purely elastic, s_{11m} decreases from its positive value to zero and then to negative values, and when it reaches the value $-c_m$, new viscous deformation starts in the opposite direction. The point where the elastic unloading ends and the inelastic unloading starts is clearly seen, for example, in the figures presented in [7]. This is the mechanism forming the hysteresis loop.

It is easy to see from the overall character of the constitutive equation (viscous part of deformation) that the stress-strain diagram depends on the rate of deformation.

Similarly it is clear from our model that if the loading process is stopped, either holding fixed deformation or fixed force, the process proceeds in the first case by relaxation and in the second case by creep. It is important to point out that in some tendons the viscous deformation of fibers cannot be neglected in this situation as it was in the short-time loading process. Here, in the long-time processes, it can play an important role.

It is known from experiments that repeated loading and unloading cycles, even if they are performed under the same conditions, do not have the same response. Repetition causes shifting of the stress-strain diagram towards higher deformations, increasing number of cycles leads to stabilization ("preconditioning"). In the same time the extent of the elastic part of unloading, mentioned above, decreases. In our concept these phenomena are modeled by a decrease of the values of H_m and c_m. With increasing number of cycles H_m approaches some minimum value, c_m approaches zero.

Figure 1 demonstrates the theoretical stress-strain diagram of the loading-unloading process and the effect of decreasing value of H_m: the change in H_m changes both the stress-strain diagram and the relaxation curves under fixed

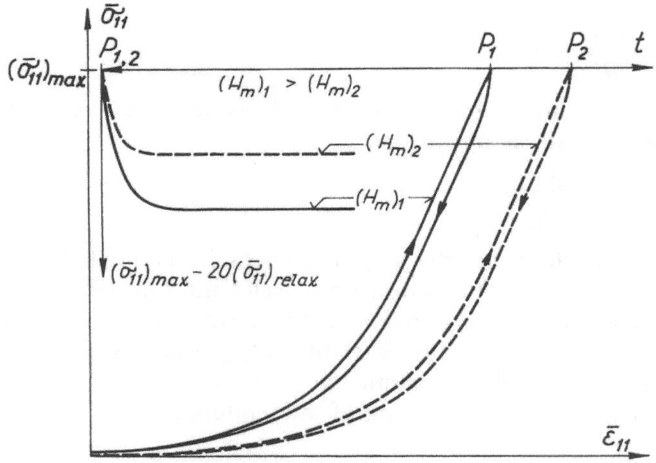

Fig. 1. Theoretical stress-strain diagram and relaxation curves of a tendon; effect of decreasing viscosity of the matrix due to repeated loading

elongation. The material parameters were chosen randomly, the fibers were assumed to remain elastic throughout the process, the input macroscopic strain rates in both the loading and unloading phases were assumed constant with the same absolute values and opposite signs, and the unloading phase follows the loading phase immediately.

Hence, from what has been demonstrated above it can be concluded that our model is able to describe all the characteristic features mentioned in the Introduction.

Specification of η_f

This task is essential, as the changing value of η_f causes the curved shape of the theoretical stress-strain diagram at the beginning of loading. It is one of the fundamental features of the mechanical behavior of tendon that the tangent modulus increases with increasing loading, which is associated with the straightening of fibers. It is natural to ask why, from the point of view of optimization of living tissues, the tendons have been optimized in this way. Our explanation is: (a) For a muscle, similarly as for any motor, the most favorable regime is to work with a constant performance. (b) On the other hand the most typical situation for the skeletal system is that during the loading process the force acting on it increases with time. (c) The tendon is assumed to have such properties that it connects the muscle with the respective skeletal part in such an optimized way that both these requirements are fulfilled (acting similarly as a gear-box in a car). Let us assume that the increase of macroscopic loading acting on the skeletal part and the respective stress in tendon is approximately linear, i.e.:

$$\frac{d\overline{\sigma}_{11}}{dt} = k_1 \tag{31}$$

The optimum regime for muscle is supposed to be working with constant specific stress power, i.e.:

$$\overline{\sigma}_{11}\frac{d(\overline{\varepsilon}_{11})_m}{dt} = k_2 \tag{32}$$

It is clear that the values of stress (force related to the cross-section of the joining tissue of tendon with muscle) and time ($\overline{\sigma}_{11}$ and t) in Eqs. 31 and 32 are identical for the tendon and the muscle, as they work in series. However, it is not the case with the variable $(\overline{\varepsilon}_{11})_m$ that relates to the muscle only. But it seems natural to assume that the contraction of the muscle is in a linear relation to the stretching of the tendon, i.e.:

$$d(\overline{\varepsilon}_{11})_m = kd(\overline{\varepsilon}_{11})_t$$

and thus using this linear relation and excluding dt from Eqs. 31 and 32 leads to the following formulae:

$$\frac{d\overline{\sigma}_{11}}{dt} = \overline{\sigma}_{11}\frac{d(\overline{\varepsilon}_{11})_m}{k_2} = \overline{\sigma}_{11}\frac{k_3 d(\overline{\varepsilon}_{11})_t}{k_2}$$

or:

$$d\overline{\sigma}_{11} = k\overline{\sigma}_{11}d\overline{\varepsilon}_{11} \tag{33}$$

where all the variables relate now to the tendon and k is some unknown constant.

Now let us turn our attention to the theoretical constitutive equation of tendon, Eq. 26. From experimental diagrams it is possible to make such approximate conclusion that the macroscopic stress $(\overline{\sigma}_{11})_L$ corresponding to the point of transition from the curved part to the linear part of the stress-strain diagram is independent of the rate of loading, of H_m and of c_m. Therefore it is assumed that η_f is function only of $\overline{\sigma}_{11}$. To model this function let us consider the simplest possible case, i.e., the m constituent as absolutely compliant. In such a case it is possible to describe the process as purely elastic deformation of fibers with voids instead of the m constituent. In Eqs. 26 and 27 this can be done by taking into account only the terms relating to elasticity and putting $\mu_m = \infty$.

The resulting formula reads:

$$\dot{\overline{\varepsilon}}_{11} = \dot{\overline{e}}_{11} = [\mu_f(2 + 3\eta_f)/(2v_f)]\dot{\overline{s}}_{11} \tag{34}$$

Furthermore, we assume that η_f reaches its zero value at the point where the curved part of the stress-strain diagram ends and the linear part begins, i.e.:

$$\eta_f = 0 \quad \text{for } \overline{\sigma}_{11} = (\overline{\sigma}_{11})_L \tag{35}$$

The relations in Eqs. 33–35 lead to the sought function for η_f :

$$\eta_f = \frac{2}{3}\left[\frac{(\overline{\sigma}_{11})_L}{\overline{\sigma}_{11}} - 1\right] \tag{36}$$

For $\overline{\sigma}_{11}$ higher than $(\overline{\sigma}_{11})_L$, η_f is defined as zero (i.e., homogeneous strain rate-field in the direction of stretching). Equation 36 was derived for $\mu_m=\infty$, but η_f is assumed to be function only of $\overline{\sigma}_{11}$ and therefore Eq. 36 is generally used.

Determination of μ_f

Let us assume that from experimental studies we know approximately the values of v_f and v_m. We now determine the value of μ_f from the gradient of the linear part of the stress-strain diagram of the tendon in question. From the preceding paragraphs we know that for a short-time loading process the viscous deformation of fibers is neglected, the experiment is performed so that the rate of deformation is constant, the linear part of the stress-strain diagram is characterized by its constant gradient and by the zero value of the structural parameter η_f. This means that *for the linear part* it holds true:

$$\beta_f = 0,\ \dot{\overline{\varepsilon}}_{11} = a,\ (d\overline{\varepsilon}_{11}/d\overline{s}_{11}) = b,\quad \dot{\overline{s}}_{11} = a/b,\ \eta_f = 0$$

where a and b are constants determinable from experiments. Using these values in Eqs. 26 and 27 leads to:

$$a = \frac{\mu_m\mu_f}{v_m\mu_f + v_f\mu_m}\left[\frac{a}{b} + v_m\beta_m(s_{11m} - c_m)/\mu_m\right] \tag{37}$$

Apart from s_{11m} Eq. 37 comprises only constants and therefore s_{11m} must be constant, too (in the extent of the linear part of the stress-strain diagram). Hence, equation (27) can be rewritten in the form:

$$\dot{s}_{11m} = 0 = [\mu_f a/b - v_f\beta_m(s_{11m} - c_m)]/(v_m\mu_f + v_f\mu_m) \tag{38}$$

The sought expression for μ_f results from Eqs. 37 and 38 in the following very simple form:

$$\mu_f = v_f b \tag{39}$$

According to this equation the gradient b of the linear part of the stress-strain diagram is independent of the rate of loading and of H_m and c_m that change in the course of the preconditioning. The experimental data that we had at our disposal verify approximately this property of our model.

Determination of H_m, η_m, and μ_m

To determine these parameters it is necessary to have two stress-strain diagrams of a "preconditioned" tendon for two different rates of loading. Then the above parameters are determined by a trial-and-error curve-fitting procedure. The dependence on η_m turns out to be very weak and according to our experience if its value is considered equal to unity, the results are fairly good. The procedure is then such that on the screen of the monitor of our calculator the two experimental diagrams for two different strain rates are depicted by two sets of separate points, some positive values are estimated for H_m and μ_m and the respective theoretical curve is plotted by our calculator following a program prepared on the basis of Eqs. 26–30, using some very small values of Δt instead of the differentials dt. At first the value of H_m is held fixed and the value of μ_m is changed so as to receive the best possible approximation of the first set of experimental points corresponding to the first strain rate. Then the other value of the strain rate is input into the program, the respective theoretical curve is plotted and compared with the other set of experimental points. If the correspondence is good, the process ends, and the values of H_m and μ_m are determined. If it is not good, the value of H_m is changed and everything is repeated up to the values of H_m and μ_m that give good correspondence for both sets of experimental points. If necessary, the whole procedure can be tried for several different values of η_m, but as mentioned above, the dependence on η_m turns out to be very weak.

Determination of H_f and c_f

These parameters relate to the viscous deformation of fibers. As mentioned above, this deformation is neglected in the short-time processes, but it can be important in the long-time processes of creep or relaxation. Usually it is easier to carry out experiments with fixed deformation and relaxation than with fixed load and creep. Therefore, below the relation of the equations to the case of relaxation are given explicitly. It is assumed that the loading process is stopped after some course of the linear part of the stress-strain relation, and then that deformation is held fixed. The formulae describing the relaxation process that follow can easily be obtained from Eqs. 26 and 27, where:

$$\dot{\varepsilon}_{11} = 0, \quad \eta_f = 0, \quad c_m = 0$$

With these values it turns out:

$$\dot{s}_{11m} = -\beta_m s_{11m}/\mu_m \tag{40}$$

and for $s_{11f} > c_f$:

$$\dot{s}_{11f} = -\beta_f(s_{11f} - c_f)/\mu_f \tag{41}$$

According to Eq. 11 it holds:

$$\bar{s}_{11} = v_m s_{11m} + v_f s_{11f} \tag{42}$$

From the above relations it is easy to see that in the case that the relaxation starts with $(s_{11f})_{start} > c_f$, the value of $\bar{s}_{11}$ which the relaxation curve approaches is

$$(\bar{s}_{11})_{lim} = v_f c_f \tag{43}$$

In this case c_f can be determined from Eq. 43 and H_f by a curve-fitting procedure using the relaxation curve (the value of H_f is subsequently changed, for every value the theoretical relaxation curve is plotted and compared with the experimental curve, the value giving the best coincidence is a sought one).

In the case that $(s_{11f})_{start} < c_f$, $\bar{s}_{11}$ approaches the value:

$$(\bar{s}_{11})_{lim} = v_f(s_{11f})_{start} = (\bar{s}_{11})_{start} - v_m(s_{11m})_{start} \tag{44}$$

and c_f and H_f cannot be determined from such an experiment (there appears no creep of the fibers).

Quantitative Comparison with Our Experiments

Our theoretical model was compared in a quantitative way with the mechanical properties of human tendon of the musculus flexor hallucis longus. The material for experimental studies was obtained from cadavers. Immediately after dissection the tendon was wrapped up in a towel wet with saline solution and frozen to the temperature –25°C. Before the experiment the tendon was thawed to the room temperature (23°C), and during the experiment the air around the sample was wetted by a special arrangement. The testing machine used was Instron 4301 adjusted to the range of 50 N. The measured length of the tendon was 50 mm, its cross-section 7.85 mm^2, and deformations were measured by the displacement of the grips of the machine. Four tendons were examined to verify the observed properties. These tendons were taken from different cadavers, and there were of course some slight differences in their parameters but not in their qualitative behavior. The data presented below correspond to one of these tendons with medium properties, not to statistical averages.

The "preconditioning" of the tendon proceeded so that it was stretched with the velocity of 1 mm/min up to the force 3 N; the machine then was stopped and the deformation held fixed for 30 min, after which the stretching proceeded again up to the force 7 N, when again the machine was stopped and the grips held fixed for 30 min. Then the sample was unloaded and in two cycles the stress-strain curves were measured for loading and unloading with the same constant strain rate up to the force 7 N. The diagrams of the two cycles were nearly identical and hence the tendon was considered "preconditioned."

The tendon parameters were determined as described abobe. Their values were:

μ_m=3.1 (MPa)$^{-1}$, η_m=1, H_m=1.667 MPa s,

μ_f=0.003726 (MPa)$^{-1}$, H_f=20000 MPa s,

$(\sigma_{11})_L$=0.373 MPa, $v_f = 1 - v_m$=0.3, c_f = 1.165 MPa

It is interesting to note that the value of μ_m is about thousand times higher than the value of μ_f. As to the order, it is in a good agreement with the relation arrived at by Ault and Hoffman [8] who described the properties of a rat tail tendon. The theoretical diagrams are compared with the experimental data in Fig. 2.

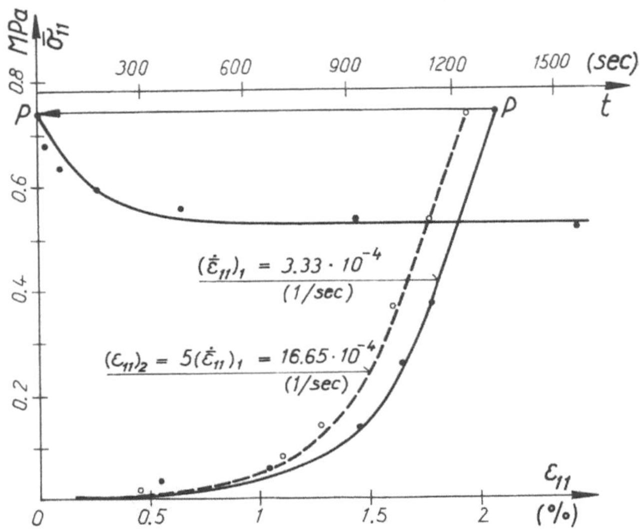

Fig. 2. Quantitative comparison of theoretical stress-strain diagrams at different rates of loading and of a relaxation curve with experimental data acquired with a sample of human tendon of musculus flexor hallucis longus.

The Relevant Clinical Experience

According to the scenario outlined above, the optimized mechanical properties of tendon are very important for the proper functioning of the respective muscle. This finding is corroborated by experience at the Orthopedic Department of the 2nd Medical Faculty of Charles University in Prague. The special fields of this Department are neurogenic disorders (cerebral palsy, poliomyelitis, myelodisplasiae), and also primary muscle disorders (myopathies). The surgery is performed principally on tendons and muscles. The number of cases of cerebral palsy reached to 1700 in the period 1970–1992.

Operations on tendons often were seen to lead to the loss of elasticity and functionality and to postoperation atrophy that improves only very slowly with rehabilitation. As a consequence the function of the respective muscles is violated, and their significant atrophy can be observed. Such atrophy of muscles is very significant especially in cases of repeated operations on one tendon, for example, in the cases of repeated prolongation or repeated shortening. Generally speaking, atrophy of muscles is significant especially in cases in which scarring of the tendons was long.

This experience led to our avoiding operations directly on tendons if at all possible and to prefering sections or transpositions of the origines musculorum or transpositions of the insertiones musculorum or aponeurotomies. Any violation of the natural function of tendon proves very unfavorable for the muscle, which is in accordance with our theoretical finding that the proper function of a tendon creates an optimum working regime for the muscle.

Conclusion

We have presented the general formulation of our mesomechanical model and an example of its application to tendon. The model proves useful not only as a tool for quantitative mathematical modeling of known phenomena but also as a means for widening out knowledge. This is shown in the discussed application to elucidating the optimized function of tendons that enable the optimum working regime of muscles.

We believe that the general model can also be successfully applied to a number of other problems not discussed here.

References

1. Kafka V (1987) Inelastic mesomechanics. World Scientific, Singapore
2. Kafka V, Jírová J (1983) A structural mathematical model for the viscoelastic anisotropic behaviour of trabecular bone. Biorheology 20:795–805
3. Kafka V (1993) Biomechanical heterogeneity. In: Valenta J (ed) Biomechanics. Elsevier, Amsterdam
4. Kafka V (1993) On hydraulic strengthening of bones. J Biomech 26:761–762
5. Kafka V, Jírová J, Smetana V (1995) On the mechanical function of tendon. Clin Biomech 10:50–56
6. Crisp JDC (1972) Properties of tendon and skin. In: Fung,YC, Ferrone N, Anliker M (eds) Biomechanics: its foundations and objectives. Prentice-Hall, Englewood Cliffs, pp 141–180
7. Fung YC (1981) Biomechanics, mechanical properties of living tissues. Springer, Berlin Heidelberg New York
8. Ault HK, Hoffman AH (1992) A composite micromechanical model for connective tissues. J Biomech Eng 114:137–146

I.5 Fundamental Considerations on Cruciate Ligament Surgery

W.K. Krudwig

Introduction

Ruptures of the anterior cruciate ligament, alone or combined, are some of the most frequent joint injuries, especially in sports [132]. At first, research concentrated almost exclusively on reconstruction of the anterior cruciate ligament. In the past emphasis has also been placed on reconstruction of the posterior cruciate ligament. However, no field in surgery has undergone such radical changes or been discussed so extensively and controversially as the treatment strategies for acute and chronic cruciate ligament tears.

The negative effects of immobilization are as known and accepted as the advantages of early functional therapy. This knowledge has thoroughly influenced treatment strategies. The number of conferences, congresses, and publications on the basic principles, indications, techniques, choice of material, early or late operation, augmentation, and clinical results of cruciate ligament surgery has risen dramatically. This is the direct consequence of the increasing number of injuries (increase of sportive activities of all classes), improvement in the diagnostic methods, and generalized surgical specialization. However, in spite of these enormous efforts the long-term results remain unsatisfactory. Also, the fundamental questions are still open to discussion [189].

By today more than 300 different reconstruction techniques for the anterior cruciate ligament have been described [191]. Autogenous, homologous, heterologous, and allogenous materials are used alone, combined, or augmented with absorbable or nonabsorbable artificial ligaments. Synthetic materials are also used. This great number of methods, however, reflects the poor long-term results and the general lack of confidence concerning the choice of an "optimal" treatment. A nationwide survey carried out in 1991, for example, observed that almost 100 different techniques are presently used in Germany to treat anterior knee instability [175].

It is astonishing and irritating that the well-known "gold standard" using the central third of the patellar tendon (B-PT-B with and without augmentation) is not as widely used as we could expect. The same can be said for the primary suture. Although many studies have confirmed that this technique does not provide sufficient healing in cases of an intraligamenteous rupture [6, 20, 52, 81, 91, 104, 118, 183, 188], this reconstruction method is still frequently used [93, 175]. Very slow revascularization prevents a stable scarring of the rupture zone. In addition, the dehiscence of the rupture stumps and the retraction of the remaining parts of the ligament are observed [138]. And even when a scarred bridging is obtained, an elongation can be expected in the long run since the collagen fibers are reduced in length compared to the natural ligament [175]. It is not clear to what extent an autogenous or alloplastic protection could prevent these physiological processes [5]. The autogenous augmentation always requires an anatomic placement. The augmentation tissue must therefore be positioned in the center of the native region of insertion of the ligament which must be augmented. Using a patellar or semitendinous tendon graft of 5–6 mm thickness, tibial and femoral tunnels of the same dimensions are required. The drilling of these holes implies, however, the destruction of central parts of that ligament which must be repaired. It is unclear then whether the eventually regained stability is obtained by the remaining parts of the native ligament or by the augmentation structure.

Only the cases of bony avulsions or ruptures directly at the insertion sites are exceptions. Reinsertion there is a possible solution and corresponds to the anatomically best form of reconstruction. The analysis of the morphology of these lesions shows, however, that this covers only 5% of all tears [183]. In case of a rupture close to the insertion with a remaining stump of only 2–3 mm this would already be related to an intraligamenteous rupture and would be treated with the same prognosis as midsubstance ruptures.

Finally, three phenomena can be found in examining the literature: (a) For each technique good and very good results are reported which could indicate that the reconstruction method itself does not affect the clinical results. (b) The repeated modifications of the different procedures in a short period of time puts in perspective the excellent results, which are often reported with a great enthusiasm after the first follow-up. (c) Other methods which do not correspond to the author's strategy are often assessed as unsuitable by repeatedly referring to a handful of references.

The main goal of this article is therefore to give an overview of today's knowledge and related scientific facts. It can also be seen as an invitation to think over critically one's own therapeutic strategy to determine an acceptable long-term compromise that could satisfy both the surgeon and the patient.

Basic Situation

The knee joint is the largest and most complicated joint of the human body and constitutes a moving link between the femur and the tibia. It is exposed to high injury risks. It is well known today that the different structures of the joints are linked in a proprioceptive net. This led to the knowledge that each element fulfills one or more functions, but that it is not isolated or independent [150]. The functional synergism of the individual structures can therefore be impaired by the loss of *only one* element, and this has complex effects on the complete coordinated movement [130].

The functional synergy of knee ligaments was noted as early as 1917 by Strasser [190] and later by Menschik [129]. Menschik considered the collateral ligaments and described the kinematics of the "four-bar system." Parts of each ligament must be under tension during the sequence of movement [90] to be able to properly control the roll-glide mechanism and to guarantee the long-term functionality of the knee joint.

Each discontinuity of one or more ligaments causes an overloading of synergistic active structures. Moreover, a load-dependent disposition of the articular surfaces results in shear stress and tensile load on the chondral surfaces, which causes exposure of the tangential fiber pattern and early destructive changes [105, 202]. This is confirmed by clinical and radiological follow-up studies of nontreated ligament insufficiencies [5, 7, 47, 52, 53, 55, 79, 96, 98, 125, 146, 158, 172, 195]. Nontreated cases with low symptoms and only slightly visible radiological changes show an objectively measured instability and an increasing rate of secondary lesions, especially in the menisci. This is generally followed by one or more surgical revisions [124, 173]. Only the functional restoration of the four-bar system can delay this process [45, 200]. The instability does not lead directly to an arthrosis but stands at the beginning and/or in the center of a pathological cycle [16, 109, 130] (Fig. 1) with the gradual destruction of the menisci and the cartilage.

The ultimate tensile strength of cruciate ligaments varies individually, depends on the level of activities and decreases with age [34, 38, 100, 144, 145, 196, 204, 205, 207, 208]. The mean maximum strength of the anterior cruciate ligaments has been found to be 480 N while that of the posterior cruciate ligament is about 940 N. It must be noted, however, that these values correspond to the bulk cross-section and therefore cannot be interpreted as absolute values.

It is through many discussions that it has been determined that isometry as such does not exist for the anterior and posterior cruciate ligament. In fact, all fibers show a change in their length pattern of minimum 1–2 mm during extension and flexion [39, 57, 60, 103]. Graf [68] therefore speaks of a "nearly isometric behavior."

The maximum reversible extensibility of the cruciate ligament is about 5% [138, 144] and greater stress results in an irreversible elongation of the liga-

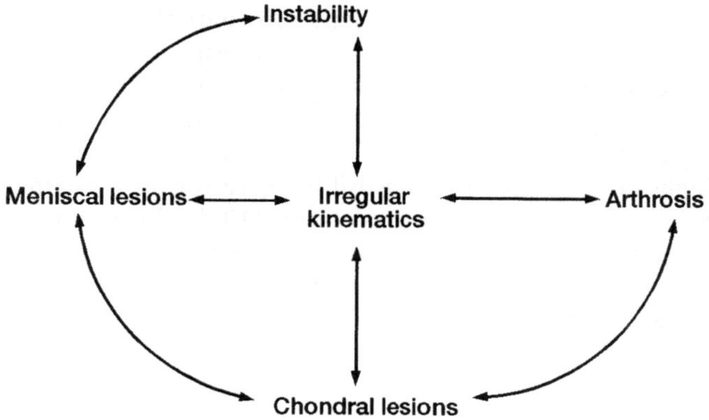

Fig. 1. Pathological cycle

ment due to interstitial ruptures. Only a few of the ligament fibers are strained, depending on the knee's flexion angle and rotation. The others are ready for use [62, 64, 113, 174]. The relatively high strength of the cruciate ligaments is therefore the sum of all fibers. In fact, the loading capacity of solitary position-dependent, load-bearing fiber bundles is much lower and can be increased if necessary by the recruitment of adjacent bundles. The load-induced fiber elongation leads, for example, in the case of the anterior cruciate ligament to proprioceptive signals which provoke a reflexlike contraction of the synergistic ischiocrural muscles. These draw back the tibial head and thereby reduce the mechanical load of the anterior cruciate ligament [13, 49, 70, 165]. However, this phenomenon can only partly be considered as a muscular protective reflex. In fact, the time of this reflex-arc (89 ms for the anterior cruciate ligament) is much longer than the impact time which causes a rapid increase of tension due to translation or rotation [70, 153].

Several studies have tried to determine the ideal "isometric" placement of tibial and femoral tunnels. These show that a deviation of even a few millimeters from the "isometric point" results in great changes of the length pattern [39, 40, 64]. The eccentric enlargement of this previously measured point (optimal insertion point) for the construction of the osseous tunnels (whose diameter is determined by the insertion region) causes an increasingly non-isometric behavior of peripheral fibers of the transplanted structure. In practice this means that the peripheral fibers of a thick graft can undergo early failure; Pelinka [161] confirms this from follow-up arthroscopies after 6 months transplantation of a 10-mm-thick patellar tendon graft. In all cases Pelinka [161] observed a circular rupture of the peripheral fibers leaving a 6- to 7-mm-thick remaining intact central bulk. For example, in cases of autogenous augmentation of the patellar tendon with the gracilis or the two or three times folded semitendinous tendon the bony tunnels have a diameter of 12 or more millimeters (at least 6 mm deviation from the central "isometric" point). The functional value of peripheral fibers of thicker grafts with regard to the load bearing ability and lifetime seems therefore questionable.

To date no method can reproduce the highly complex and complicated anatomy and function of cruciate ligaments [64, 109, 153]. Each reconstruction is confined to the replacement of a partial function and can therefore fulfill only the demand of a partial repair (especially from the point of view of the proprioception). "Today's surgical techniques do not allow the operator to duplicate the ligament's anatomy. The aim must be the reconstruction of the ligament's function" [57]. Fundamental doubts are being expressed increasingly concerning the possibility of restoring a normal knee function by the existing reconstruction procedures [193]. This opinion is supported by a study by Anetzberg et al. [8], which shows that despite clinically successful reconstructions of the anterior cruciate ligament a disposition of the articular surfaces remains with clearly higher loading of the dorsal tibial plateau.

Basic Principles

Proprioception

The scientific research in neuroanatomic and neurophysiological fields leads today to a better understanding of the nerve supply of the joint and the neuromuscular reflex-arc for the control and the coordination of the stabilizing muscles (see Chap. I.2). Even in last century studies were published on the nerve structures in the knee joint [51, 142, 163], but only recently new investigations put this subject at the center of attention. The knee joint is now understood as a sensorimotor complex where the cruciate ligaments have both a passive, mechanical stabilizatory and a sensory function with a neuromuscular feedback [13, 15, 16, 56, 63, 69, 70, 74, 94, 99, 102, 106, 162, 180, 181, 187, 210, 212]. The femoral nerve, obturator nerve, tibial nerve, and peroneal nerve innervate the capsule of the knee joint and the cruciate ligaments. These nerval units guarantee through different receptors and neurotransmitters a continuous data exchange. In 1944 Abbott et al. [1] studied the nervous supply of the collateral and cruciate ligaments and described it as the outgoing point of a "kinetic chain."

In the literature several types of topographically more or less numerous sensory nerve endings can be found: Pacinian corpuscles, which pass on information about joint position, Ruffini's end-organs, which react slowly to changes in tension and acceleration, Golgi tendon apparatus as a motion receptor, and free nerve endings as pain receptors. In addition to the above nervous structures present in the capsule and the cruciate ligaments, mechanoreceptors can also be observed in the menisci [11, 199].

To summarize, we can establish that the proprioception is responsible for the undisturbed physiological motion and loading capability of the knee joint. Each disorder or loss of function of one or more of these structures leads to a breakage of the kinetic chain with increasing pathologies. An initially satisfactory and subjectively sufficient compensation by intact structures is synonymous with overload of these structures and therefore only temporary [61]. For example, the synergistic effect of the meniscal dorsal horns are increasingly under pressure as the anterior translation of the tibial head increases. Secondary meniscal lesions appear within a short time [14, 52, 54, 77, 86, 87, 168, 209].

Operative Trauma

If we accept the knee joint as a complex organ and not only the sum of topographically adjacent individual structures, the question arises to what extent the surgical treatment results in further damages of previously intact structures. If we simoultaneously consider the basic biological principle stating maximum efficiency with minimum material, this leads us to following conclusion: each loss of an autogenous periarticular structure result in an additional weakening of the already predamaged knee joint. This again means an overloading of the synergistic functional structures. Every incision causes, depending on its length, a clinically more or less relevant destruction of effective proprioceptive areas. The operative transposition of complete structures (extraarticular surgery and reinforcement) including capsular tissues leads to an extensive scarring with disruption of the nervous communication. The positive effect of such a method is estimated to be small [95]. Today arthroscopic and minimally invasive methods are therefore preferred, and this is the direct consequence of these findings.

More and more patients coming in for operative revision have been operated on several times, and the question remains whether the renewed or already existing pain and limitation of function are the consequences of the primary injury or the previous surgical procedures. The sum of the iatrogenic damages due to technically perfect operations often lead to a shift of the clinically dominant problem. Further interventions are then frequently needed.

It is obvious that the functional value of individual structures or their parts must be differentiated with regard to the knee joint, and this depends on the judgment of each operator. On one hand, untreated instabilities result in an early degeneration and destruction of compensating structures and ultimately in the arthrosis of the knee joint. On the other hand, the reconstructive operations require the maximum possible care of intact tissues. The proposed compromise must, however, take into account following statement: after the sacrifice of an intact autogenous structure for the reconstruction of an anterior or posterior cruciate ligament, the outgoing situation of the knee joint is fundamentally worse in case of an operative failure or a renewed trauma.

The degree of proprioceptive disorder depends on the extent of the primary injury *and* the surgical technique. Kumar [110] summarized the problem at the annual meeting of the international PCL study group: "Don't rob Peter to pay Paul."

The recommendation has occasionally been made that the transplant should be removed from the healthy side. This certainly stems from the idea that the already injured knee should not be more traumatized. However, it neglects the possibility of problems at the donor site, the always existing operative risks (infection, thrombosis), and the possibility of further injuries of both knees.

Materials

Autografts

Patellar Tendon

The autogenous patellar tendon graft (midthird bone-tendon-bone) is one of the most widely used and extensively studied grafts for the reconstruction of both cruciate ligaments. Despite differences in the fine structure [2, 4, 25, 27, 28] the patellar tendon graft is morphologically the most ligamentlike autogenous tissue. Its biomechanical properties are similar to those of cruciate ligaments [33, 147]. This is the reason why it is commonly accepted to be the best suitable graft for cruciate ligament reconstructions [33, 71, 73, 147]. The primary anchorage of the patellar tendon is without doubt the most stable one. The rapid revascularization of the bone blocks leads within a few weeks to the bony ingrowth and the revascularization of the tendinous segment [9, 19, 23, 72, 94, 197]. The disadvantage of all autografts lies in the rapid avitality of the tendon and the protracted transformation [9, 10, 12, 24, 25, 27, 35, 37, 40, 101, 127, 141, 151, 197]. Amiel et al. [3] termed this process "ligamentization." Agreement can also be found on the weakening of every autograft after 6 weeks of transplantation. During this period the autograft has a relative tensile strength under 20% and gains strength only slowly. Authors disagree merely on the time of the transformation (necrosis, revitalization, collagen synthesis, and remodeling). This makes it difficult to plan the period where protection is necessary [104, 148]. While some authors assume that grafts have a relative tensile strength of 50% after 30 weeks [3], others think that this is reached at the earliest after 1 year [25, 28, 97]. Still, the morphology corresponds neither to that of a tendon nor to that of a cruciate ligament [97]. Special attention must be given the study by Noyes et al. [147], which noted that during normal daily activities the ligament is loaded to up to 25% of its maximum strength. During effort the load can increase up to 50% of the ligament's maximum strength. Chen et al. [36] established stress values for daily activities which were between 67 and 458 N. Normally the ligament is loaded up to 500 N, and in extreme situations the value can reach 2000 N [207]. This means that an unprotected graft reaches its functional stress capability after only 6 weeks. Further loading or small accidents can lead to an acute overtaxing of the mechanical properties of the graft with recurrent microtraumas and resulting elongation or even to the total rupture of the transplant.

Surgeons normally expect that tendinous grafts are transformed into an approximately functional ligament. This can be confirmed by good macroscopic views obtained with control arthroscopies. Numerous studies show, however, that this is probably not the case [23, 35, 40]. In fact, they confirm that even after a few years the graft consists increasingly of collagen fiber of type III (in the natural ligament the predominant collagen type is type I), and that necrotic areas are coming through its central part [26, 29, 111, 126]. This phenomenon corresponds to an insufficient adaptation resulting in diminishing mechanical properties. This could explain why short-term results are good while long-term results decline with time. In fact, a continuous disproportion exists between mechanical strength of the transplant and its loading. The intention to increase the mechanical properties by using a thicker graft leads to an additional weakening of the donor site. Moreover, problems appear with regard to the isometry and the space in the notch.

The hope that the transplant could possibly be linked into the proprioceptive system is still not obvious. Even if the ingrowth of nerve fibers could be found in grafts it does not prove that these nerves have been integrated into the sensory system [59, 179].

On one hand, the removal of the patellar tendon means an intrusion upon the proprioception and the neuromuscular balance of the knee joint. On the other hand, the complications related to the donor site (fracture of the patellar or quadriceps tendon, restricted range of extension) and the symptoms, especially in the femoropatellar joint, cannot be ignored [21, 22, 48, 76, 80, 95, 122, 123, 136, 146, 149, 159, 169, 171, 182, 184, 192].

Since the number of load-bearing fibers is reduced to one-third, the remaining two-thirds of the patellar tendon must take over this part of the load. The defect of the patellar tendon, adapted or not, can result in a lowering of the patella (patella baya) [31, 116]. The pathophysiological consequences of this phenomenon are not to be calculated. Sonographic follow-ups show an enormous swelling of the patellar tendon after removal of the central third and must be interpreted as a result of an inflammatory reaction and a chronic overloading [17, 22, 178]. The defect of the patellar tendon is filled with a nonoriented collagen scar tissue.

Semitendinous Tendon Graft

The muscle group of the pes anserium, which is positioned medially on the tibial head, is composed of the semitendinous, sartorius, semimembranous, and gracilis muscle. These oppose the external rotation of the tibia and support the internal rotatory stability of the knee. The semitendinous tendon is responsible for 20% of the internal rotatory stability [131]. The harvest of this tendon leads to a weakening of this functional unity and results in a permanent external rotation of the tibia combined with the corresponding incorrect rotatory position of the tibial plateau [117, 137]. This incorrect positioning has been confirmed by the gait analysis conducted by Eichhorn (personal communication).

The primary anchorage of the two or three times folded tendon is much more difficult to achieve and less stable compared to that of the primary fixation of the patellar tendon. We have noted above the problems related to the large thickness of the grafts; difficulties in obtaining an isometric placement and the limited space in the notch have been mentioned above. The decreased strength of the semitendinous graft is identical to that of the patellar tendon graft and is not significantly improved by doubling this structure [40, 170].

Other Transplants

Only in individual cases are free or flapped grafts of the fascia lata, quadriceps, biceps, gracilis, or plantar tendon used. Grafts can also be taken from the menisci, skin, and periosteum. Fundamentally, all these grafts are subject to the above structural changes.

Xenografts

Xenogenous materials (for example, lyophylizated dura, tendons of the kangaroo and calf) have shown very poor short-term results. They are no longer used due to their related biological problems.

Allografts

Allografts are being used increasingly, especially in United States and Japan [89, 167, 185]. Their advantages are obvious: unlimited availability, absence of the problems related to the donor site, reduction in the proprioceptive disorder, shorter operating time, and they are biological structures. In addition to various clinical results [75] about the incorporation of allogenous tissues it must be noted that the use of such material is heavily criticized since the HIV problem appeared [32]. Secure disinfection of the allograft is extremely difficult; on the one hand, the fundamental properties must be preserved (elasticity and stability), and, on the other, the bone blocks must also be disinfected. A procedure which disinfects these different structures without altering one of them is not known to date [65]. The current irradiation dose is 25–30 kGy, and it is not proven that this provides an effective virus inactivation [119]. A higher radiation inevitably leads to structural changes of the biological tissue and to the uselessness of the graft for transplantation [65]

The applied radiation therefore introduces an unpredictable risk. In addition, the allograft must remodel as an autograft and can therefore be considered to have similar mechanical properties [50, 88, 112, 143, 186]

Augmentation

The augmentation is defined as the increase in the effective loaded cross-section of biological structures. This can be achieved with a coaxial augmentation (in-line augmentation) or with a parallel augmentation in contact with the autogenous or allogenous structure (on-line augmentation) [105]. Its aim is a load protection of the present transplant during "ligamentization" and should permit early functional therapy [148].

Autogenous and Allogenous Tissues

If it is accepted that all autogenous and allogenous structures undergo a structural transformation with a long period of biomechanical weakness, the question arises as to how allogenous/autogenous augmentation is able to do justice to this task. In fact, from the moment that the transplant loses its mechanical strength the autogenous/allogenous structure undergoes the identical period of weakness. This is why the desired protective function of an allogenous or autogenous augmentation is only partly fulfilled. The problem of the isometric positioning of such a thick structure has been discussed in detail above. Moreover, we must consider the following facts: further removal of autogenous tissues leads to a further intervention into the proprioception and threatens the stability of the knee.

Absorbable Materials

Most absorbable materials used for ligament grafts augmentation loose their mechanical properties after 4-6 weeks. Since the functional weakness of transplants is known to occur around the sixth week, these materials seem unsuitable. In fact, the protection fails the moment that it is most needed [84, 85, 164, 194]. Compared to autogenous and allogenous tissues these grafts have the advantage only that they are less thick.

Nonabsorbable Materials

The already discussed healing and remodeling periods of autogenous and allogenous transplants lead us to the conclusion that only permanent alloplastic materials are suitable for augmentation. Alloplastic materials should demonstrate a strength close to that of the transplant. At 5% elongation it should possess an equivalent tensile load. The resultant stress sharing of both components should reach 50% [45, 105, 177]. In case of further stress, strength should increase in order to protect the biological structure from overloading.

Synthetic Grafts

Since the long-term results with autogenous reconstructions are unsatisfactory, the use of synthetic grafts merits consideration for such an operation. Synthetic materials have been clinically used since the seventies after animal trials [157]. The theoretical advantages of alloplastic grafts are obvious: (a) unlimited availability, (b) care of intact autogenous structures, (c) minimization of the operative trauma and reduction of the operation time, (d) technically easier surgical procedure, and (e) immediate functional stability and shorter healing process. Nevertheless, most studies showed a high proportion of failure with partly catastrophic clinical outcomes [92, 120, 121, 155, 160, 166, 198]. However, this should not lead to the conclusion that synthetic grafts are generally unsuitable [83, 92].

The success or the failure of synthetic ligaments depends on the mechanical characteristics of the material, graft design, biocompatibility, operating technique, healing, and anchoring.

Material Requirements

The material used for ligament prostheses must be a synthetic fatigue resistant material. It must be free from industrial contamination and soluable additives and must be chemically stable and insensitive to immunological reactions of human tissue and fluid. They must furthermore demonstrate a linear stress-strain behavior, good fatigue resistance to torsion, and low elasticity [41, 45, 140].

Design

Round grafts show a variable cross-sectional stress distribution when they are led around edges and curves. With greater loads this leads to an increased wear of the inner fiber structure with primary abrasion and consequently early failure. With oval or flat shapes this risk is reduced with regard to the influence of the design. The mechanical resistance of synthetic grafts is also largely influenced by its shape.

Woven textile shapes seem to be particularly suitable for long-term implantation [45, 46]. In the case of braided materials the fibers are squeezed under the load, especially if the load is not applied axially. This increases the inherent dynamic wear.

The graft should have a porosity which encourages ingrowth of fibroblasts. They improve the quality of the anchorage and isolate the implant from the physiological environment. The fibroblasts have a diameter of 15–20 μm. The granulating clot is, however, much larger. Therefore a porosity of 200–250 μm is examined as optimal and permits ingrowth of granulating tissue and capillary supply [108].

Biocompatibility

The biocompatibility must be guaranteed both intraarticularly and extraarticularly. The human body shows a consistent foreign body reaction [41, 43, 45]. When a material dissolution takes place, the substances are phagocytosed. If the foreign body remains stable, it is isolated from the rest of the environment by a fibrous tissue. This means that the biocompatibility of a foreign body depends on its stability against the body's defense mechanism. Suitable materials for ligament reconstruction should therefore neither undergo primary changes in their molecular structure nor lose their consistency or load capacity.

Several studies have tested the biocompatibility of different materials and their particular biological reactions [38, 42, 45, 46, 176].

Operative Technique

For the operating technique the first concern should be to avoid sharp edges around which the graft is led. The intraligamenteous consequences of this have are described above. In addition, compression or friction of the synthetic graft reduces its mechanical lifetime. The release of small synthetic particles can result in synovial inflammatory reactions. These considerations are also valid in autogenous and allogenous reconstructions. With the exception of the patellar tendon graft (primary bony anchorage) the implant, especially synthetic ones, should be positioned such that edges are avoided. In the case of the natural anterior cruciate ligament and when the femoral insertion of a synthetic ligament is parallel to the knee's flexion axis (see Chap. II.3), the flexion angle of the knee is equivalent to the torsion in the ligament. Bending and torsion of the synthetic ligament vary with its placement in the tunnel [211]. Therefore proper placement of the synthetic ligament is very important (see Chap. II.3). Autogenous and allogenous tendinous transplants require, in addition to the above protection during remodeling, a sufficient protection to permit a stable reinsertion without elongation.

From these findings and to limit the high failure rate the classical over-the-top technique emerged. The replacing structure was led through the dorsal capsule while the tibial tunnel remained at the insertion plane of the native ligament. As a result of this technique a high rupture rate was observed. The reason was the large deviation from the isometric behavior leading to an elongation of more than 10 mm of the graft during knee flexion [78]. However, it is possible to achieve a nearly isometric behavior by using two different methods: (a) by leading the graft through a dorsal groove in the over-the-top passage (more precisely "through-the-top") [82, 128, 152, 154] and (b) by a dorsomedial displacement of the tibial tunnel behind the native tibial anterior cruciate insertion (this corresponds to a parallel and dorsal displacement of both insertion planes [105, 109, 201, 203].

The synovitis observed with artificial ligaments were always the consequence of an important release of particles due to wear or rupture. Using synthetic grafts means therefore a meticulous implantation technique which minimizes mechanical irritations. Operative failures inducing wear or squeezing of the synthetic structures become visible through synovial reactions while mistakes in reconstructions using biological grafts are clinically less easy to detect. In both cases the reconstruction, however, failed. On the other hand, the absence of clinical reactions in case of synthetic ligaments means that the reconstruction is successful.

Healing and Anchorage

Healing and anchorage are two factors which are initially related and which depend on one another. The earlier a stable primary anchorage of the implant takes place, the quicker is the secondary anchorage: the reacting collagen tissue in the insertion site improves and supports the primary anchorage. For autogenous and allogenous transplants a strong insertion is of prime importance. This is obtained by creating a load bearing area between the tendon and the bone (in case of the patellar tendon this part is totally transplanted). Each mechanical disturbance can lead to microtraumas with bridged scarring of the defect, elongation of the transplant, and finally a renewed instability. Autogenous and absorbable augmentation cannot solve this problem, as we noted above. Only the use of a permanent augmentation can improve the situation in protecting the graft from traumatic overloading and in guaranteeing a mechanically sufficient stability. Prosthetic ligaments can be fixed by staples or screws. The additional connective tissue anchorage reduces the loads applied on the primary fixation.

To summarize, a suitable alloplastic cruciate ligament must have following properties: fatigue resistance, chemical stability, woven textilelike constructions with linear load behavior, low elasticity, and good torsional fatigue resistance [45, 46].

In terms of these basic design criteria it is not surprising that some prostheses show poor results. Stiff, ropelike, synthetic grafts with a low torsional fatigue resistance and which are implanted without any care for isometry cannot fulfill the expected demands. The increasing discussion about isometric or nearly isometric positioning of the replacing structure and the use of guiding devices lead to better results in case of biological reconstructions. On the other hand, synthetic grafts are no longer used because of negative experience with them.

At this point a fundamental question arises: was it always only the synthetic graft that failed? In fact, synthetic ligaments have been implanted with identical techniques to early autografts, and it was precisely the failure of the latter which encouraged the surgeons to use synthetic ligaments. In the case of biological materials the surgical procedure was declared to be responsible for failure. The negative results obtained with alloplastic ligaments were explained by the failure of the synthetic material, which were therefore abandoned. This means that alloplastic materials did not even have the chance to demonstrate their intrinsic functional utility.

Looking back, we must certainly admit that the negative results obtained using synthetic ligaments were due to a combination of a improper material choice, unsuitable designs, *and* an incorrect operative procedure. The use of alloplastic materials does not mean that the technique is easy. On the contrary, it implies a meticulous implantation [66]. It has been thought that the biological healing process would forgive and correct slight operative deficiencies in allogenous and autogenous transplantations, therefore making them less sensitive to an isometry. The general good results of biological replacements have, however, been explained above. This must not be taken as a reproach. At that time most of the experiences and scientific findings known today and generally accepted were still unknown or only beginning to enter research. However, the further developments were only implemented for the biological procedures.

Recent follow-ups demonstrate results for reconstruction or augmentation with synthetic materials comparable to those obtained with biological grafts [67, 114, 115, 134, 135]. The operating technique plays a decisive role here. In ruptures of the synthetic ligament an important foreign body reaction is observed. This is due to the release of synthetic wear particles. In ruptures resulting from renewed trauma no important synovial reaction is observed. Histological studies on explanted ligaments show that the amount of foreign body reaction is directly related to the stressed segments. More severe reaction are found only in the surrounding area of the rupture or near the wear surfaces. At the mechanically less constrained part of the ligament (prosthetic body) the reactions were found to be much lower [30, 58, 107]. Such a cellular reaction is, however, the natural reaction of the human body on implants [42].

Admittedly, it must be considered that, contrary to the biological transplants, synthetic ligaments do not adapt functionally with time, since they are inorganic. Therefore they undergo fatigue from the first moment of loading [44]. Consequently, alloplastic reconstructions have a time limited functionality, and the procedure must take into account their possible replacement [45]. A sufficient functional adaptation of biological transplants is, however, still doubtful. That is the reason why revisions should also be considered in primary biological reconstruction strategies.

Revision

The question still remains as to what type of structure should be used in case of a renewed instability. Some reports pretend that the third part of the patellar tendon can be again removed for reconstruction after 3 years from the first operation [18]. This assumes that the primary defect heals not only as a collagen scar tissue but turns out to be a differentiated tendinous tissue. However, this contradicts normal collagen tissue healing. Studies show in fact that scarring and remodeling of the tissue take place, but that they do not achieve the homogeneity of the normal patellar tendon [17–19]. Generally the defect caused by the harvest of an autogenous structure is not compensated, and every incision corresponds to the disorder of the proprioception. It is questionable whether it makes sense to increasingly harvest and weaken primary intact structures in order to reconstruct the central restraints, especially since this also submits them to higher loading.

This was observed when we considered the history of a knee which had been operated on several times. Unidirectional instabilities changed to multidirectional, and the reconstruction prognosis was increasingly negative. When the use of a synthetic ligament is then considered as a salvage procedure, it would be overloaded from the beginning, therefore limiting its lifetime. It can be expected that the graft will fail, and this is the result of the general functional disorder of the joint.

Use of a suitable, correctly implanted synthetic ligament as a primary reconstruction allows stabilization of the articulation with conservation of functionally important structures, and this again allow their own relief. At the same time the autogenous structures are at the surgeon's disposal in case of revision (for example, after failure of the prosthesis). By then a long-term satisfactory method for the reconstruction with autografts will perhaps exist.

Summary

Despite innumerable conferences and publications, no standard concept for the reconstruction of acute cruciate ligament tears or chronic knee instabilities has been established. In 1984 Noyes et al. [147] wrote: "An intraarticular cruciate substitution is a rather complex operative procedure, with many factors still to be studied. It requires selection of a graft with appropriate mechanical properties, meticulous surgical technique, correct fixation sites, correct adjustment of graft tension, postoperative protection allowing time for remodeling, and a careful and detailed rehabilitation program.... To what extent all of this occurs is presently unknown." In the meantime fundamental research has been conducted. Optimization of the procedures with regard to isometry and choice of material and to reduction of the operative trauma by using arthroscopic and minimal invasive methods have led to better short- and medium-term results. Long-term results remain unsatisfactory. In 1995 Soslowsky and Goldstein [189] summarized the situation: "While many of these injuries require repair and/or reconstructive procedures, the optimal choice of techniques, devices, or repair strategies remains unresolved.... While much is known about the pros and cons of various ligament reconstructions, even more remains unknown at this time."

Agreement has been found concerning the necessity of operative reconstruction in order to avoid secondary lesions. In fact, conservative treatments are only partly able to achieve a sufficient long-term muscular compensation [153]. Authors also agree on the isometric or nearly isometric placement of the graft. This is one of the fundamental assumptions for the ligament reconstruction. The proprioception of the joint is also widely acknowledged as well as the fact that each injury leads to a lasting disorder of the neuromuscular function. Moreover, no doubts remain on the fact that at best a partial functional recovery can be achieved. Transformation (ligamentization) of biological transplants with first enormously decreased load capacity is also well known: simply the time of each phase and the moment of sufficient strength are still under discussion.

The literature in recent years shows a preference for the midthird patellar tendon as autogenous or allogenous transplant. The number of different techniques is not to be overlooked. The uniformity of good results, on the one hand, and equally bad results, on the other, is irritating. After numerous follow-up studies, the primary suture for the healing of intraligamenteous cruciate ligament ruptures was found to be useless.

The disadvantages of all allogenous and autogenous grafts lie in the time needed for biological transformation (ligamentization) resulting in a reduced mechanical load bearing capacity and a slow regaining of its strength. Consequently, the need of a protective augmentation by nonabsorbable alloplastic materials before overloading is obvious. Autogenous structures are submitted to the same structural changes as the transplant itself and weakens therefore at the same moment. They lead to a thickening of the whole transplant which increases the deviation from the necessary isometric position. Absorbable materials loose their stability after a few weeks and thereby their protective function. They are consequently just as ineffective.

The midthird patellar tendon (B-PT-B) is known to be *the* transplant since it is morphologically and from the view of the anchorage similar to the natural ligament. The semitendinous tendon is a part of the pes anserium and therefore has an important function for the internal rotatory stability. Its harvest leads to the displacement of the tibial plateau resulting in an uneven loading of the articular surfaces.

After a temporary loss of strength autogenous grafts all have the disadvantage of the associated proprioceptive disorder of the joint. Primary intact and functionally important structures of the injured knee are partly or completely sacrificed. Therefore, synergistic and compensatory useful tissues are submitted to lasting overloading. The resulting proprioceptive disturbance increases to a degree depending on the iatrogenic part. The theoretical idea that biological transplants are connected to the proprioceptive control after some time by the ingrowth of nerve receptors has not been confirmed. All the possible complications and complaints described cannot be put aside. Passler et al. [156] summarized this in a few words: "The cruciate ligament surgery brings quite a lot of complications with itself."

The use of allogenous transplants, which avoid the complication at the donor site and reduce the amount of proprioceptive disorder, must be discouraged. In fact, no disinfection procedure can guarantee a secure viral disinfection without altering the structure of the graft.

The use of alloplastic material must be carefully considered. Complicated results with early failure were in the past the result of a improper material selection, an unsuitable design, and/or an incorrect operative technique. The unsatisfactory results of biological reconstruction procedures, without the consideration of isometric positioning, led to the use of synthetic ligaments. Their advantages were evident: unlimited availability, care of autogenous structures, minimization of operative trauma with shortening of the operating time, easier operating techniques, and the immediate functional stability and shorter healing process.

However, the synthetic grafts have been implanted in the same way as biological structures and could therefore not fulfill the expected demands. The mis-loading which was responsible for the discrete and slow failure of biological structures led in the case of synthetic ligaments to massive synovitis with clinically dramatic results, stimulated by the release of synthetic wear particles. In case of the biological grafts, their misplacement (with regard to the isometry) was considered to be the main reason for failures. In case of synthetic ligaments, only the quality of the material was considered to be unsuitable and was therefore abandoned. The synthetic grafts generally did not even enjoy the chance to prove their efficiency by using them with improved reconstruction procedures. Today, however, reports confirm that synthetic ligaments can achieve equivalent results compared to the biological graft.

With respect to the previously known facts and experiences we can summarize the problem as following: the cruciate ligament surgery is still seeking an optimal procedure. Since the allografts are not secure (because of the virus activity [119], only two different procedures seem to fulfill the material, mechanical, and biological demands: (a) Reconstruction with the midthird patellar tendon, augmented with a permanent alloplastic material for the whole period of ligamentization. The subsequent removal of the synthetic ligament must be possible [67]. (b) Synthetic ligament with great demands on the quality of the material, the design and especially the operative technique.

"The problems related to the use of synthetic ligaments are still not solved. But we are now at the very beginning of the research, just as the arthoplastic surgery was thirty years ago. Therefore we can count on further improvements and good long-term results. We can assume that this technique will be successful at least in one of the fields of ligament surgery." [139]

References

1. Abbott LC, Saunders JB, Bost FC, Anderson CE (1944) Injuries to the ligaments of the knee joint. J Bone Joint Surg 26:503–521
2. Amiel D, Frank C, Harwood F, Fronek J, Akeson WH (1984) Tendons and ligaments: a morphological and biochemical comparison. J Orthop Res 1:257–265
3. Amiel D, Kleiner JB, Akeson WH (1986) The natural history of the anterior cruciate ligament autograft of patellar tendon origin. Am J Sports Med 14:449–462
4. Amiel D, Kuiper S (1990) Experimental studies on anterior cruciate ligament Grafts. In: Daniel D, Akeson WH, O'Connor J (eds) Knee ligaments: structure, function, injury, and repair. Raven, New York, pp 379–388
5. Andersson C, Odenstein M, Good L, Gillquist J (1989) Surgical or non-surgical treatment of acute ruptue of the anterior cruciate ligament. J Bone Joint Surg Am 71:965–973
6. Andersson C, Odenstein M, Gillquist J (1991) Knee function after surgical or nonsurgical treatment of acute ruptures of the anterior cruciate ligament: a randomized study with along-term follow-up period. Clin Orthop 264:255–263
7. Andersson C, Gillquist J (1992) Treatment of acute isolated and combined ruptures of the anterior cruciate ligament. A long-term follow-up study. J Bone Joint Surg Am 20:7–12
8. Anetzberger H, Müller-Gerbl M, Scherer MA, Metak G, Blümel G, Putz R (1994) Veränderung der subchondralen Mineralisierung nach Rekonstruktion des vorderen Kreuzbandes beim Schaf. Unfallchirurg 97:655–660
9. Arnoczky SP, Tarvin GB, Marshall Jl (1982) Anterior cruciate ligament replacement using patellar tendon: an evaluation of graft revascularization in the dog. J Bone Joint Surg Am 64:217–224
10. Arnoczky SP (1983) Anatomy of the anterior cruciate ligament. Clin Orthop 172:19–25
11. Assimakopoulos AP, Katonis PG, Agapitos MV, Exarchou EI (1992) The innervation of the human meniskus. Clin Orthop 275:232–236
12. Ballock RT, Woo SL-Y, Jyon RM, Hollis JM, Akeson WH (1989) Use of patellar tendon autograft for anterior cruciate ligament reconstruction in the rabbit: a long-term histological and biomechanical study. J Orthop Res 7:474–485
13. Baratta R, Solomonow M, Zhou BH, Letson D, Chuinard R, D'Ambrosia R (1988) Muscular coactivation, the role of antagonist musculature in maintaining knee stability. Am J Sports Med 16:113–122
14. Barber FA (1987) Meniscus repair results of an arthroscopic technique. Arthroscopy 3:25–30
15. Barrack RL, Skinner HB, Buckley SL (1989) Proprioception in the anterior cruciate deficient knee. Am J Sports Med 17:1–6
16. Barrack L, Skinner HB (1990) The sensory function of knee ligaments. In: Daniel D, Akeson WH, O'Connor J (eds) Knee ligaments: structure, function, injury, and repair. Raven, New York, pp 95–114
17. Benedetto KP, Sperner G, Glötzer W, Klima G, Ebner W (1991) Die Entwicklung des Lig. Patellae nach Transplantatentnahme – Sonographische Verlaufskontrolle. In: Poigenfürst J, Pelinka H (eds) Rekonstruktion des vorderen Kreuzbandes – Methodenvergleich. Enke, Stuttgart, pp 78–84
18. Benedetto KP (1995) The gold standard in the reconstruction of the anterior cruciate ligament. Chirurg 66:1061–1070
19. Bergljung L (1970) Vascular reactions in tendon healing. A stereomicroangriogaphic study of tendon suturing techniques and tendon transplantation. Angiology 21:375–384
20. Blatter G, Tissi R (1991) Is suture of the ruptured anterior cruciate ligament without augmentation sensible? Unfallchirurgie 17:232–235
21. Blazina ME, Fox JM, Carlson GJ, Jurgutis JJ (1975) Patella baja. J Bone Joint Surg Am 57:1027
22. Bonamo JJ, Krinick RM, Sporn AA (1984) Rupture of patellar tendon after use of the central third for anterior cruciate reconstruction. J Bone Joint Surg Am 66:1294–1297
23. Bosch U, Decker B, Möller HD, Kasperczyk WJ, Oestern HJ, Tscherne H (1989) Healing at the graft fixation site under functional conditions in posterior cruciate ligament reconstruction. A morphological study in sheep. Arch Orthop Trauma Surg 108:154–158

24. Bosch U, Decker B, Kasperczyk W, Oestern HJ, Tscherne H (1989) Biological aspects of long-term failure of autografts after cruciate ligament replacement. Arch Orthop Trauma Surg 108:368–272
25. Bosch U, Kasperczyk WJ (1992) Healing of the patellar tendon autograft after posterior cruciate ligament reconstruction – a process of ligamentization? Am J Sports Med 20:558–566
26. Bosch U, Kasperczyk WJ, Decker B, Nerlich A, Oestern HJ, Tscherne H (1991) Heilungsphasen nach experimentellem Kreuzbandersatz mittels Patellarsehnentransplantat – Morphologische Aspekte. In: Poigenfürst J, Pelinka H (eds) Rekonstruktion des vorderen Kreuzbandes – Methodenvergleich. Enke, Stuttgart, pp 1–5
27. Bosch U, Möller H, Iburg T, Decker B, Kasperczyk WJ, Oestern HJ (1993) Morphologische Untersuchungen zur Veränderung des Kollagenfibrillendurchmessers in einem Patellarsehnentransplantat nach Kreuzbandersatz. Langenbecks Arch Chir Forum 319–324
28. Bosch U, Kasperczyk WJ, Oestern HJ, Tscherne H (1994) Biology of posterior cruciate ligament healing. Sports Med Arthrosc Res 2:88–99
29. Bosch U, Kasperczyk WJ (1995) Ligamentisation des Patellabandes? In: Hasselbach C, Scherer MA (eds) Wiederherstellende Chirurgie des vorderen Kreuzbandes. Demeter, Gräfelfing, pp 51–56
30. Boszotta H, Helperstorfer W, Pflanzl W (1993) Foreign body synovitis – a limiting factor in use of the Trevira ligament in cruciate ligament surgery? Unfallchirurgie 19:138–143
31. Brülhart K, Kieser C (1991) Veränderung der Patellahöhe nach Entnahme eines Transplantats aus der Patellarsehne. In: Poigenfürst J, Pelinka H (eds) Rekonstruktion des vorderen Kreuzbandes – Methodenvergleich. Enke, Stuttgart, pp 85–88
32. Buck RE, Malinin, Brown MD (1989) Bone transplantation and human immundeficieny syndrome (AIDS). Clin Orthop 240:129–136
33. Butler D, Noyes F, Grood E, Miller E, Malek M (1979) Mechanical properties of transplants for anterior cruciate ligament. Orthop Trans 3:180–192
34. Butler DL, Kay MD, Stouffer DC (1986) Comparison of material properties in fascicle-bone units from human patellar tendon and knee ligaments. J Biomech 19:425–432
35. Butler DL, Grood ES, Noyes FR, Olmstead ML, Hohn RB, Arnoczky SP, Siegel MG (1989) Mechanical properties of primate vascularized vs. nonvascularized patellar tendon graft; changes over time. J Orthop Res 7:68–79
36. Chen EH, Black J (1980) Materials design analysis of the prothetic anterior cruciate ligament. J Biomed Mater Res 14:567–586
37. Chiroff RT (1975) Experimental replacement of the anterior cruciate ligament. J Bone Joint Surg Am 57:1124–1127
38. Claes L (1983) Biomechanische Eigenschaften humaner Bänder. In: Burri C, Claes L (eds) Aktuelle Probleme in Chirurgie und Orthopädie: alloplastischer Bandersatz. Bd 25. Huber, Bern, pp 12–19
39. Claes L, Kiefer H, Dürselen L (1991) Biomechanische Untersuchungen zur funktionellen Belastbarkeit ligamentärer Ersatzplastiken am Kniegelenk. In: Weise K, Weller S (eds) Kapsel-Band-Verletzungen des Kniegelenks. Springer, Berlin Heidelberg New York, pp 45–60
40. Clancy WG, Narechania RG, Rosenberg TD, Gmeiner JG, Wisnefske DD, Lange TA (1981) Anterior and posterior cruciate ligament reconstruction in rhesus monkeys. J Bone Joint Surg Am 63:1270–1284
41. Contzen H (1962) Materialtechnische Voraussetzungen und biologische Gewebereaktion bei der Implantation von Kunststoffen. Bruns Beitr Klin Chir 104:179
42. Contzen H, Straumann F, Paschke E (1967) Grundlagen der Alloplastik mit Metallen und Kunststoffen. Thieme, Stuttgart
43. Contzen H (1969) Biologische Grundlagen der Alloplastik. In: Gohrband, Gabka, Berndorfer (eds) Handbuch der plastischen Chirurgie, vol 1. Gruyter, Berlin
44. Contzen H (1970) Unbelebte Hilfsmittel (Allenthesen) in der operativen Medizin. In: Breitner (ed) Chirurgische Operationslehre I; Ergänzung 9. Urban and Schwarzenberg, Munich

45. Contzen H (1985) Material requirements and biological conditions for the alloplastic knee ligament substitution. Unfallchirurgie 11(5):242–246
46. Contzen H (1986) Materialtechnische Voraussetzungen und biologische Grundlagen für den alloplastischen Kniebandersatz (am Beispiel des Treviabandes). Hefte Unfallheilkd 181:837–834
47. Cross MJ, Powell JF (1984) Long-term followup of posterior cruciate ligament rupture: a study of 116 cases. Am J Sports Med 12:292–297
48. DeLee JC, Craviotto DF (1991) Rupture of quadrizeps tendon after a central third patellar tendon anterior cruciate ligament reconstruction. Am J Sports Med 19:415–416
49. Draganich LF, Andiacchi TP, Andersson GBJ (1987) Interaction between intrinsic knee mechanics and the knee extensor mechanism. J Orthop Res 5:539–547
50. Drez DJ Jr, DeLee J, Holdon JP, Arnoczky SP et al. (1991) Anterior cruciate ligament reconstruction using bone-patellar tendon-bone allografts. A biological and biomechanical evaluation in goats. Am J Sports Med 19:256–263
51. Ellis GV (1839) Extract from a communication of the nerves of the knee joint. Lancet 1:595–597
52. Feagin JA, Curl WC (1976) Isolated tear of the anterior cruciate ligament: five-year follow-up study. Am J Sports Med 4(3):95–100
53. Fetto JF, Marshall JL (1980) The natural history and diagnosis of anterior cruciate ligament insufficiency. Clin Orthop 147:29–38
54. Finsterbush A, Frankl U, Matan Y (1990) Secondary damage to the knee after isolated injury of the anterior cruciate ligament. Am J Sports Med 18:475–479
55. Fowler PJ, Regan WD (1987) The patient with symptomatic chronic anterior cruciate ligament insufficiency – results of minimal arthroscopic surgery and rehabilitation. Am J Sports Med 15:321–324
56. Freeman MAR, Wyke BD (1966) Articular contribution to limb muscle reflexes: the effects of partial neurectomy of the knee. Br J Surg 53:61–69
57. Friederich NF, O'Brien WR (1990) Zur funktionellen Anatomie der Kreuzbänder. In: Jakob RP, Stäubli H-U (eds) Kniegelenk und Kreuzbänder. Springer, Berlin Heidelberg New York, pp 80–95
58. Friedman MJ, Ferkel RD (1988) Prothetic ligament reconstruction of the knee. Saunders, Philadelphia
59. Fromm B (1995) Die Nervenversorgung des vorderen Kreuzbandes und des Kreuzband-Allotranplantates. In: Hasselbach C, Scherer MA (eds) Wiederherstellende Chirurgie des vorderen Kreuzbandes. Demeter, Gräfelfing, pp 57–62
60. Fuchs M, Schabus R, Kwasny O (1991) Gibt es die Isometrie des vorderen Kreuzbandes. In: Poigenfürst J, Pelinka H (eds) Rekonstruktion des vorderen Kreuzbandes – Methodenvergleich. Enke, Stuttgart, pp 19–22
61. Furman W, Marshall JL, Girgis FG (1976) The anterior cruciate ligament. A functional analysis based on postmortem studies. J Bone Joint Surg Am 58:179–189
62. Fuss FK (1991) Anatomie des vorderen Kreuzbandes. In: Poigenfürst J, Pelinka H (eds) Rekonstruktion des vorderen Kreuzbandes – Methodenvergleich. Enke, Stuttgart, pp 9–10
63. Gardner E (1948) The innervation of the knee joint. Anat Rec 101:109–130
64. Gaudernak T, Schmiedmayer B (1991) 2D-Computersimulation des vorderen Kreuzbandes. In: Poigenfürst J, Pelinka H (eds) Rekonstruktion des vorderen Kreuzbandes – Methodenvergleich. Enke, Stuttgart, pp 12–18
65. Gibbons MJ, Butler DL, Grood ES (1989) Dosage dependent effects of gamma irradiation on material properties of frozen bone-patellar-bone allografts. Trans Orthop Res Soc 14:513–521
66. Gillquist J (1993) Cruciate ligament protheses. Techniques, results and perspectives. Orthopade 22:381–385
67. Gotzen L, Petermann J (1994) Rupture of the anterior cruciate ligament in athletes. Chirurg 65:910–919
68. Graf B (1987) Isometric placement of substitutes for the anterior cruciate ligament. In: Jackson DW, Drez DJ (eds) The anterior cruciate deficient knee. Mosby, St. Louis, pp 102–113

69. Grigg P, Hoffmann AH (1984) Ruffini mechanoreceptors in isolated joint capsule: response corrolated with strain energy density. Somatosens Res 2:149–162
70. Grüber J, Wolter D, Lierse W (1986) Anterior cruciate ligament reflex (LCA-reflex). Unfallchirurg 89:551–554
71. Hackenbruch W (1991) Arthroskopisch unterstützte VKB- Plastik mit autologem Lig. Patellae. In: Poigenfürst J, Pelinka H (eds) Rekonstruktion des vorderen Kreuzbandes – Methodenvergleich. Enke, Stuttgart, pp 57–67
72. Hackenbruch W, Schmiedebusch H, Müller W, Staubesand J (1986) Arthroscopy and electron microscopy following anterior cruciate reconstruction. In: Trickey EL, Hertel P (eds) Surgery and arthroscopy of the knee. Springer, Berlin Heidelberg New York, pp 192–200
73. Hackenbruch W, Schmiedebusch H (1990) Morphologische Ultrastruktur nach Rekonstruktion. In: Jakob RP, Stäubli H-U (eds) Kniegelenk und Kreuzbänder. Springer, Berlin Heidelberg New York, pp 268–271
74. Halata Z, Haus J (1989) The ultrastructure of sensory nerve endings in the human anterior cruciate ligament. Anat Embryol (Berl) 179:415–421
75. Harner CD, Olson EI, Fu FH, Irrgang JI, Maday MG (1992) The use of fresh frozen allograft tissue in knee ligament reconstruction: indications, techniques, results and controversies. Scientific exhibit, American Academy of Orthopaedic Surgeons. Annual Meeting, Washington, pp 20–25
76. Harner CD, Irrgang JJ, Paul J, Dearwater S, Fu FH (1992) Loss of motion after anterior cruciate ligament reconstruction. Am J Sports Med 20:499–506
77. Hart JAL (1982) Meniscal injury associated with acute and chronic instability of the knee joint. J Bone Joint Surg Br 64:119–127
78. Hassenpflug J, Blauth W, Rose D (1985) Tension characteristics of transplants replacing the anterior cruciate ligament, together with a contribution to the criticism of the "over-the-top" technic. Unfallchirurg 88:151–158
79. Hawkins RJ, Misamore GW, Merritt TR (1986) Follow-up of the acute nonoperated isolated anterior cruciate ligament tear. Am J Sports Med 14:205–210
80. Hefti F, Gächter A, Jenny H, Morscher A (1982) Replacement of the anterior cruciate ligament. A comparative study of four different methods of reconstruction. Arch Orthop Trauma Surg 100:83–94
81. Hefti FL, Kress A, Fasel J, Morscher EW (1991) Healing of the transected anterior cruciate ligament in the rabbit. J Bone Joint Surg Am 73(3):373–383
82. Hefzy MS, Grood ES, Noyes ER (1989) Factors affecting the region of most isometric femoral attachments. II. The anterior cruciate ligament. Am J Sports Med 17:208–216
83. Henche HJ (1990) Einbau eines Kreuzbandes aus Kunststoff – ein Arthrosemodel? Arthroskopie 3:1
84. Hertel P, Lais E, Bernard M (1991) Nachbehandlung: das Berliner Modell. In: Weise K, Weller S (eds) Kapsel-Band-Verletzungen des Kniegelenks. Springer, Berlin Heidelberg New York, pp 72–77
85. Hoffmann R, Lobenhoffer C, Krettek C, Tscherne H (1989) PDS (Polydioxanon)-Augmentation der vorderen Kreuzbandrekonstruktion. Hefte Unfallheilkd 207:247
86. Indelicato PA, Bittar ES (1985) A perspective of lesions associated with ACL insufficiency of the knee. Clin Orthop 198:77–80
87. Irvine GB, Glasgow MMS (1992) The natural history of the meniscus in anterior cruciate ligament insufficiency. Arthroscopic analysis. J Bone Joint Surg Br 74:403–405
88. Jackson DW, Grood ES, Arnoczky SP, Butler DL, Simon TM (1987) Freeze dried anterior cruciate ligament allografts: preliminary studies in a goat model. Am J Sports Med 15:295–303
89. Jackson DW, Windler GE, Simon TM (1990) Intraarticular reaction associated with the use of freeze-dried, ethylene oxide-sterilized bone-patellar tendon bone allografts in the reconstruction of the anterior cruciate ligament. Am J Sports Med 18:1–11
90. Jäger M, Wirth CJ (1978) Kapselbandläsionen. Thieme, Stuttgart
91. Järvinnen M, Kannus P (1985) Clinical and radiological long-term results after primary ligament surgery. Arch Orthop Trauma Surg 104:1–6

92. Jensen KU, Klein W (1990) Problems and complications with artificial ligaments. Arthroskopie 3:15–23
93. Jerosch J, Drescher H, Schröder M, Lewejohann B (1994) Current treatment modalities in acute and chronic ACL insufficiencies in Germany – results of a nationwide survey. Sportmedizin 95:48–59
94. Johansson H, Sjölander P, Sojka P (1991) A sensory role for the cruciate ligaments. Clin Orthop 268:161–178
95. Johnson RJ, Beynnon BD, Nichols CE, Renstrom PAFH (1992) The treatment of injuries of the anterior cruciate ligament. J Bone Joint Surg Am 74:140–151
96. Kannus P, Järvinen M (1987) Conservatively treated tears of the anterior cruciate ligament. Long-term results. J Bone Joint Surg Am 69:1007–1012
97. Kasperczyk WJ, Bosch U, Oestern HJ, Tscherne H (1991) Heilungsphasen nach experimentellem Kreuzbandersatz mittels Patellarsehnentransplantat: biomechanische Aspekte. In: Weise K, Weller S (eds) Kapsel-Band-Verletzungen des Kniegelenks. Springer, Berlin Heidelberg New York, pp 32–37
98. Keller PM, Shelbourne KD, Carroll JRM, Rettig AC (1993) Nonoperatively treated isolated posterior cruciate ligament injuries. Am J Sports Med 21:132–136
99. Kennedy JC, Weinberg HW, Wilson AS (1974) The anatomy and function of the anterior cruciate ligament as determined by clinical and morphological studies. J Bone Joint Surg Am 56:223–235
100. Kennedy JC, Hawkins RJ, Willis RB, Dannylchuck KD (1976) Tension studies of human knee ligaments: yield point, ultimate failure, and disruption of the cruciate and tibial collateral ligaments. J Bone Joint Surg Am 58:350–355
101. Kennedy JC, Roth JH, Mendenhall HV, Sanford JB (1980) Intraarticular replacement in the anterior cruciate deficient knee. Am J Sports Med 8:1–8
102. Kennedy JC, Alexander IJ, Hayes KG (1982) Nerve supply of the human knee and its functional importance. Am J Sports Med 10:329–335
103. Kiefer H, Dürselen L, Claes L (1992) Experimentelle Untersuchungen zur Biomechanik des Kniebandapparates. Springer, Berlin Heidelberg New York
104. Kleiner JB, Roux RD, Amiel D, Woo SL-Y, Akeson WH (1986) Primary healing of the ACL. Trans Orthop Res Soc 11:13–55
105. Knaepler H, Krudwig W, Witzel U (1994) Reflections upon different therapeutic conceps of the insufficiency of the anterior cruciate ligament – basics and clinical choises. Aktuelle Traumatol 24:188–194
106. Kocher MS, Fu FH, Harner CD (1994) Neuropathophysiology. In: Fu FH, Harner CD, Vince KG (eds) Knee surgery, vol 1. Williams and Wilkins, Baltimore, pp 231–249
107. Kock HJ, Stürmer KM, Letsch R (1991) Biological tissue response to prothetic replacement of the anterior cruciate ligament with PET implants (Trevira Hochfest). Unfallchirurg 94:594–602
108. Konikoff J (1974) Development of a single stage active tendon prothesis. J Bone Joint Surg Am 56:848
109. Krudwig W, Kerschen J, Witzel U, Draznin N (1992) Alloplastic replacement of the anterior cruciate ligament. Material, technical and biomechanical requirements, operative technique. Aktuelle Traumatol. 22:170–175
110. Kumar R (1995) Results of the West Herts two bundle PCL reconstruction. IIIrd Int Symp of the PCL Study Group, 1721 October, Dijon
111. Kurosaka M, Abe S, Iguchi T, Fujita N, Yoshiya S, Hirohata K (1990) Light and electron microscopic study of remodeling and maturation of the autogenous graft for anterior cruciate ligament reconstruction. Abstracts, 4th Congress of the European Society of Knee Surgery and Arthroscopy. Stockholm, pp 42–43
112. Lane JG, McFadden P, Bowden K, Amiel D (1993) The ligamentization process: a 4 year case study following ACL reconstruction with a semitendinosus graft. Arthroscopy 9:149–153
113. Lang J, Wachsmuth W (1972) Bein und Statik. Springer, Berlin Heidelberg New York
114. Letsch R, Stürmer KM, Kock HJ, Schmid-Neuerburg KP (1993) Primary repair of acute tears of the anterior cruciate ligament and protection by synthetic augmentation (Trevira

Hochfest). Indication, technique, and results of a five-year-study. Unfallchirurg 96:499-507
115. Letsch R, Stürmer KM, Kock HJ, Wissing H, Schmit-Neuerburg KP (1994) Implantation of a PET-ligament (Trevira Hochfest) as an anterior cruciate ligament prothesis in salvage cases of the knee. Medium to long-term results of a clinical study. Unfallchirurgie 20 (6):293-301
116. Linclau L (1984) Iatrogenic patella "baja." Acta Orthop Belg 50:75-80
117. Lipscomb AB, Anderson AF (1986) Tears of the anterior cruciate ligament in adolescents. J Bone Joint Surg Am 68:19-28
118. Lobenhoffer P, Tscherne H (1993) Rupture of the anterior cruciate ligament. Current status of treatment. Unfallchirurg 96:150-168
119. Lobenhoffer P, Lattermann C (1995) Cruciate ligament allograft reconstruction - clinical perspective. Chirurg 66:1071-1078
120. Lopez-Vazques E, Juan JA, Vila E, Debon J (1991) Reconstruction of the anterior cruciate ligament with a dacron prothesis. J Bone Joint Surg Am 73:1294-1300
121. Macnicol MF, Penny ID, Sheppard L (1990) Early results of the Leeds-Keio anterior cruciate ligament replacement. J Bone Joint Surg Br 73:377-380
122. Marder RA, Raskind JR, Caroll M (1991) Prospective evaluation of arthroscopically assisted anterior cruciate ligament reconstruction. Patellar tendon versus semitendinosus and gracilis tendon. Am J Sports Med 19:497-484
123. Marks PH, Harner CD, Fu FH (1994) Complications of knee ligament surgery. In: Fu FH, Harner CD, Vince KG (eds) Knee surgery, vol 1. Williams and Wilkins, Baltimore, pp 897-910
124. McDaniel WJ, Dameron TB (1980) Untreated ruptures of the anterior cruciate ligament. J Bone Joint Surg Am 62:698-705
125. McDaniel WJ, Dameron TB (1983) The untreated anterior cruciate ligament rupture. Clin Orthop 172:158-163
126. McLean I, Deacon O, Oakes B (1992) The final morphology of anterior cruciate ligament grafts. Abstracts, First World Congress of Sports Trauma. Palma de Mallorca, pp 132-133
127. McPherson GK, Mendenhall HV, Gibbons DF et al. (1985) Experimental, mechanical and histological evaluation of the Kennedy Ligament Augmentation Device. Clin Orthop 196:186-195
128. Melhorn JM, Henning CE (1987) The relationship of the femoral attachment site to the isometric tracking of the anterior cruciate ligament graft. Am J Sports Med 15:539-542
129. Menschik A (1987) Biometrie. Springer, Berlin Heidelberg New York
130. Metak G, Scherer MA (1995) Beanspruchungswandel der Menisken bei Läsion des vorderen Kreuzbandes. In: Hasselbach C, Scherer MA (eds) Wiederherstellende Chirurgie des vorderen Kreuzbandes. Demeter, Gräfelfing, pp 18-27
131. Meystre J-L (1990) Rekonstruktion mittels Semitendinosussehne. In: Jakob RP, Stäubli H-U (eds) Kniegelenk und Kreuzbänder. Springer, Berlin Heidelberg New York, pp 385-392
132. Miasaka KC, Daniel D, Stone ML, Hirshman P (1991) The incidence of knee ligament injuries in the general population. Am J Knee Surg 4/1:3-9
133. Mockwitz J, Contzen H (1982) Treatment of inveterate knee ligament lesions with alloplastic material. Unfallchirurgie 8:176-179
134. Mockwitz J, Contzen H (1984) Alloplastische Korrektur chronischer Kniebandinstabilitäten mit einem Band aus Polyaethylenterephthalat In Rettich H (ed) Biomaterialien und Nahtmaterial. Springer, Berlin Heidelberg New York
135. Mockwitz J, Rau B (1988) Ideal indications for the use of a textile ligament of high strength Trevira in alloplastic replacement of the anterior cruciate ligament in chronic isolated rupture. Unfallchirurgie 14:276-282
136. Mohtadi NGH, Webster-Bogaert S, Fowler PJ (1991) Limitation of motion following anterior cruciate ligament rconstruction. Am J Sports Med 19:620-625
137. Mott HW (1983) Semitendinosus anatomic reconstruction for cruciate ligament insufficiency. Clin Orthop 172:90-92
138. Müller W (1983) The knee: form, function and ligament reconstruction. Springer, Berlin Heidelberg New York

139. Munzinger U (1990) Heutiger Stand des künstlichen Ersatzes. In: Jakob RP, Stäubli H-U (eds) Kniegelenk und Kreuzbänder. Springer, Berlin Heidelberg New York, pp 529–542
140. Neugebauer R, Burri C, Claes L, Piehler J (1983) Vergleichende histologische und biomechanische Untersuchungen zum Bandersatz mit verschiedenen alloplastischen Materialien. In: Schreiber HW (ed) Chirurgisches Forum '83 für experimentelle und klinische Forschung. Springer, Berlin Heidelberg New York, pp 115–119
141. Newton PO, Shuji H, Woo SL-Y (1990) Experimental studies on anterior cruciate ligament autografts and allografts: mechanical studies. In: Daniel D, Akeson WH, O'Connor J (eds) Knee ligaments: structure, function, injury, and repair. Raven, New York, pp 389–399
142. Nikoladoni C (1873) Untersuchungen über die Nerven aus der Kniegelenkkapsel des Kaninchens. Med Jahrb 56:401–414
143. Nikolaou PK, Seaber AV, Glisson RR, Ribbeck BM, Bassett FH (1986) Anterior cruciate ligament allograft transplantation. Long-term function, histology, revascularization, and operative technique. Am J Sports Med 14:348–360
144. Noyes FR, Torvic PJ, Hyde WB, De Lucas JL (1974) Biomechanics of ligament failure. II. An analysis of immobilization. Exercise and reconditioning effects in primates. J Bone Joint Surg Am 56:1406–1418
145. Noyes FR, Grood FS (1976) The strength of the anterior cruciate ligament in humans and rhesus monkeys. J Bone Joint Surg Am 58:1074–1082
146. Noyes RF, Mooar PA, Metthews DS, Butler DL (1983) The symptomatic anterior cruciate-deficient knee. I. The long-term functional disability in athletically active individuals. J Bone Joint Surg Am 65:154–174
147. Noyes FR, Butler DL, Grood ES, Zernicke RF, Hefzy MS (1984) Biomechanical analysis of human ligaments grafts used in knee ligament repair and reconstruction. J Bone Joint Surg Am 66:344–352
148. Noyes FR, Barber SD (1992) The effect of ligament-augmentation device on allograft reconstruction for chronic ruptures of the anterior cruciate ligament. J Bone Joint Surg Am 73(7):960–973
149. O'Brien SJ, Warren RF, Pavlov H, Panariello R, Wickiwicz TL (1991) Reconstruction of the chronically insufficient anterior cruciate ligament with the central third of the patellar ligament. J Bone Joint Surg Am 73:278–286
150. O'Connor J, Shercliff T, FitzPatrick D, Biden E, Goodfellow J (1990) Mechanics of the knee. In: Daniel D, Akeson WH, O'Connor J (eds) Knee ligaments: structure, function, injury, and repair. Raven, New York, pp 201–237
151. O'Donoghue DH, Rockwood CA, Frank GR, Jack SC, Kenyon R (1966) Repair of the anterior cruciate ligament in dog. J Bone Joint Surg Am 48:503–519
152. Paar O (1988) Management of fresh and older ruptures of the anterior cruciate ligament with the Kennedy modified polypropylene ligament augmented tendoplasty. Chirurg 59:788–792
153. Paar O, Magin MN (1995) Principles and limits of conservative treatment in anterior cruciate deficiency. Chirurg 66:1050–1060
154. Paar O, Magin MN, Witzel U (1995) On the femoral attachment site of prothetic ligament in anterior cruciate-insufficient knees. Experimental study on cadaver knees in a modified over-the-top technique. Chirurg 66:1115–1119
155. Parson JR, Bhayani S, Alexander H, Weiss AB (1985) Carbon fibre debris within the synovial joint. Clin Orthop 196:69–76
156. Passler JM, Schippinger G, Schweighöfer F, Fellinger M, Seiber F-J (1995) Complications of 283 cases using a free patellar tendon graft in ACL deficient knees. Unfallchirurgie 21(5):240–246
157. Pattee GA, Snyder SJ (1988) Prothetic reconstruction of the anterior cruciate ligament: historical overview. In: Friedman MJ, Ferkel RD (eds) Prothetic ligamentreconstruction of the knee. Saunders, Philadelphia, pp 29–33
158. Pattee GA, Fox JM, del Pizzo W, Friedman MJ (1989) Four to ten years follow-up of unreconstructed anterior cruciate ligament tears. Am J Sports Med 17:430–435
159. Paulos LE, Rosenberg TD, Drawbert J, Manning J, Abbott P (1987) Infrapatellar contracture syndrome. Am J Sports Med 15:331–341

160. Paulos LE, Rosenberg TD, Grewe SR, Tearse DS, Beck CL (1992) The Gore-Tex anterior cruciate ligament prothesis. Am J Sports Med 20:246–252
161. Pelinka H (1993) Erfahrungen mit dem Laser in der Arthroskopie. Second Augsburger Forum "Arthroskopie - aktuell," Augsburg
162. Rameharan JF, Wyke BD (1972) Articular reflexes at the knee joint: an electromyographic study of the articular nerves in the knee joint. Am J Physiol 223:1276–1280
163. Rauber A (1874) Über die Vaterschen Körper der Gelenkkapsel. Centr Med Wiss 12:305–306
164. Rehm KE, Schultheis KH (1985) Ligament replacement with polydioxanone (PDS). Unfallchirurgie 11:264–273
165. Renström P, Arms SW, Stanwyck TS, Johnson RJ, Pope MH (1986) Strain within the anterior cruciate ligament during hamstring and quadriceps activity. Am J Sports Med 14:83–87
166. Richmond JC, Manseau CJ, Patz R, McConville O (1992) Anterior cruciate reconstruction using a Dacron ligament prothesis. Am J Sports Med 20:24–28
167. Roberts TS, Drez D, McCarthy W, Paine R (1991) Anterior cruciate ligament reconstruction using freeze dried, ethylene oxide-sterilized, bone-patellar tendon-bone allografts, Am J Sports Med 19:35–41
168. Rosenberg TD, Scott SM, Coward DB (1986) Arthroscopic meniscal repair evaluated with repeat arthroscopy. Arthroscopy 2:14–20
169. Rosenberg TD, Franklin JL, Baldwin GN, Nelson KA (1992) Extensor mechanism function after patellar tendon graft harvest for anterior cruciate ligament reconstruction. Am J Sports Med 20:519–526
170. Roth JH, Kennedy JC (1980) Intraarticular reconstruction of the anterior cruciate ligament in rabbits. Orthop Trans 4:399
171. Sachs RA, Daniel DM, Stone ML, Garfein RF (1989) Patello-femoral problems after anterior cruciate ligament reconstruction. Am J Sports Med 17:760–765
172. Sandberg R, Balkfors B, Nilsson B, Westlin N (1987) Operative versus non-operative treatment of recent injuries to the ligaments of the knee. J Bone Joint Surg Am 69:1120–1126
173. Satku K, Kumar VP, Ngoi SS (1986) Anterior cruciate ligament injuries. J Bone Joint Surg Br 68:458–461
174. Schabus R (1988) Die Bedeutung der Augmentation für die Rekonstruktion des vorderen Kreuzbandes. Acta Chir Austr [Suppl 76]
175. Scherer MA (1992) Experimentelle und klinische Untersuchungen zur autogenen Rekonstruktion des vorderen Kreuzbandes. Habilitationsschrift. Technical University, Munich
176. Scherer MA, Ascherl R, Siebels W, Hipp E (1992) Basic experiments on allplastic reconstructive procedures on the anterior cruciate ligament. Dtsch Z Sportmedizin 43:380–390
177. Scherer MA, Früh H-J, Ascherl R, Mau H, Siebels W, Blümel G (1992) Kinetics of resorption of different suture materials depending on the implantation site and the species. In: Planck H, Dauner M, Renardy M (eds) Degradation phenomena on polymeric biomaterials. Springer, Berlin Heidelberg New York, pp 77–96
178. Scherer MA, Früh HJ, Ascherl R, Siebels W (1993) Biomechanical studies of change in the patellar tendon after transplant removal. Aktuelle Traumatol 23:129–132
179. Scherer MA, Metak G (1995) Nachweis nervaler Strukturen nach autogener Kreuzbandrekonstruktion. In: Hasselbach C, Scherer MA (eds) Wiederherstellende Chirurgie des vorderen Kreuzbandes. Demeter, Gräfelfing, pp 67–69
180. Schultz RA, Miller DC, Kerr C, Micheli L (1984) Mechanoreceptors in human cruciate ligaments. J Bone Joint Surg Am 66A:1072–1076
181. Schutte MJ, Dabezies EJ, Zimny ML, Happel LT (1987) Neural anatomy of the human anterior cruciate ligament. J Bone Joint Surg Am 69A:243–247
182. Schwarz W, Böck W, Franzreb M, Gerngroß H (1995) Das Zyklopssyndrom – Arthrofibrose des Kniegelenkes nach Ersatzplastik des vorderen Kreuzbandes. In: Hasselbach C, Scherer MA (eds) Wiederherstellende Chirurgie des vorderen Kreuzbandes. Demeter, Gräfelfing, pp 146–153

183. Seiler H, Frank HR (1993) Suture of the anterior cruciate ligament – what is the real value of this method? Unfallchirurg 96:443–450
184. Shelbourne KD, Wilckens JH, Mollabashy A, DeCarlo M (1991) Arthrofibrosis in acute anterior cruciate ligament reconstruction. Am J Sports Med 19:332–336
185. Shino K, Inoue M, Horibe S, Nagano J, Ono K (1988) Maturation of allograft tendons transplanted into the knee. J Bone Joint Surg Br 70:556–560
186. Shino K, Inoue M, Horibe S, Hamada M, Ono K (1990) Reconstruction of the anterior cruciate ligament using allogenic tendon. Long-term follow-up. Am J Sports Med 18:457–465
187. Solomonow M, Baratta R, Zhou BH, Shoji H, Bose W (1987) The synergistic action of the anterior cruciate ligament and thigh muscles in maintaining joint stability. Am J Sports Med 15:207–213
188. Sommerlath K, Lysholm K, Gillquist J (1991) The long-term course after treatment of anterior cruciate ligament ruptures. Am J Sports Med 19:156–162
189. Soslowsky L, Goldstein S (1995) Many unknowns in ACL repair. The AAOS Bulletin
190. Strasser H (1917) Lehrbuch der Muskel- und Gelenkmechanik. Bd III. Springer, Berlin Heidelberg New York
191. Strobel M, Eichhorn J (1995) Present state and surgical technique in the reconstruction of the anterior cruciate ligament. Aktuel Chir 30:210–220
192. Sturm GM, Friedman MJ, Fox JM, Ferkel RD, Dorey FH, Del Pizzo W, Snyder SJ (1989) Acute anterior cruciate ligament reconstruction – analysis of complications. Clin Orthop 253:184–189
193. Swenson TM, Fu FH (1993) Anterior cruciate ligament reconstruction: long-term results using autograft tissue. Clin Sports Med 12:709–722
194. Tiling T, Ure B, Rixen D (1991) Die funktionelle Belastbarkeit des Kniegelenks nach Band-Kapsel-Nähten. In: Weise K, Weller S (eds) Kapsel-Band-Verletzungen des Kniegelenks. Springer, Berlin Heidelberg New York, pp 39–45
195. Torg JS, Barton TM, Pavlov H, Stine R (1989) Natural history of the posterior cruciate ligament-deficient knee. Clin Orthop 246:208–216
196. Trent PS, Walker PS, Wolf B (1976) Ligament strength, patterns strength and rotational axes of the knee joint. Clin Orthop 117:263–270
197. Wentzensen A, Drobny T, Perren S (1991) Heilungsvorgänge bei ligamentären Strukturen. In: Weise K, Weller S (eds) Kapsel-Band-Verletzungen des Kniegelenks. Springer, Berlin Heidelberg New York, pp 11–25
198. Wilk RM, Richmond JC (1993) Dacron ligament reconstruction for chronic anterior ligament insufficiency. Am J Sports Med 21:374–380
199. Wilson AS, Legg PG, McNeur JC (1969) Studies of the innervation of the medial meniscus in the human knee joint. Anat Res 165:485–492
200. Witzel U (1988) Biomechanische Untersuchungen des Kniegelenkes im Hinblick auf die Belastungscharakteristik künstlicher Kreuzbänder. Second Meeting of "Alloplastischer Bandersatz aus Trevira-hochfest," Unfallklinik, Frankfurt
201. Witzel U, Paar O, Krudwig W (1990) Biomechanische Untersuchungen zu isometrischen Zwei- und Einkanaltechniken beim vorderen und hinteren Kreuzbandersatz. Eleventh 11. Vortragsreihe "Künstlicher Bandersatz," DVM, Berlin
202. Witzel U (1992) Knorpel- und Knochendefekte des Kniegelenks als Folge pathologisch oder traumatisch bedingter Artikulationsstörungen. In: Ittel TH, Sieberth HG, Matthiaß HH (eds) Aktuelle Aspekte der Osteologie. Springer, Berlin Heidelberg New York
203. Witzel U (1993) Biomechanik des Kniegelenks. In. Kus H (ed) Polymere in der Medizin. T XXIII, no 1–2. CHZ Ars Polonia, Warsaw, pp 19–37
204. Woo SL-Y, Gomez NA, Seguchi Y, Endo CM, Akeson WH (1983) Measurement of mechanical properties of ligament substance from a bone-ligament-bone-preparation. J Orthop Res 1:22–20
205. Woo SL-Y, Inoue M, McGurk-Burleson E, Gomez Ma (1987) Treatment of the medial collateral ligament injury II. Am J Sports Med 15:22–29
206. Woo SL-Y, Adams DJ (1990) The tensile properties of human anterior Cruciate Ligament (ACL) and ACL Graft Tissues. In: Daniel D, Akeson WH, O'Connor J (eds) knee ligaments: structure, function, injury, and repair. Raven, New York, pp 279–289

207. Woo SL-Y, Hollis JM, Adams DJ, Lyon RM, Takai S (1991) Tensile properties of the human femur-anterior cruciate ligament-tibia complex. Am J Sports Med 19:217–225
208. Woo SL-Y, Smith BA, Johnson GA (1994) Biomechanics of knee ligaments. In: Fu FH, Harner CD, Vince KG (eds) Knee surgery, vol 1. Williams and Wilkins, Baltimore, pp 155–172
209. Woods GW, Chapman DR (1984) Repairable posterior menisco-capsular disruption in anterior cruciate ligament injuries. Am J Sports Med 12:381–385
210. Zimny ML, Schutte M, Dabezies R (1986) Mechanoreceptors in the human anterior cruciate ligament. Anat Res 214:204–209
211. Gely P, Drouing, Thiry PS, Tremblay GR (1984) Torsion and bending imposed on a new anterior cruciate ligament prosthesis during knee flexion: an evaluation method. J Biomech Eng 106:285–94
212. Yahia L'H, Newman N, St George M (1992) Innervation of the canine cruciate ligaments. A neurohistochemical study. Anat Histol Embryol 21(1):1–8

Part II
Artificial Ligaments: Materials and Designs, Mechanical Performance, Lifetimes, Fixation Techniques, Biofunctionality

II.1 The History of Ligament Substitutes

C. Chaput and N. Duval

Introduction

Interest in the relationship between the type of injury, the type of knee ligament lesion, and the stability of articulation dates from the nineteenth century. Hyrtl in 1847 [112] and Hönigschmied in 1893 [110] are reported to have realized the first studies. For the first time in 1885 Mayo Robson [164] described a ligament repair to the level of the knee joint. Since this first publication on ligament lesions and their treatment, studies on knee ligament injuries and descriptions of the various repair and reconstruction procedures have continued to develop. Especially over the past 20 years this literature increased tremendously due to the development of new devices for injured ligaments.

Today a retrospective study of data on ligament ruptures and their treatment shows the important diversity of repair concepts and the great variety of surgical procedures and ligament substitutes. Indeed, the torn ligaments repair is realized by sutures, while the reconstruction or the replacement of these ligaments uses autogenous biological structures (autologous substitution), collagen bioprostheses, biological allograft transplantations, and finally synthetic structures used as biological scaffolds, stents, or ligament prosthesis.

Biological Substitutes

Autologous Substitutions

Autologous substitution generally refers to an autologous structure transfer, usually a noncritical tendon surrounding the knee joint. However, this should not alter the functionality and the global viability of the articulation of the knee.

First Autologous Substitutions

In 1917 Hey Groves [105] described the first reconstruction of the anterior cruciate ligament (ACL) of the knee with the transference of autogenous tendinous structures. He used a proximal fascia lata band running through the femur, articulation, and tibia. He also performed in this same way a reconstruction of the posterior cruciate ligament (PCL) using the semitendinosus tendon (posterior Hey Groves technique). In 1920 he used a distal fascia lata band to perform a similar reconstruction (Hey Groves' intraarticular plasty [106]). This autologous substitution was introduced by Smith [221] for the reconstruction of the ACL and lateral ligaments (LL).

In 1926 Bennet [22] carried out an extraarticular procedure to reconstruct an ACL-deficient unstable knee using a free band of fascia along the medial line of the articulation. He relied upon the fact that the articulation of the knee can remain stable without ACL when the other ligaments, particularly LL, are fully operational. This opinion was also expressed by others, including Cotton and Morrison [56] and Bosworth and Bosworth [30].

In 1936 Campbell [40, 41] performed an intraarticular reconstruction using a strip of the patellar and quadriceps tendons. During the same period Palmer [195] in Sweden carried out important work on ligament lesions (definition of the "drawer sign") and on reconstruction following the technique described by Hey Groves [105]; he emphasized the care in realizing the bony tunnels. Later the semitendinosus tendon [104, 158] and portions of the patellar tendon [102] were used as autologous substitution for ligament reconstructions (ACL). The semitendinosus and rectus femoris tendons are reported by Merle D'Aubigné [174, 175] for the reconstruction of posteromedial and anteromedial ligaments. O'Donoghue [189–191] recommended a modified Groves procedure for a chronic knee instability and did not recommend dynamic reconstructions presented by Augustine [17] (Augustine's technique) and Helfet [104]. The meniscus, an avascular fibrocartilaginous structure, was also proposed by Lindstrom [153] for the reconstruction of the ACL.

The Development of Reconstruction Procedures

In 1967 Lemaire [146] presented an extraarticular reconstruction using a band of fascia lata that he had used since 1961, the Lemaire anterolateral extraarticular reconstruction. The aim of this technique was to control the rotary instability and to prevent a anterior subluxation of the lateral tibial plateau. The fascia lata band was left attached distally to Gerdy's tubercle, crossed a lateral transcondylar tunnel, and returned to itself. The use of the patellar tendon intensified with the work of Jones [128, 129] and Lam [143]. Jones carried out a transplant with the central third of the patellar tendon (Jones intraarticular plasty) while Lam modified the site of tibial attachment of the medial third of the same tendon. This procedure, with or without modification, was studied by several surgeons, including Brückner [32], Brostöm et al. [31], Gillquist et al. [88], Alm and Gillquist [8], Eriksson (Eriksson's intraarticular plasty [74]), and later Clancy [52] and Johnson et al. [127].

Transplantation of the gracilis was performed in 1950 by Lindeman [152] in Germany using an intercondylar tunnel and a transtibial bony tunnel. This procedure, initially proposed by Lindeman (Lindeman's intraarticular plasty), was later reported by Dutoit [69]. Another procedure was developed after the technique of Lindeman using the gracilis and the semitendinosus tendons, the activopassive Lindeman [35].

Slocum et al. [220] defined the notion of the rotary instability and its relationship with ligament lesions (ACL, PCL, LL). He proposed the transposition of the pes anserinus tendons as an extraarticular reconstruction. Results of this procedure were published by O'Donoghue [191], Nicholas [183], and Chick et al. [44]. Important developments were made during the 1970s. McIntosh et al. [165] proposed the "pivot shift test" to diagnose the instability related to a ruptured ACL. Torg et al. [233] described advantages of the "Lachman test" over the "anterior drawer test." Benefiting from these new developments, instabilities of the knee joint were submitted to classification [111, 135, 137]. Various other autogenous fascial substitutions [70, 115, 184, 215] and tendinous substitutions [169, 231] were also applied on extraarticular reconstructions.

In 1972 McIntosh et al. [165] used a distal fascia lata strip that they used to control the anterolateral instability and developed the "over-the-top" technique (McIntosh intra- and extraarticular plasty). Ellison [70] used a band of fascia lata with its distal bony insertion; he passed it under the lateral collateral ligament, then returned it to its original insertion (Ellison's extraarticular plasty). According to him, this procedure of ACL reconstruction is dynamic. Using this procedure, Kennedy obtained disappointing results, and the technique was modified by Ellison himself [71] in an attempt to improve stability. Insall [115] also used the fascia lata with a distal bone bloc passing over the lateral condyle, intraarticular and then fixed on the tibial surface just in front of the anterior tibial spine.

The semitendinosus tendon was used by Cho [45], Lipscomb et al. [154, 155], and Zaricznyj [253]. The patellar tendon was also proposed for combined intra- and extraarticular ligament reconstruction by Drez [68], Clancy et al. [52, 53], and Ritter et al. [204]. Reconstructions using a meniscal substitution that have been introduced by Lindstrom et al. [153] were reported by Collins et al. [54], Tillberg [232], and Walsh [241].

During the 1980s authors [251] presented results of reconstruction procedures proposed by Hey Groves [107], Jones [127–129], McIntosh [165], Lemaire [146], Insall [114, 115], Slocum [220] or Helfet [104]. Biga et al. [26] reported their results with the procedure of Insall [114, 115], and Dejour et al. [65] reported their results using the medial third of the patellar tendon and a Lemaire anterolateral extraarticular reconstruction. In Switzerland Jakob et al. [119] proposed a intraarticular reconstruction using the medial third of the patellar tendon with a Hoffa pediculate plasty, and a systematic extraarticular plasty. The PCL reconstruction is studied by Courvoisier [57] and Insall [114] using the lateral and the medial gastrocnemius muscle.

Lerat et al. [148, 149] reported on a patellar tendon reconstruction according to procedures derived from those of Jones [128, 129] and McIntosh [165]. It is also used with the quadriceps tendon for a "plastie quadricipitale bicroisée" [148]. This technique allows simultaneous reconstruction of the PCL and ACL following dislocation of the knee. Kennedy et al. [138] and Noyes et al. [186–

188] published studies on the mechanical properties of human ACL and various autogenous tendon strengths including the semitendinosus, gracilis, patellar tendon, fascia lata, and iliotibial band. Various evaluations have shown modifications of the initial mechanical properties of these autogenous tissues related to their remodeling and revascularization. The autologous ligament substitution or biological augmentation of the injured ligaments of the knee is still the most commonly performed surgery despite the introduction of new concepts of ligament reconstruction.

Autologous Skin Transplants

Anderson [13] presented his results of skin transplants for the reconstruction of LL and ACL. He compared his results to those related to fascia lata reconstructed ligaments. Gibbens [87] also reported on the use of a skin autotransplant for ACL substitution in a canine model. Swart [228] and Delchef [66] reported the use of the skin in orthopedic surgery, especially for ligament and tendon reconstruction. Autologous skin transplant as biological substitute of the ACL was published by Leighton [145], Forster et al. [77], and Padovani [193]. Nevertheless, the development of skin ligament transplants remained limited in comparison with others materials.

Allografts

The allografts represent an exciting alternative to the autografts. Indeed, it is possible to store an important number of bone-tendon-bone and bone-ligament-bone units of various sizes and shapes for surgical needs. For ACL reconstruction deep-frozen or cryopreserved fascia lata, Achille's tendon, bone-patellar tendon-bone, and bone-ACL-bone units are possible considerations. Gluck in 1881 and Rehn in 1910 [213] are the firsts to report on tendon allograft transplantation in humans. Noyes and Butler [185], Sheh et al. [217], Paulos et al. [198, 199] and Butler et al. [36] studied the biomechanical behavior of various fresh deep-frozen transplants compared to other collagenic transplants. Webster and Werner [242] used a flexor tendon allograft for ACL reconstruction in animals; Shino et al. [218, 219] used an allogenic tendinous transplant and Curtis et al. [58] used cryopreserved fascia lata for the same purpose. Vasseur et al. [236] studied the bone-ligament-bone complex as an ACL allograft in dogs.

The harvesting, sterilization, and conservation of the allografts are factors that influence performances of such biological transplants. Jackson et al. [116–118] studied bone-patellar tendon-bone and bone-ACL-bone allografts sterilized by ethylene oxide and preserved by cryopreservation. Others studied factors related to the behavior of such ACL allografts [79, 212]. These evaluations showed that the revascularization takes place between 6 and 12 weeks after implantation. On the other hand, the matrix architecture remained normal with viable cells. Eight months after transplantation the biomechanical strength of the allograft was 30% less than normal specimens. Finally, an inflammatory reaction was observed for deep-frozen allografts, but this reaction was of the same magnitude of those described for fresh autografts and less marked than those described for fresh allografts. However, the stability of the articulation was not always provided by these allografts.

General concerns have been recently stressed for the direct transmission of pathogens, virus, or infection agents through transplantation of biological fluids, grafts, or derived products. Allogenic biological substitutes may appear as potential transmitters for pathogens or infectious agents. This limits the attractiveness of allografts in tissue repair or reconstruction.

Heterologous Bovine Collagen Xenografts

In 1982 a collagen bioprosthesis of bovine origin was presented by Abbink [1], and later by Berg et al. [24, 25]. This bovine xenograft, the ProCol CL bioprosthesis (Xenotech, Irvine, Calif., USA) has been treated in 2% glutaraldehyde in an effort to "reticulize" its collagen fibers content and to render it more stable and more resistant.

McMaster et al. [166–168] pioneered that biomaterial inspired by techniques under way in the field of cardiac valves bioprostheses. Thomas et al. [230] and Turner et al. [235] considered this xenograft as unsatisfactory material for an ACL reconstruction following in vivo animal studies. McMaster [166] reported bony tunnel incorporation and good short-term results with the bovine xenograft. Abbink [2] presented satisfactory clinical results using the bovine xenograft reconstruction for the chronic ACL deficient knee. Collagen particles were observed during degradation or rupture of the graft; they were found in the synovium and may be responsible of synovitis reactions. Abbink [2] and Whipple [245] reported such observations and insisted on technical details during the implantation of these bioprosthesis. Allen et al. [6] expressed the opinion that bovine xenograft bioprostheses were not satisfactory in the long-term replacement of the ACL. Other problems reported with xenografts are antigenicity and permanent strain which are probably related to acellularity and remodeling.

Sutures

Ligament repair by sutures is probably the simplest surgical procedure. It is also the oldest synthetic material introduction, since initiated by Mayo Robson in 1885 [164]. Attempts to realize a synthetic reconstruction of a ligament, such as the ACL, are intimately linked to this type of material. Smith [221] presented an ACL replacement using several sutures of silk. He referred to an ACL synthetic replacement performed previously by Corner using steel wires. A severe synovitis reaction was observed in the presence sutures of silk and required their extraction. O'Donoghue [190, 191] performed such repairs with natural suture materials (silk and cotton). DePalma [194] and O'Donoghue

[190, 191] contributed greatly to the development of ligament repair techniques. In 1960 Johnson [126b] introduced braided Nylon (polyamide) in canine ACL replacement. Vaughn [237] presented a similar type of work with Nylon. Allman [7] reported resorbable polyglycolic acid (PGA) Dexon application for collateral ligaments and ACL repair. In animals Saidi et al. [214] studied Ti-Cron, Teflon, Tevdek, Ethicon, and a steel suture, but these first artificial ligaments yielded disappointing results. Many authors described surgical repair techniques by these various sutures during the 1970s [63, 64, 73, 94, 162, 163, 247]. Polylactide sutures for ligaments and tendons are still commercially available at the present time [103]. Although these sutures failed for the replacement of the ligaments of the knee, they have allowed the development of the more complex synthetic structures that we know today as prosthetic substitutes of the various ligaments.

Table 1. Textile trademarks

Synthetic fiber trademarks	Companies
Nylon, Dacron, Teflon, Lycra, Nomex, Kevlar	Du Pont of Nemours, USA
Ti-cron, Dexon	Davis & Geck, Cyanamid Canada, Canada
Tedvek	Deknatel, USA
Ethicon, PDSR	Ethicon, USA
Silastic	Dow-Corning, USA
Terylene	Imperial Chemical, UK
Tersuisse, Kermel	Rhône-Poulenc Fibres, France
Trevira	Telos, Germany
Dacrilène	Sciences & Médecine, France
Gore-Tex	Gore, USA
Impra	Impra, USA
Spectra	Allied-Signal, USA
Marlex	Phillips Petroleum, USA
PHP	Cendis Medical, France
Hercules	HIMONT (Hercules-Montedison), USA
Graphil	Courtaulds, UK
Plastifil	A.W. Showell, UK
Rigidor	Serofin, France
Sigraphil	Sigry, Germany

Synthetic Implants

The first attempts to replace the ACL with polymeric substitutes appeared in the beginning of the 1960s. In 1959 Gort and Rostrup [92] used a Teflon (polytetrafluoroethylene) structure to replace the ACL in a canine model. Various prosthetic substitutes made of Teflon, woven or braided, are then experimented in animals [37, 72]. Johnson [126b] and Vaughn [237] studied fine braided structures of Nylon (polyamide) for ACL replacement. Gupta et al. [97] reported on silicone coated Dacron (aromatic polyester; polyethylene terephthalate) braided prostheses in dogs. Lavsan (polyester) prostheses have been studied by several authors in Eastern Europe [27, 86, 98, 179, 239, 240]. (See Tables 1, 2.)

Table 2. Prosthetic ligaments and artificial ligament augmentation devices

Developed/commercialized synthetic ligaments	Biomedical companies
ProCol CL bioprosthesis	Xenotech, Irvine, Calif., USA
PolyFlex	Richard, Memphis, Tenn., USA
Cruciate Ligament Stent, Proplast	Vitek, Houston, Tex., USA
Integraft	Hexcel Medical, Dublin, Calif., USA
Lafil	Braun Melsungen, Melsungen, Germany
Biocarb	Société Européenne de Propulsion, Claret, France
Gore-Tex ACL Ligament	Gore, Flagstaff, Ariz., USA
PHP Ligaments	Cendis Medical, Paris, France
Spectra Polyethylene Ligament	Consumer Products, New Brunswick, N.J., USA
Kennedy LAD	3M, St. Paul, Minn., USA
Stryker Ligament	Stryker, Kalamazoo, Mich., USA
Surgigraft ABC prostheses	Surgigraft, Leeds, UK
Leeds-Keio ligament	Howmedica, USA
SEM Ligament	Sciences et Médecine, Paris, France
Ligastic and Ligapro	Orthomed, Dijon, France
LARS	LARS, Arc-sur-Tille, France
Lygeron	Implants, Bordeaux, France (Distributors)
MBL 500 and MBL 200	MATCO, Avignon, France (Distributors)
Proflex Prostheses	Protek, Bern, Switzerland
PDS band	Ethicon, Somerville, N.J., USA
Westminster	MATCO, Avignon, France (Distributors)

Kennedy [138] and Grood and Noyes [96] reported on polyethylene substitutes. They studied the ultrahigh molecular weight (UHMW) polyethylene Hercules 1900 prosthesis, known as PolyFlex prosthesis (Richard, Memphis, Tenn., USA). The fixation was carried out with polymethylmethacrylate bone cement and was not built according to the textile technology. Further biomechanic studies by Beauchamp et al. [19] and by Chen and Black [43] showed nonsatisfactory performances of this prosthesis for ACL replacement. Winston [246] and Meyers et al. [176] reported high-density polyethylene fibers, polypropylene filaments (Marlex), and polyester fibers (Dacron) as an LL scaffold in animals. These porous synthetic materials did support the ingrowth of collagenic fibrous tissues, particularly for the porous Marlex, but Meyers et al. [176] concluded that none of these materials (polypropylene and polyester) was satisfactory for ACL. Fatigue failure of synthetic ligaments was observed in dogs.

Several other implants were developed during the same period. A composite synthetic structure (cruciate ligament stent, Proplast, Vitek, Houston, Tex., USA) made of a nucleus of aramide fibers Nomex, of FEP (fluoroethylene propylene) polymer and a coating of carbon fibers and polytetrafluoroethylene was presented by James et al. [120] with a 50% success rate for ACL replacement and a 80% success rate for PCL replacement. Woods [252] reported on this same ligament in 1979. The surgical technique of Proplast used the medial third of the patellar tendon and a femoral and tibial transosseous tunnel fixation. Tremblay et al. [234] designed a composite ligament strong enough to meet the requirements of an ACL prosthesis.

Jenkins [121, 125, 126] pioneered animal studies using carbon fibers (Graphil HMS) to stimulate the development of neotendons and neoligaments. Carbon fibers were initially found as very biocompatible in extraarticular environment. Many authors including Forster et al. [76], Tayton et al. [229], Wolter [250], and Mendes et al. [171, 172] demonstrated the biocompatibility of carbon fibers. The extreme fineness, high tensile strength, and low elongation rate of carbon fibers were viewed of great interest in ligament and tendon surgery. Carbon fiber tows were used first, and provided highly filamentous structures with almost good inductive properties for the formation of collagenic tissue. Nevertheless, Rushton et al. [211] and Mendes et al. [171, 172] showed that carbon fibers must keep their integrity to preserve their biological and biomechanical compatibility. Rushton et al. [211] reported fragmentation of carbon fibers and observed many cases of synovitis following ligament reconstructions in humans. Béjui et al. [21], Goutallier et al. [95], Goodship et al. [90], Amis et al. [9, 11], Mendes et al. [170–172], and Cooke et al. [55] all reported the tendency of carbon fibers to fragment in particles with a strong migratory behavior.

The first carbon fiber ligament reconstructions have been realized with naked fibers. The application of these naked carbon fibers seemed inappropriate for ACL reconstruction because they were unable to withstand strong mechanical constraints. To solve this problem, carbon fiber ligaments have been protected with a coating or a protective envelope. Many synthetic polymers and copolymers have been tried for this purpose: polylactic acid

(PLA) [4a, 5, 15, 227, 249], PLA copolymer and PGA [23], and PLA copolymer and polycaprolactone [243]. Carbon fibers have a very low strain to rupture, generally less than 2%. To improve the elasticity of carbon fiber ligaments some authors have tried a flat braided structure [48, 50, 180–182]. Claës and Neugebauer [51] reported that the invasion of the carbon braid by biological tissue (collagen) increases its elasticity and its resistance to rupture. The presence of collagen is then provided by an envelope of fascia lata or by the torn portions ligament to repair [180, 225].

Many evaluations of carbon ligaments have been presented in humans and in animals, but results are very controversial. In animal studies Claës et al. [51] documented ruptures of carbon ligaments for ACL reconstruction. Gleason et al. [89], Amis et al. [9], and Goodship et al. [90] reported fragmentation of carbon ligaments in numerous particles and migration of these particles. Collagen penetration of the carbon grafts did not seem satisfactory during animal studies [9, 159, 160, 172, 173] and in human studies [151, 211]. Strum et al. [227], Thomas et al. [230], Amis et al. [9, 11] found no benefit in the use of carbon fibers for ligament reconstruction. However, Aragona et al. [14], Neugebauer et al. [181, 182], and Claës and Neugebauer [51] in animals and Burri and Neugebauer [34], Jenkins [124], Strover and Firer [225], Witvoët and Christel [249], and Mendes et al. [171] in humans reported encouraging results for these reconstructions.

Numerous experiment and applications of carbon fibers have been carried out for ligament reconstruction of the shoulder [34], hip [81, 141, 150], and knee [4a, 5, 10, 11, 15, 20, 21, 33, 34, 59, 89, 91, 93–95, 100, 122, 123, 181, 182, 210, 224, 225, 227, 248, 250]. Carbon fiber ligaments have been marketed under various forms. Jenkins, CFS, and Integraft (Hexcel Medical, Dublin, Calif., USA) are longitudinal tows of oriented carbon fibers. The Integraft is protected with a coating of PLA or polylactide-cocaprolactone [4a, 5, 210]. Lafil (Braun Melsungen, Melsungen, Germany), and Biocarb (SEP, Claret, France) are more complex flat-braided ribbons. Carbon fiber ligaments have been rejected due to complications related to the material and to its filamentous form, and they are generally no longer marketed.

In 1985 the Dacron prosthesis (polyester) was experimented in dogs for ACL reconstruction [196]. Peterson et al. [200] presented a composite segmented polyurethane prosthesis. Rey et al. [202, 203] presented an aramide fibers prosthesis. Claës et al. [47] and Bolton et al. [28, 29] presented a Gore-Tex (microporous polytetrafluoroethylene) braided filaments prosthesis. All were presented as ACL substitutes. Recent developments in the area of artificial ligaments have concentrated on synthetic materials such as polyester, polyethylene, polypropylene, microporous polytetrafluoroethylene (Gore-Tex) and the polyaramides (types I, II, and III).

The Gore-Tex prosthesis was a braided ligament composed of three multifilament beams (Gore, Flagstaff, Ariz., USA). Bolton et al. [28], Ahlfeld et al. [3], Ferkel et al. [75] and Indelicato et al. [113] presented encouraging preliminary results with a certain number of ruptures and the presence of intraarticular particles of Gore-Tex after rupture. Stonebrook et al. [223], Markolf [161], Fox

[78], Vireday et al. [238] reported on in vitro and in vivo biomechanic performances of the Gore-Tex ligament.

Winston [246] reported on polyethylene fibers for LL reconstruction. New ligament prostheses made of braided or knitted structures of high molecular weight polyethylene (PHP fibers) were introduced by Medical Cendis (Paris, France). In 1989 several other polyethylene ligament prostheses were developed. One of these (USA patent 4713910), was made of high-tenacity polyethylene fibers Spectra 900 or 1000 (Allied-Signal, New York, N.Y., USA) of two multifilament strands placed under independent tension. The two strands have a diameter of 4–6 mm and are protected by a synthetic envelope (polymer or textile) in the bony tunnels. This ligament, proposed by Consumer Products (New Brunswick, N.J., USA), was offered as an ACL /PCL substitute. Richards (Memphis, Tenn., USA) proposed an UHMW polyethylene/polypropylene ligament. A central braid of polyethylene or UHMW polypropylene protected by a tear-resistant envelope of unknown composition is available in cylindrical, elliptical, or flat shape.

Kennedy [134, 136] used a polyethylene cousin, the polypropylene fiber, to support a autogenous biological transplant. The Kennedy ligament augmentation device (3M, St. Paul, Minn., USA) allows a biological and synthetic substitution of the ACL. Kennedy [136] and Roth et al. [206, 207] presented numerous reports on this ligament.

Several comparisons of carbon fibers and polyester fibers have been experimented for the repair of ligaments in animals [9, 11]; They reported the best results with polyester fibers when considering the resistance to fragmentation, the collagen penetration and the level of the bony fixation of the ligament. Thomas et al. [230] and Turner et al. [235] evaluated various ligaments made of carbon (Modmor 115 fiber), polyester (Terylene), and composite (carbon/polyester). They extensively presented the Westminster composite ligament developed by Aichroth in the United Kingdom (Westminster Hospital, London [4]). This prosthesis is composed of a multifilamentous carbon core and a braided polyester envelope.

Polyester fibers have been known in orthopedic surgery for many years. The Lavsan (polyester) was employed by several surgeons for ligament reconstructions [86, 239], and the Dacron was initially introduced in this area by Gupta et al. [97]. Rubin et al. [208, 209] reported on a knitted polyester structure (Ethicon Mersilene Mesh RM53), and a 5-mm braided band (Ethicon Mersilene RS22) while Meyers et al. [176] reported on woven and knitted polyester structures USCI DeBakey 6 mm and USCI Savage external velour 4 mm as well as a knitted Marlex mesh. Arnold et al. [16], Kapparkas et al. [130], and Park et al. [196] employed polyester fibers for acromioclavicular dislocation surgery. Frazier [80], King et al. [139, 140] and Amstutz et al. [12] all proposed making polyester artificial tendons. Polyester fibers have been largely and constantly applied for the surgery of ligaments and tendons. However, their use has been eclipsed by the adventure of carbon fiber ligaments, especially for the reconstruction or the replacement of ligaments of the knee. Their incorporation in artificial ligament structures was nevertheless generalized in the late 1980s.

Hinko [108] experimented a knitted and woven polyester fiber prosthesis in animals. Park et al. [196] proposed this prosthesis for ACL reconstruction. The Stryker prosthesis (Stryker, Kalamazoo, Mich., USA) was composed of narrow bands of woven polyester incorporated in a velour knitted polyester envelope (Dacron). For Lukianov et al. [157], the Stryker prosthesis was rather a prosthetic substitute than a tutor or a support for a growing neoligament. Lemons et al. [147] published the biomechanical characteristics of this substitute. Strover et al. [226] proposed a series of braided and double-thread strand ligament prostheses made of polyester/polyester and polyester/carbon composites: the Surgigraft ABC prostheses (Surgigraft, Leeds, UK). Another concept was developed by the University of Leeds (UK) and the University of Keio (Japan): a treillis constituted by an open woven structure of polyester. The Leeds-Keio ligament was presented as a scaffold to ease the growth of a neoligament. Seedhom [216] and Fujikawa [82–85] presented their results with this ligament.

The Leed-Keio ligament provides attractive opportunities for bony fixations at the extremities and for collagenous tissue ingrowth in the intraarticular portion. However, up to 1988 the clinical data did not provide the expected success [255].

In Europe several polyester implants have been marketed as artificial ligaments or prosthetic reinforcements for ACL reconstruction: the SEM (Sciences et Médecine, Paris, France), Ligastic, and Ligapro (Orthomed, Dijon, France), Lygeron (Implants, Bordeaux, France), MBL 500 and MBL 200 (MATCO, Avignon, France), and Proflex (Protek, Bern, Switzerland). Few literature reports on the properties and performances in vitro and in vivo in animals are available for these various prostheses. Recently two new polyester (Dacron) ligaments were released: Apex and LARS. The Apex is a simple multifilament structure, while the LARS is a strengthened knitted structure.

Polyaramide fibers were proposed for ligament substitution by James [120] and Woods [252] in a composite prosthesis (Proplast ligament). Dahhan et al. [60, 61] then experimented it under a textile form. Later Kevlar (type I: poly-*p*-phenylene terephthalamide) was evaluated and tested in the articulation of the knee by Wening et al. [244] and Passuti et al. [197], while Nomex (type II: poly-*m*-phenylene isophtalamide) was studied by Rey et al. [202, 203] and Leandri et al. [144]. Prothaid (Paris, France) manufactures the Nomex and Ligaid ligaments.

Resorbable and Tissue-Engineered Ligaments

Migliaresi and Nicolais [178] proposed a resorbable/nonresorbable structure made of Dacron polyester fibers in a polyhydroxyethylmethacrylate (pHEMA) matrix. The biodegradable portion of the artificial tendon was represented by this pHEMA matrix that reproduces the natural structure of a tendon and receives the collagen ingrowth [177, 1787]. Stol et al. [222] presented the biological and mechanical performances of a composite collagen/pHEMA implant, and Cifkova et al. [46] evaluated this implant in animals with limited success. There has been no clinical application of any collagen/pHEMA implant.

Cabaud et al. [38, 39] proposed a prosthesis obtained from three strands of ten Dexon sutures no 2.0 (PGA, Davis & Geck, River, N.J., USA) incorporated within a braided structure having the Y shape. Rodkey et al. [205] presented a woven ligament composite of polyester Dacron and PGA Dexon whose envelope is a 50%/50% mix and the core is a 80%/20% mix. Claës et al. [49] carried out an animal comparative study of several ligament prostheses and proposed a woven fine band of polydioxanone suture (PDS; Ethicon, Somerville, N.J., USA). Ligaments of PDS have been used for ligament reconstructions in Germany by Rehm and Schultheis [201], Hohlbach et al. [109], Diehl et al. [67], Haupt and Duspiva [101], Lobenhoffer et al. [156], Haas and Blauth [99], and Oberbillig et al. [192].

Recently, biodegradable polymers, for example, PLA and reconstituted collagen, have been used to develop new artificial ligaments. An experimental model of a poly-L-lactic acid biodegradable artificial ligament was proposed in Europe by Laitinen et al. [142]. Davis et al. [62] evaluated in vitro and in vivo a composite hydrogel of poly-2-hydroxyethylmethacrylate/polycaprolactone and PLA in rabbits. Other synthetic resorbable polymers studies include PGA by Zieren et al. [254], Vicryl (a copolymer of PLA and PGA) by Diehl et al. [67] and polyglyconates such as polytrimethylene carbonate. Structural architectures of these ligaments are not really complex and consist essentially of braided threads or woven bands.

Chapvil et al. [42] evaluated a cross-linked collagen braided structure in vivo in goats as a temporary support for ACL reconstruction. Kato and Silver [133] developed the collagen fiber ligament, and Kato et al. [131, 132] evaluated it. The main objective of theses studies was to manufacture a reconstituted collagen fiber ligament.

As for all soft tissues, bioabsorbable polymeric biomaterials have been proposed recently to develop temporary templates or scaffolds for tissue reconstruction. Direct implantation of bioabsorbable ligaments or reimplantation of cell-seeded and tissue-engineered ligament implants are the main ways for this purpose. However, there are few advanced or clinical applications performed up to now for a controlled and genuine ligament regeneration.

Conclusion

As for many soft tissues, ligament and tendon replacement began mainly with biological grafts, autogenous structures and allografts. However, site donor morbidity, possible transmission of virus, and pathogens, and storage and sterilization problems limit the attractiveness of biological grafts, particularly when the success rate is not as high as expected. Alloplastic substitutes, nonbioabsorbable and inert, were then proposed as therapeutic alternatives for a rapid and complete tissue function recovery. However, in recent decades ligamentoplasty with synthetic ligaments has often been associated with controversy and implant failures. It does not seem to ensure a good treatment for a long-term reconstruction of ligaments. Logically, bioabsorbable materials, bioprocessed structures and tissue-engineered ligaments are now being introduced and promoted for ligament reconstruction and regeneration.

What can be expected for ligament substitutes? The surgical treatment of ligament injuries is a complex process with multiple interactive factors, each of which is crucial for the clinical success. Many studies, developments, and evaluations have been performed on ligament substitutes and implants. Some progresses have been registered, but ligament failures still occur with the latter generation of synthetic ligaments. There is no miracle biomaterial for the ligament prosthetics or implants. More precise and complete definition, understanding, and control over the multiple factors governing the failure or success of the ligament replacement or reconstruction must be achieved. Still a lot of work must be done, so stay tuned!

References

1. Abbink EP (1986) Prosthetic ligament reconstruction of the knee. Presented at the third annual course for the American Academy of Orthopedic Surgeons. Scottsdale, AZ
2. Abbink EP (1988) Clinical experience in correction of chronic anterior cruciate ligament deficiency with bovine xenografts: a 5-year study. In: Friedman MJ, Ferkel RD (eds) Prosthetic ligament reconstruction of the knee." Saunders, London, pp 101–111
3. Ahlfeld SK, Larson RL, Collins HR (1987) Anterior cruciate reconstruction in the chronically unstable knee using an expanded polytetrafluoroethylene (PTFE) prosthetic ligament. Am J Sports Med 15(4):326–330
4. Aichroth PM, Jones CB, Thomas NP (1986) Carbon fibre and Dacron composites in the prosthetic reconstruction of the anterior cruciate ligament. J Bone Joint Surg Br 64:841

4a. Alexander H, Weiss AB, Parsons JR (1983) Absorbable polymer-filamentous carbon composites – a new class of tissue scaffolding materials. Aktuelle Probl Chir Orthop 26:78–91

5. Alexander H, Weiss AB, Parsons JR (1983) Resorbable polymer-carbon fiber compound materials – a new method of support and leading frame for joint reconstruction. Aktuelle Probl Chir Orthop 25:83–97
6. Allen PR, Amis AA, Jones MM, Heatley FW (1987) Evaluation of preserved bovine tendon xenografts: a histological, biomechanical and clinical study. Biomaterials. 8(2):146–152
7. Allman FL Jr (1973) Polyglycolic acid suture in routine sports injury surgical practice. Surg Gynecol Obstet 136(4):607–610
8. Alm A, Gillquist J (1974) Reconstruction of the anterior cruciate ligament by using the medial third of the patellar ligament. Acta Chir Scand 140:289–296
9. Amis AA, Campbell JR, Kempson SA, Miller JH (1984) Comparison of the structure of neotendons induced by implantation of carbon or polyester fibres. J Bone Joint Surg Br 66:131–139
10. Amis AA (1988) The strength of artificial ligament anchorages. A comparative experimental study. J Bone Joint Surg Br 70(3):397–403
11. Amis AA, Kempson SA, Campbell JR, Miller JH (1988) Anterior cruciate ligament replacement. Biocompatibility and biomechanics of polyester and carbon fibre in rabbits. J Bone Joint Surg Br 70(4):628–634
12. Amstutz HC, Coulson WF, David E (1976) Reconstruction of the canine Achilles and patellar tendons using Dacron mesh silicone prosthesis. I. clinical and biocompatibility evaluation. J Biomed Mater Res 10:47–59
13. Anderson RL (1956) The repair of knee ligaments by cutis-graft transplants. J Bone Joint Surg Am 38:1369–1378

14. Aragona J, Parsons JR, Alexander H, Weiss AB (1981) Soft tissue attachment of a filamentous carbon-absorbable polymer tendon and ligament replacement. Clin Orthop 160:268–278
15. Aragona J, Parsons JR, Alexander H, Weiss AB (1983) Medial collateral ligament replacement with a partially absorbable tissue scaffold. Am J Sports Med 11(4):228–233
16. Arnold JA, Stuard DT, Harrison WE, Nelson CL, Weber ER (1977) Dacron repair of acromio-clavicular separations. Sports Medicine Meeting, Memphis
17. Augustine RW (1956) The unstable knee. Am J Surg 92:380–388
18. Balkfors B (1982) The course of knee-ligament injuries. Acta Orthop Scand [Suppl] 198(53):7–14
19. Beauchamp P, Laurin CA, Bailon JP (1979) A study of the tensile strength of cruciate ligaments with regard to the possibilities of prosthetic replacement. Rev Chir Orthop Reparatrice Appar Mot 65(4):197–207
20. Béjui J, Perot F, Perault F, Vignon E, Béjui-Thivolet F, Hartmann D (1984) Etude expérimentale des fibres de carbone intra-articulaires dans le genou du chien avec et sans conservation du ligament croisé antérieur. Rev Chir Orthop [Suppl II] 70:27–29
21. Béjui J, Tabutin J, Perot F, Patricot LM, Vignon E, Arlot M (1982) Prothèse ligamentaire en fibres de carbone. Rev Chir Orthop 68:97–102
22. Bennett GE (1926) The use of fascia for the reinforcement of relaxed joints. Arch Surg 13:655–666
23. Bercovy M, Goutallier D, Voisin MC, Geiger D, Blanquaert D, Gaudichet A, Patte D (1985) Carbon-PGLA prostheses for ligament reconstruction. Experimental basis and short-term results in man. Clin Orthop (196):159–168
24. Berg WS, Starhurski TM, Moran JM (1983) Mechanical properties of bovine xenografts. Orthop Trans 7:279–280
25. Berg WS, Starhurski TM, Moran JM (1984) High loading rate and oscillatory evaluation of bovine xenografts. Trans Annu Meet Orthop Res Soc 9:381
26. Biga N, Judlin D, Thomine JM, Léger H (1986) Résultat de l'intervention d'Insall dans le traiement des laxités chroniques antérieures du genou. Rev Chir Orthop 72:115–120
27. Bogutskaia EV (1971) Regeneration of cruciate ligaments on the base of Lavsan implant in the human joint. Ortop Travmatol Protez 32:55–57
28. Bolton CW, Bruchman WC (1983) Mechanical and biological properties of the Gore-Tex expanded polytetrafluoroethylene prosthetic ligament. Aktuelle Probl Chir Orthop 26:40
29. Bolton CW, Bruchman WC (1985) The Gore-Tex expanded polytetrafluoroethylene prosthetic ligament. Clin Orthop 186:202–213
30. Bosworth DM, Bosworth BM (1936) Use of fascia lata to stabilize the knee in cases of ruptured crucial ligaments. J Bone Joint Surg Am 18: 178–179
31. Brostom L, Gillquist J, Liljedahl SO, Lindvall N (1968) Behandling av inveterarad ruptur av frme korsbandet. Lakartidningen 65:4479–4487
32. Brückner H (1966) Eine neue Methode der Krutzbandplastik. Chirurg 37:413–414
33. Burri C, Neugebauer R (1981) Technique of the alloplastic ligament substitution with carbon fibres. Unfallchirurgie 7(6):289–297
34. Burri C, Neugebauer R (1985) Carbon fiber replacement of the ligaments of the shoulder girdle and the treatment of lateral instability of the ankle joint. Clin Orthop (196):112–117
35. Buscayret C (1987) Le Lindeman activo-passif, non-isométrique: technique de traitement des laxités antérieures du genou. Rev Chir Orthop [Suppl II] 73:205–208
36. Butler DL, Noyes FR, Waltz KA, Gibbons MJ (1987) Biomechanics of human knee ligament allograft treatment. Trans Orthop Res Soc 12:128
37. Butler HC (1964) Teflon as a prosthetic cruciate ligament in repair of ruptured anterior cruciate ligaments. Am J Vet Res 25:55–60
38. Cabaud HE, Feagin JA, Rodkey WG (1982) Acute anterior cruciate ligament injury and repair reinforced with a biodegradable intra-articular ligament. Experimental studies. Am J Sports Med 10(5):259–265
39. Cabaud HE, Feagin JA, Rodkey WG (1985) Acute anterior cruciate ligament injury and repair reinforced with a biodegradable material device for repair and reconstruction of injured tendons. Am J Sports Med 13:242–247

40. Campbell WC (1936) Repair of the ligaments of the knee. Surg Gynecol Obstet 61:964–968
41. Campbell WC (1939) Reconstruction of the ligaments of the knee. Am J Surg 43:473–480
42. Chapvil M, Speer DP, Holubec H, Chapvil TA, King DH (1993) Collagen fibers as a temporary scaffold for replacement of ACL in goats. J Biomed Mater Res, 27:313–325
43. Chen EH, Black J (1980) Material design analysis of the prosthetic anterior cruciate ligament. J Biomed Mater Res 14:567–586
44. Chick RP, Collins RH, Rubin BD (1981) The pes anserinus transfer. A long term follow up. J Bone Joint Surg Am 63:1449–1452
45. Cho KO (1975) Reconstruction of the anterior cruciate ligament by semitendinosus tenodesis. J Bone Joint Surg Am 57:608–612
46. Cifkova I, Stol M, Holusa R, Adam M (1987) Calcification of poly(2-hydroxyethyl methacrylate)-collagen composites implanted in rats. Biomaterials 8:30–34
47. Claës L, Burri C, Neugebauer R, Piehler P, Mohr W (1983) Animal experimental studies on the comparison of various alloplastic materials for ligament replacement. Aktuelle Probl Chir Orthop 25:109–115
48. Claës L, Burri C, Neugebauer R, Wolter D, Rose P (1983) The influence of various carbon fibre braiding techniques and methods of fixation on the extensibility of ligament prostheses. Biomaterials 4(2):134–136
49. Claës L, Dürselen L, Kiefer H, Mohr W (1987) The combined anterior cruciate and medial collateral ligament replacement by various materials: a comparative animal study. J Biomed Mater Res 21:319–343
50. Claës L, Neugebauer R (1983) Mechanical and biomechanical properties of ligament replacement with carbon fibers. Aktuelle Probl Chir Orthop 25:55–59
51. Claës L, Neugebauer R (1985) In vivo and in vitro investigation of the long-term behavior and fatigue strength of carbon fiber ligament replacement. Clin Orthop (196):99–111
52. Clancy WG Jr (1983) Anterior cruciate ligament functional instability. A static intra-articular and dynamic extra-articular procedure. Clin Orthop 172:102–106
53. Clancy WG Jr, Nelson DA, Rider RA (1982) Anterior cruciate ligament reconstruction using one-third, augmented by extra-articular tendon transfers. J Bone Joint Surg Am 64:352–359
54. Collins HR, Hughston JC, DeHaven KE (1974) The meniscus as a cruciate ligament substitute. Am J Sports Med 2:11–21
55. Cooke PH, Goodship AE (1986) Biomechanical compatibility of cruciate ligament prostheses. Eng Med 15(2):95–98
56. Cotton FJ, Morrison GM (1934) Artificial ligaments at the knee: technique. New Engl J Med 210:1331
57. Courvoisier E (1979) La réfection active du croisé postérieur dans la laxité chronique du genou. Rev Chir Orthop [Suppl II] 65:51–53
58. Curtis R, Delee J, Drez D (1985) Reconstruction of the anterior cruciate ligament with freeze-dried fascia-lata allografts in dogs. Am J Sports Med 13:408
59. Dahhan P, Benazet JP, Rey P, Leandri J (1981) Réparation du ligament croisé antérieur par fibres de carbone. J Chir (Paris) 118:275–280
60. Dahhan P, Benazet JP, Rey P, Leandri J (1982) Carbon fiber ligament prosthesis: present problems and future prospects. Int J Artif Organs 5(3):195–198
61. Dahhan P, Léandri J, Rey P (1984) Prothèses ligamentaires et tendineuses base d'une fibre nouvelle. Biomat 84. Cahiers du Codernac, Bordeaux, pp 135–143
62. Davis PA, Huang SJ, Ambrosio L, Ronca D, Nicolais L (1991) A biodegradable composite artificial tendon. J Mater Sci Mater Med 3:359–364
63. Dejour H (1983) Les résultats du traitement des laxités antérieures du genou. Introduction matériel d'étude-méthodologie. Rev Chir Orthop 69(4):255–257
64. Dejour H (1983) Les résultats du traitement des laxités antérieures du genou. Rev Chir Orthop 69:253–302
65. Dejour H, Walch G, Deschamps G, Chambat P (1987) Arthrose du genou sur laxité chronique antérieure. Rev Chir Orthop 73:157–170

66. Delchef J, Delchef Jr J, Falla G (1959) La peau en chirugie orthopédique. Acta Orthop Belg 25(1):29–46
67. Diehl K, el-Ahmad M, Franzl K (1987) Capsular ligament surgery of the knee joint with resorbable materials. 5-year clinical experience with Vicryl and PDS ligaments. Z Orthop Ihre Grenzgeb 125(5):467–472
68. Drez D (1978) Modified Ellison procedure for chronic anterior cruciate instability. Orthopaedics 147:51–55
69. uToit GT (1967) Knee joint cruciate ligament substitution: the Lindeman (Heidelberg) operation. S Afr J Surg 5:25–30
70. Ellison AE (1979) Distal iliotibial-band transfer for antero-lateral rotatory instability of knee. J Bone Joint Surg Am 61:330–337
71. Ellison AE (1980) The pathogenesis and treatment of anterolateral rotatory instability. Orthop Clin North Am 147:51–55
72. Emery MA, Rostrup O (1960) Repair of the anterior cruciate ligament with 8 mm tube Teflon in dogs. Can J Surg 4:111–115
73. England RL (1976) Repair of the ligaments about the knee. Symposium on ski trauma and skiing safety. Orthop Clin North Am 7(1):195–204
74. Eriksson E (1976) Reconstruction of the anterior cruciate ligament. Symposium on ski trauma and skiing safety. Orthop Clin North Am 7:167–179
75. Ferkel RD, Fox JM, Wood D, Del Pizzo W, Friedman MJ, Snyder SJ (1989) Arthroscopic "second look" at the Gore-Tex ligament. Am J Sports Med 17(2):147–153
76. Forster IW, Rlis ZA, McKibbin B, Jenkins DH (1978) Biological reaction to carbon fiber implants: the formation and structure of a carbon-induced "neo-tendon." Clin Orthop 131:299–308
77. Forster WJ, Imhoff RK, Cordell JT (1963) "Close-joint" repair of anterior cruciate ligament rupture in the dog. J Am Vet Med Assoc 163:281–283
78. Fox JM (1988) Arthroscopic insertion of the prosthetic Gore-Tex ligament. In: Friedman MJ, Ferkel RD (eds) Prosthetic ligament reconstruction of the knee. Saunders, London, pp 165–179
79. France EP, Paulos LE, Rosenberg TD, Harner CD (1988) The biomechanics of anterior cruciate allografts. In: Friedman MJ, Ferkel RD (eds) Prosthetic ligament reconstruction of the knee. Saunders, London, pp 180–185
80. Frazier CH (1981) Tendon repairs with Dacron vascular graft suture. Orthopedics 4:539–540
81. Freedman LS, Jenkins AI, Jenkins DH (1988) Carbon fibre reinforcement for chronic lateral ankle instability. Injury 19(1):25–27
82. Fujikawa K (1985) Reconstruction of the cruciate ligament of the knee using an artificial ligament. Kango Gijutsu. Jpn J Nursing Art 31(1):99–101
83. Fujikawa K (1988) Clinical study of anterior cruciate ligament reconstruction with the Leeds-Keio artificial ligament In: Friedman MJ, Ferkel RD (eds) Prosthetic ligament reconstruction of the knee. Saunders, London, pp 132–139
84. Fujikawa K (1989) Clinical study on anterior cruciate ligament reconstruction with the scaffold type artificial ligament (Leeds-Keio). Nippon Seikeigeka Gakkai Zasshi J Jpn Orthop Assoc 63(8):774–788
85. Fujikawa K, Iseki F, Seedhom BB (1989) Arthroscopy after anterior cruciate reconstruction with the Leeds-Keio ligament. J Bone Joint Surg Br 71(4):566–570
86. Gegechkori IUA (1969) Restoration of injured ligaments of the knee joint with the use of Lavsan prosthetics. Voen Med Zh 8:85
87. Gibbens R (1957) Patelectomy and a variation of Paatsama's operation on the anterior cruciate ligament of a dog. J Am Vet Med Assoc 15:557–558
88. Gillquist J, Liljedahl SO, Lindvall H (1971) Reconstruction for old rupture of the anterior cruciate ligament: a follow up study. Injury 2:371–278
89. Gleason TF, Barmada R, Ghosh L (1984) Can carbon fiber implants substitute for collateral ligament? Clin Orthop 191:274–280
90. Goodship AE, Wilcock SA, Shah JS (1985) The development of tissue around various prosthetic implants used as replacements for ligaments and tendons. Clin Orthop 196:61–68

91. Gorecki A, Kus W, Pykao R, Bazewicz S, Powroznik A (1985) Experimental studies on the usefulness of carbon fibers in the reconstruction of the anterior cruciate ligament of the knee. Chirurgia Narzadow Ruchu I Ortopedia Polska 50(3):181–186
92. Gort J, Rostrup O (1959) Teflon fabric for ligament reconstruction: an experimental study. Can J Surg 3:75–78
93. Goutallier D, Bercovy M (1987) Use of carbon fibers in surgery of recent and old lesions of the antero-external cruciate ligament of the knee. Experimental studies in sheep and initial results in human practice. Acta Orthop Belg 53(3):342–347
94. Goutallier D, Bercovy M, Blanquaert D, Voisin MC, Gaudichet A, Trevisan D, Geiger G, Rouveix B (1983) Replacement of the anterior cruciate ligament by carbon fibers. An experimental study. Rev Chir Orthop Reparatrice Appar Mot 69(3):187–194
95. Goutallier D, Sterkers Y, Normand X (1983) Analyse critique des résultats du traitement des laxités fraiches. Déductions thérapeutiques. Le problème des sutures du ligament croisé antérieur. Les résultats du traitement des laxités antérieures du genou. Rev Chir Orthop 69:266–268
96. Grood ES, Noyes FR (1976) Cruciate ligament prosthesis: strength, creep, and fatigue properties. J Bone Joint Surg Am 58(8):1083–1088
97. Gupta BN, Brinker WO (1969) Anterior cruciate ligament prosthesis in the dog. J Am Vet Med Assoc 154:1057–1061
98. Gurgenidze NI, Shakhbazov ET (1977) Remote results of plastic reconstruction of the knee joint ligaments with Lavsan. Ortop Travmatol Protez 8:68–69
99. Haas N, Blauth M (1989) Injuries of the acromio- and sternoclavicular joint – surgical or conservative treatment? Orthopade 18:234–245
100. Harilainen A, Myllynen P (1987) Treatment of fresh tears of the anterior cruciate ligament. A comparison of primary suture and augmentation with carbon fibre. Injury 18(6):396–400
101. Haupt PR, Duspiva W (1988) Polydioxanone augmentation-plasty in injuries of the cruciate ligament. Unfallchirurg 91(3):97–105
102. Hauser EDW (1947) Extra-articular repair for ruptured collateral and cruciate ligaments. Surg Gynecol Obstet 84:339–345
103. Lancaster R (1995) Clinical and in vitro performance of an absorbable sort-tissue repair device. Proceedings of the Medical Textile Conference, Professional Development. Clemson University, 6–7 September 1995, Clemson, SC
104. Helfet AJ (1948) Function of the cruciate ligament of the knee joint. Lancet 1:665–667
105. Hey Groves EW (1917) Operations for the repair of the crucial ligaments. Lancet 11:674
106. Hey Groves EW (1920) The crucial ligaments of the knee joint. Their function, rupture and operative treatment of the same. Br J Surg 7:505–515
107. Hey Groves EW (1980) Operation for repair of the crucial ligaments. Clin Orthop 147:4–6
108. Hinko PJ (1981) The use of a prosthetic ligament in repair of the torn anterior cruciate ligament in the dog. J Am Anim Hos Assoc 17:563
109. Hohlbach G, Schildberg FW, Stalling F, Hoppner W (1987) The PDS tensionband – a new method of early functional treatment of surgically repaired ligament rupture of the upper ankle joint. Chirurg. 58(11):769–774
110. Hönigschmied J (1893) Leichenexperimente uber die Zerreissungen der Bänder in Kniegelenk. (Cité Bengt Balkfors, 1982) Dtsch Z Chir 36:587–620
111. Hughston JC, Andrews JR, Cross MJ (1976) Classification of knee ligament instabilities. Parts I and II. J Bone Joint Surg Am 58:159–179
112. Hyrtl J (1847) Handbuch der Topographischen Anatomie und ihrer Praktisch Medizinish-Chirurgishen Anwendungen. (Cité Bengt Balkfors, 1982) JV Wallishausser, Vienna
113. Indelicato PA, Pascale MS, Huegel MO (1989) Early experience with the Gore-Tex polytetrafluoroethylene anterior cruciate ligament prosthesis. Am J Sports Med 17(1):55–52
114. Insall J, Hood R (1982) Bone block transfer of the medial head of the gastrocnemius for posterior cruciate insufficiency. J Bone Joint Surg Am 64:691–9
115. Insall J, Joseph DM, Aglietti P (1981) Bone block iliotibial-band transfer for anterior cruciate insufficiency. J Bone Joint Surg Am 63:560–569

116. Jackson DW, Grood ES, Arnoczky SP (1987) Freeze-dried anterior cruciate ligament allografts. Preliminary studies in a goat model. Am J Sports Med 15:295–303
117. Jackson DW, Grood ES, Arnoczky SP (1988) The effects of processing techniques on the mechanical properties of bone-anterior ligament-bone allograft. An experimental study in goats. Am J Sports Med 16:101–5
118. Jackson DW, Windler GE, Simon TM (1990) Intra-articular reaction associated with the use of freeze-dried, ethylene oxide-sterilized bone-patella tendon-bone allografts in the reconstruction of the anterior cruciate ligament. Am J Sports Med 18(1):1–11
119. Jakob RP, Kipfer W, Klaue K, Staübli HU, Gerber C (1988) Etude critique de la reconstitution du ligament croisé antérieur du genou par la plastie pédiculée sur le Hoffa partir du tiers médian du tendon rotulien. Rev Chir Orthop 74:44–51
120. James SL, Woods GW, Homsy CA, Prewitt JM, Slocum DB (1979) Cruciate ligament stents in reconstruction of the unstable knee. Clin Orthop 143:90–6
121. Jenkins DH (1978) The repair of cruciate ligaments with flexible carbon fibre. A longer term study of the induction of new ligaments and of the fate of the implanted carbon. J Bone Joint Surg Br 60(4):520–522
122. Jenkins DH (1983) Alloplastic ligament replacement with carbon fibers – clinical use and results. Aktuelle Probl Chir Orthop 25:130–2
123. Jenkins DH (1983) Ligament replacement in experiments on animals. Aktuelle Probl Chir Orthop 26:92–95
124. Jenkins DH (1985) Ligament induction by filamentous carbon fiber. Clin Orthop 196:86–7
125. Jenkins DH, Forster IW, McKibbin B, Ralis ZA (1977) Induction of tendonand ligament formation by carbon implants. J Bone Joint Surg Br 59(1):53–57
126. Jenkins DH, McKibbin B (1980) The role of flexible carbon-fibre implants as tendon and ligament substitutes in clinical practice. A preliminary report. J Bone Joint Surg Br 62(4):497–499
126b. EJohnson FL (1960). Use of braided Nylon as a prosthetic anterior cruciate ligament of the dog. J Am Vet Med Assoc 137:646–7
127. Johnson RJ, Eriksson E, Haggmark T, Pope MH (1984) Five to ten year follow up evaluation after reconstruction of the anterior cruciate ligament. Clin Orthop 183:122–140
128. Jones KG (1963) Reconstruction of the anterior cruciate ligament. A technique using the central one-third of the patellar ligament. J Bone Joint Surg Am 45:925–932
129. Jones KG (1970) Reconstruction of the anterior cruciate ligament using the central one third of the patellar ligament. J Bone Joint Surg Am 52:1302–1308
130. Kappakas GS, McMaster JH (1978) Repair of acromioclavicular separations using a Dacron prosthesis graft. Clin Orthop 131:247
131. Kato YP, Christiansen DL, Hahn RA, Shieh S-J, Goldstein JD, Silver FH (1989) Mechanical properties of collagen fibres: a comparison of reconstitued and rat tail tendon fibres. Biomaterials 10:38–42
132. Kato YP, Dunn MG, Tria AJ, Zawadsky JP, Silver FH (1991) Preliminary assessment of a collagen fiber ACL prosthesis. 16th Annual Meeting of the Society for Biomaterials, Charleston, May 20–23
133. Kato YP, Silver FH (1990) Formation of continuous collagen fibres: evaluation of biocompatibility and mechanical properties. Biomaterials 11:169–175
134. Kennedy JC (1983) Application of prosthetics to anterior cruciate ligaemnt reconstruction and report. Clin Orthop 172:125–128
135. Kennedy JC, Fowler PJ (1971) Medial and anterior instability of the knee. An anatomical and clinical study using stress machines. J Bone Joint Surg Am 53:1257–1270
136. Kennedy JC, Roth JH, Mendenhall HV, Sanford JB (1980) Intra-articular replacement in the anterior cruciate ligament-deficient knee. Am J Sports Med 8:1–8
137. Kennedy JC, Stewwart R, Walker DM (1978) Anterolateral rotatory instability of the knee joint. J Bone Joint Surg Am 60:1031–1039
138. Kennedy JC, Weinberg HW, Wilson AS (1974) The anatomy and function of the anterior cruciate ligament as determinated by clinical and morphological studies. J Bone Joint Surg Am 56:223–235

139. King RN, Dunn HK, Bolstad KE (1976) A single unit digital flexor tendon prosthesis. J Biomed Mater Res 6:157–165
140. King RN, McKenna GB, Statton WO (1977) Novel uses of fibers as tendons and bones. J Appl Polym Sci Appl Polym Symp 31:335–350
141. Kinzl L, Burri C, Neugebauer R (1983) Carbon fiber ligament replacement in the treatement of chronic instability of the external joint ligament. Zentralbl Chir 108:1192–1197
142. Laitinen O, Tormälä P, Taurio R, Skutnabb K, Saarelainen K, Iivonen L, Vainionpää S (1992) Mechanical properties of biodegradable ligament augmentation device of poly(L-actide) in vitro and in vivo. Biomaterials 13:1012–1016
143. Lam SJS (1968) Reconstruction of the anterior cruciate ligament using the Jones procedure and its Guy's hospital modification. J Bone Joint Surg Am 50:1213–1224
144. Léandri J, Dahhan P, Tarragano O, Cerol M, Geiger D, Rey P (1988) Long term follow up implantation and new designed prosthesis for anterior cruciate ligament replacement. Third World Biomaterials Congress. Kyoto, Japan
145. Leighton RL (1961) Repair of ruptured anterior cruciate ligament with whole thickness skin. Small Anim Clin 1:246–259
146. Lemaire M (1967) Ruptures anciennes du ligament croisé antérieur du genou. J Chir (Paris) 102:311–320
147. Lemmons J, Devrnja R, Rigney D, Bhate A (1988) Fatigue analyses of Stryker Dacron prosthetic ligament. Abstract. Third World Biomaterials Congress, Kyoto, Japan
148. Lerat JL, Dupré-Latour L, Herzberg G, Moyen B (1987) Revue de 100 patients opérés pour laxité chronique antérieure du genou par un procédé dérivé des méthodes de K. Jones et de D McIntosh. Rev Chir Orthop [Suppl II] 73:201–203
149. Lerat JL, Dupré-Latour L, Tarquini C, Dumont P (1986) Remplacement des deux ligaments croisés avec un transplant unique provenant du système extenseur. Ligamentoplastie quadricipale bicroisée. Rev Chir Orthop 72:223–226
150. Leyshon RL (1982) Chronic lateral ligament instability of the ankle – bilateral repair using flexible carbon fibre in an international sportsman. Br J Sports Med 16(3):172–173
151. Leyshon RL, Channon GM, Jenkins DHR, Rlis ZA (1984) Flexible carbon fibre in late ligamentous reconstruction for instability of the knee. J Bone Joint Surg Br 66(2):196–200
152. Lindeman K (1950) ber den Plastichen Ersatz Kreutzbänder durch gestielte Sehnenverpflanzung. Z Orthop Ihre Grenzgeb 79:316–334
153. Lindstrom N (1959) Cruciate ligament plastics with meniscus. Acta Orthop Scand 29:150–151
154. Lipscomb AB, Johnston RK, Southwick WO (1982) Evaluation of hamstring strength following use of semitendinosis and gracilis tendons to reconstruct the anterior cruciate ligament. Am J Sports Med 10:340–342
155. Lipscomb AB, Johnston RK, Synder RB (1981) The technique of cruciate ligament reconstruction. Am J Sports Med 9:77–81
156. Lobenhoffer P, Blauth M, Tscherne H (1988) Resorbable augmentationplasty and functional after care in fresh rupture of the anterior cruciate ligament. An improved treatment concept. Z Orthop Ihre Grenzgeb 126:296–299
157. Lukianov AV, Richmond JC, Barrett GR, Gillquist J (1989) A multicenter study on the results of anterior cruciate ligament reconstruction using a Dacron ligament prosthesis in salvage cases. Am J Sports Med 17:380–386
158. Macey HB (1939) A new operative procedure for repair of ruptured cruciate ligaments of the knee joint. Surg Gynecol Obstet 69:108–109
159. Mäkisalo S, Paavlainen P, Holmström T, Skutnabb K (1989) Carbon fiber as a prosthetic anterior cruciate ligament. A biochemical and histological analysis in pigs. Am J Sports Med 17(4):459–462
160. Mäkisalo S, Skutnabb K, Holmstrom T, Gronblad M, Paavolainen P (1988) Reconstruction of anterior cruciate ligament with carbon fiber. An experimental study on pigs. Am J of Sports Med 16(6):589–593
161. Markolf KL (1988) Clinical and laboratory studies with the Gore-Tex ligament at UCLA. In: Friedman MJ, Ferkel RD (eds) Prosthetic ligament reconstruction of the knee. Saunders, London, pp 149–155

162. Marshall JL, Olsson SE (1971) Instability of the knee. A long-term experimental study in dogs. J Bone Joint Surg 53:1561
163. Marshall JL, Warren RF, Wickiewicz TL, Reider B (1979) The anterior cruciate ligament: a technique of repair and reconstruction. Clin Orthop 143:97–105
164. Mayo Robson AW (1903) Ruptured crucial ligaments and their repair by operation. Ann Surg 37:716–718
165. McIntosh DL, Darby TA (1976) Lateral substitution reconstruction. Proceedings of the Canadian Orthopaedic Association. J Bone Joint Surg (Br) 58:142
166. McMaster WC (1985) A histological assessment of canine anterior cruciate substitution with bovine xenograft. Clin Orthop (196):196-201
167. McMaster WC (1986) Mechanical properties and early clinical experience with xenograft. Biomaterials. Bull Hosp Jt Dis Orthop Inst 46(2):174–184
168. McMaster WC (1988) Biomechanics profile of Procol cruciate bioprosthesis. In: Friedman MJ, Ferkel RD (eds) Prosthetic ligament reconstruction of the knee. Saunders, London, pp 89–94
169. McMaster WC, Weinert CR, Scranton P (1974) The diagnosis and management of isolated anterior cruciate tears. A preliminary report on reconstruction with the gracilis tendon. J Trauma 14:230–235
170. Mendes DG, Iusim M, Angel D, Rotem A, Mordohohovich D, Roffman M, Lieberson S, Boss J (1986) Ligament and tendon substitution with composite carbon fiber strands. J Biomed Mater Res 20(6):699–708
171. Mendes DG, Iusim M, Angel D, Rotem A, Roffman M, Grishkan A, Mordohohovich D, Boss J (1985) Histological pattern of biomechanic properties of the carbon fiber-augmented ligament tendon. A laboratory and clinical study. Clin Orthop 196:51–60
172. Mendes DG, Soudry M, Angel D, Grishkan A, Roffman M (1985) Carbon reconstruction in the anterior cruciate deficient knee. Orthopedics 8(10):1244–1248
173. Mendes DG, Soudry M, Roffman M, Boss J (1988) Maturation of composite ligament. Clin Orthop 234:291–295
174. Merle d'Aubigné R (1950) Traitement des laxités du genou. Journées orthopédiques. Rev Orthop 36:297–300
175. Merle d'Aubigné R, Ramadier JO (1964) Ruptures ligamentaires du genou. Mem Acad Chir 24(25):726–729
176. Meyers JF, Grana WA, Lesker PA (1979) Reconstruction of the anterior cruciate ligament in the dog. Am J Sports Med 7(2):85–90
177. Migliaresi C, Nicodemo L, Nicolais L (1987) Hydrogels for artificial tendons. In: Peppas NA (ed) Hydrogels in med and pharmacy, vol 2. CRC, Boca Raton, pp 65–93
178. Migliaresi C, Nicolais L (1980) Composite materials for biomedical applications. Int J Artif Organs 3:114–118
179. Nazaretskii AS, Vendenoz VI, Smolianoz LA, Zherebtsov SN, Fiderov AA (1975) Use of Lavsan in orthopedic-traumatological practice. Ortop Travmatol Protez 1:58–61
180. Neugebauer R, Burri C (1985) Carbon fiber ligament replacement in chronic knee instability. Clin Orthop 196:118–123
181. Neugebauer R, Claës L (1983) Reaction of tissues to carbon fibers as ligament prostheses of the sheep knee. Aktuelle Probl Chir Orthop 25:103–108
182. Neugebauer R, Claës L (1983) The biological reaction of the tissues to carbon fibre ligament prosthesis in sheep-knees. Aktuelle Probl Chir Orthop 26:96–100
183. Nicholas JA (1973) The five-one reconstruction for anteromedial instability of the knee. J Bone Joint Surg Am 55:899–922
184. Nicholas JA, Minkoff J (1978) Iliotibial band transfer through the intercondylar notch for combined anterior instability. Am J Sports Med 6:341–353
185. Noyes FR, Butler DL, Grood ES, Zernicke RF, Hefzy MS (1984) Biomechanical analysis of human ligament grafts used in knee-ligament repairs and reconstructions. J Bone Joint Surg 66:344–352
186. Noyes FR, Butler DL, Paulos LE (1983) Intra-articular cruciate reconstruction, part I: perspective on graft strength, vascularization and immediate motion after replacement. Clin Orthop 172:71–77

187. Noyes FR, Matthews DS, Mooar PA (1983) The symptomatic anterior cruciate-deficient knee. II. the results of rehabilitation, activity modification, and counseling on functional disability. J Bone Joint Surg Am 65:163–174
188. Noyes FR, Mooar PA, Matthews DS (1983) The symptomatic anterior cruciate-deficient knee. I. the long-term functional disability in athletically active individuals. J Bone Joint Surg Am 65:154–162
189. O'Donoghue DH (1950) Surgical treatment of fresh injuries to the major injuries of the knee. J Bone Joint Surg Am 32:721–738
190. O'Donoghue DH (1955) An analysis of end results of surgical treatment of major injuries to ligaments of the knee. J Bone Joint Surg Am 37:1–13
191. O'Donoghue DH (1963) A method for replacement of the anterior cruciate ligament of the knee. J Bone Joint Surg Am 45:905–924
192. Oberbillig C, Kirschner P (1989) Postoperative continuous passive motion following augmentation of anterior cruciate ligament ruptures with polydioxanone ligaments. Early results of a prospective study. Unfallchirurgie 15(3):145–151
193. Padovani M (1964) Traitement des laxités post-traumatiques du genou. Mem Acad Chir 24/25:730–732
194. De Palma (1954) Diseases of the knee. Lippincott, Philadelphia
195. Palmer I (1938) On the injuries of the ligaments of the knee joint. Acta Chir Scand [Suppl] 53:1–282
196. Park JP, Grana WA, Chitwood JS (1985) A high-strength Dacron augmentation cruciate ligament reconstruction. Clin Orthop 196:175–185
197. Passuti N, Daculsi G, Guoin F, Martin S, Vigneron M (1989) Ligamentoplastie du ligament croisé antérieur: Etude expérimentale de fibers aramides intra-articulaires chez le chien. Rev Chir Orthop 75:353–359
198. Paulos LE, France EP, Rosenberg TD (1987) Comparative material properties of allograft tissues for ligament replacement: effects of type, age, sterilization and preservation. Trans Orthop Res Soc 12:129
199. Paulos LE, Rosenberg TD, Gurley WD (1988) Anterior cruciate ligament allografts. In: Friedman MJ, Ferkel RD (eds) Prosthetic ligament reconstruction of the knee. Saunders, London, pp 186–192
200. Peterson CJ, Donachy JH, Kanelak A (1985) A segmented polyurethane composite prosthetic anterior cruciate ligament. In vivo study. J Biomed Mater Res 19:589–594
201. Rehm KE, Schultheis KH (1985) Ligament replacement with polydioxanone (PDS). Unfallchirurgie 11(5):264–273
202. Rey P, Léandri J, Dahhan P, Tanageno O, Carol ML, Geiger D (1988) Tendon ligament prosthesis. The Third World Biomaterials Congress. April, Kyoto
203. Rey P, Dahhan P, Léandri J, Wierzbicki Z (1985) New fiber for ligament and tendon prosthesis. Trans Am Soc Artif Int Organs 31:305–307
204. Ritter MA, Leaming ES, McCarroll JR (1983) Preliminary report on the Jones Ellison, Slocum repair for symptomatic anterior cruciate deficient knees. Am J Sports Med 11:89–94
205. Rodkey WG, Cabaud HE, Feagin JA, Perlik PC (1985) A partially biodegradable material device for repair and reconstruction of injured tendons. Experimental studies. Am J Sports Med 13(4):242–247
206. Roth JH, Kennedy JC, Anderson C (1988) Intra-articular reconstruction of the anterior cruciate ligaments in rabbits. Transplant Implant Today 5:59–65
207. Roth JH, Kennedy JC, Lockstadt H, McCallum CL, Cunning LA (1985) Polypropylene braid augmented and nonaugmented intraarticular anterior cruciate ligament reconstruction. Am J Sports Med 13(5):321–336
208. Rubin RM, Marshall JL (1975) Porous hydrophilic polymer: good and bad news in the orthopedic application of cruciate ligament substitution. J Biomed Mater Res 9:375–380
209. Rubin RM, Marshall JL, Wang J (1975) Prevention of knee instability. Clin Orthop 113:212–236
210. Rusch RE, Wells ER (1986) The intra-articular use of bio-degradable polymer coated carbon filament as an adjunct to repair or reconstruction of cruciate ligaments. Acta Orthop Belg 52(4):564–573

211. Rushton N, Dandy DJ, Naylor CPE (1983) The clinical, arthroscopic and histological findings after replacement of the anterior cruciate ligament with carbon fibre. J Bone Joint Surg Br 65(3):308–310
212. Sabiston P, Frank C (1987) Ligament allografts: a biological solution to chronic ligament deficiency. Transplant Implant Today 4:43–47
213. Sabiston P, Frank C, Lam T, Shrive N (1990) Allograft ligament transplantation. A morphological and biochemical evaluation of a medial collateral complex in a rabbit model. Am J Sports Med 18(2):160–168
214. Saidi K, Beauchamp P, Laurin CA (1976) Prosthetic replacement of the anterior cruciate ligament in dogs. Can J Surg 19(6):547–549
215. Scott WN, Schosheim PM (1983) Intra-articular transfer of the iliotibial muscle-tendon unit. Clin Orthop 172:97–101
216. Seedhom BB (1988) The Leeds-Keio ligament: biomechanics. In: Friedman MJ, Ferkel RD (eds) Prosthetic ligament reconstruction of the knee. Saunders, London, pp 118–131
217. Sheh M, Butler DL, Stouffer DC (1986) Mechanical and structural properties of the human cruciate ligament and patellar tendon. Trans Orthop Res Soc 11:236
218. Shino K, Inove M, Horibe S, Hamada M, Ono K (1990) Reconstruction of the anterior cruciate ligament using allogeneic tendon. Am J Sports Med 18(5):457–465
219. Shino K, Kawasaki T, Hirose H, Gotoh I, Inove M, Ono R (1984) Replacement of the anterior cruciate by an allogeneic tendon graft. J Bone Joint Surg Br 66:672
220. Slocum DB, Larson RL (1968) Rotary instability of the knee. Its pathogenesis and a clinical test to demonstrate its presence. J Bone Joint Surg 50:211–225
221. Smith A (1918) The diagnosis and treatment of injuries of the crucial ligaments. Br J Surg 6:176–189
222. Stol M, Tolar M, Adam M (1985) Poly(2-hydroxyethyl methacrylate) - collagen composites which promote muscle cell differentiation in vitro. Biomaterials 6:193–197
223. Stonebrook SN, Berman AB, Bruchman WC, Bain JR (1988) Functional biomechanics of the Gore-Tex cruciate-ligament prosthesis: effects of implant tensioning. In: Friedman MJ, Ferkel RD (eds) Prosthetic ligament reconstruction of the knee. Saunders, London, pp 140–148
224. Strover AE (1983) Technical advances in the reconstruction of knee ligaments using carbon fibre. Aktuelle Probl Chir Orthop 26:127–134
225. Strover AE, Firer P (1985) The use of carbon fiber implants in anterior cruciate ligament surgery. Clin Orthop 196:88–98
226. Carson-Pullicino VN, McCall JW, Strover AE (1994) MRI of the knee following prosthetic anterior cruciate ligament reconstruction. Clin Radiol 49(2):89–99
227. Strum GM, Larson RL (1985) Clinical experience and early results of carbon fiber augmentation of anterior cruciate reconstruction of the knee. Clin Orthop 196:124–138
228. Swart HA (1952) The use of cutis in orthopedic operations. South Med J 45:87–91
229. Tayton K, Phillips G, Rlis Z (1982) Long-term effects of carbon fibre on soft tissues. J Bone Joint Surg Br 64(1):112–114
230. Thomas NP, Turner IG, Jones CB (1987) Prosthetic anterior cruciate ligaments in the rabbit. A comparison of four types of replacement. J Bone Joint Surg Br 69(2):312–316
231. Thompson SK, Calver R, Monk CJE (1978) Anterior cruciate ligament repair for rotatory instability. The Lindeman dynamic muscle-transfer procedure. J Bone Joint Surg Am 60:917–920
232. Tillberg B (1977) The late repair of torn cruciate ligaments using menisci. J Bone Joint Surg Br 59:15–19
233. Torg JS, Conrad W, Kalen V (1976) Clinical diagnosis of anterior cruciate ligament instability in the athlete. Am J Sports Med 4:84–91
234. Tremblay GR, Laurin CA, Drovin G (1980) The challenge of prosthetic cruciate ligament replacement. Clin Orthop 147:88–92
235. Turner IG, Thomas NP (1986) A preliminary histological study of the fixation of four prosthetic anterior cruciate ligaments in the rabbit. Clin Mat 1:259–267
236. Vasseur PB, Rodrigo JJ, Stevenson S, Clark G, Sharkey N (1987) Replacement of the anterior cruciate ligament with a bone-ligament-bone anterior cruciate allograft in dogs. Clin Orthop 219:268–277

237. Vaughn LC (1963) A study of the replacement of the anterior cruciate ligament in the dog by fascia, skin and nylon. Vet Rec 75:537–541
238. Vireday C, Settlage R, Bain JR (1988) Tissue response to a porous ePTFE prosthetic ligament: correlation of laboratory and clinical findings. Society for Biomaterials, Symposium on Retrieval Analysis of Surgical Implants Biomaterials. Snowbird, USA
239. von Mironova ZS (1978) Spätresultate der Rekonstruktion des Bandapparates des Kniegelenks mit Lawsan. Zentralbl Chir 103:432
240. von Mironova ZS, Bogutskaia EV (1975) Remote results of alloplastic reconstruction of cruciform ligaments of the knee joint. Vestn Khir Im I I Grek 115(8):85–88
241. Walsh JJ Jr (1972) Meniscal reconstruction of the anterior cruciate ligament. Clin Orthop 89:171–177
242. Webster DA, Werner FW (1981) Freeze-dried flexor tendons in anterior cruciate ligament reconstruction. Clin Orthop 181:238–243
243. Weiss AB, Blazina ME, Goldstein AR, Alexander H (1985) Ligament replacement with an absorbable copolymer carbon fiber scaffold - early clinical experience. Clin Orthop 196:77–85
244. Wening JV, Langendorff U, Delling G, Marquardt H, Hoffman M, Jungbluth KH (1989) Initial results of the biocompatibility, cytotoxicity and genotoxicity of Aramid. Unfallchirurgie 15(5):215–220
245. Whipple TL (1988) Arthroscopic anterior cruciate ligament reconstruction with Procol xenograft bioprosthesis. In: Friedman MJ, Ferkel RD (eds) Prosthetic ligament reconstruction of the knee. Saunders, London, pp 112–117
246. Winston LA, Tercy AF, Jardine JH, Parrish FF (1978) The result of replacement of partial or total collateral ligaments with Marlex mesh in the knees of dogs. Clin Orthop 137:287–290
247. Witvoet J (1983) Traitement des laxités fraiches. Les résultats du traitement des laxités antérieures du genou. Rev Chir Orthop 69:258–263
248. Witvoet J, Christel P (1985) Reinforcement of plastic surgery of the anterior cruciate ligament using carbon fiber. A propos of 44 cases. Rev Chir Orthop Reparatrice Appar Mot [Suppl II] 71:121–125
249. Witvoet J, Christel P (1985) Treatment of chronic anterior knee instabilities with combined intra- and extra-articular transfer augmented with carbon-PLA fibers. Clin Orthop 196:143–153
250. Wolter D (1983) Biocompatibility of carbon fibers and carbon microparticles. Aktuelle Probl Chir Orthop 25:30–38
251. Woods GW (1985) Synthetics in anterior cruciate ligament reconstruction: A review. Symposium on the anterior cruciate ligament, part II. Orthop Clin North Am 16:227–235
252. Woods GW, Homsy CA, Prewitt JM, Tullos HS (1979) Proplast leader for use in cruciate ligament reconstruction. Am J Sports Med 7(6):314–319
253. Zaricznyj B (1983) Reconstruction of the anterior cruciate ligament using free tendon graft. Am J Sports Med 11:164–176
254. Zieren HU, Holzmuller WA, Rosenberg J, Rehm KE (1991) Are patellar tension band reconstructions with absorbable materials possible? An animal experiment. Unfallchirurg 94:634–639
255. Moyen B (1988) Ligaments artificiels pour le remplacement du ligament croisé antérieur du genou. Cahiers d'enseignements de la SOFCOT. Clinique Chirurgicale Orthopédique, Paris, pp 149–178

II.2 A Classification of Prosthetic Ligament Failures

N. Duval and C. Chaput

Introduction

The vast majority of ligament implants have been constructed of the synthetic textiles. They are therefore composed of artificial polymeric materials and manufactured in fibrous structures organized according to the defined models. Since the studies of Voorhes et al. [16] and Harrison [4] which brought textiles into the implantable synthetic material field, more complex porous textile structures have been used for prostheses and surgical implants. Cardiovascular surgery was certainly the first area to benefit of artificial prostheses built from polyester-type (polyethylene terephthalate, polyester PET; Dacron) or poly-tretrafluoroethylene-type (polytetrafluoroethylene, PTFE; Teflon) textiles (1952–1957). The first synthetic ligament prosthesis, a textile implant made of PTFE, was introduced by Gort and Rostrup [3] and Emery and Rostrup [2].

A specific point of ligament substitutes lies in a demanding and precise function, and the triple (material-structure-function) relationship. Indeed, the choice of a synthetic material and the selection of an implant structure are totally dependent on the biological and biomechanical needs related to the exact final function: reconstruction, reinforcement, or replacement. This multiplies the conceptual choices and possible options for realizing a ligament implant and renders more complex the area of artificial ligament substitutes.

The understanding of this point suggests the answer to the question of why artificial knee ligaments fail. The reasons for failure can be divided into three major groups: bad material, bad structure, bad surgical technique. These groups can be considered separately, but one must realize that most of the time a failed ligament is related to a combination of these groups to some degree. This chapter describes these three groups of factors and their significance to the orthopedic surgeon who wishes to work with artificial ligaments.

Artificial Ligament Materials

Diversity of Synthetic Materials

Many artificial materials have been experimented with in ligament prostheses. The preliminary experience of cardiovascular surgery probably allowed the first selection. The synthetic polymers chosen therefore consist of PTFE, polyethylene terephthalate (polyester PET), and polyolefines (polyethylene, PE; polypropylene, PP). These synthetic materials were products of postwar industrial chemistry (1947–1950). PTFE, PET, PE, and PP are considered relatively inert, resistant to biological attacks and to sterilization. They have established themselves as mechanically reliable. PET and PTFE have demonstrated a certain success as constitutive materials of arterial prostheses. Teflon (PTFE), Marlex, Hercules, Spectra, PHP (polyolefines), Lavsan, Dacron, Terylene, Tersuisse, and Dacrilène (PET) are commercial brands of polymeric synthetic ligament implants.

The materials available at present can be divided into six classes: (a) materials of carbon fiber or mineral origin, (b) polyesters, (c) PE-PP, (d) PTFE, (e) others, and (f) composites.

Carbon Fiber and Mineral-Origin Materials

Most of the works published on ligament prostheses or implants made of mineral origin fiber relates to carbon fiber. The carbon fiber consists of atoms of carbon, more or less organized. It is obtained in controlled atmosphere by progressive thermal decomposition of a precursory thread. This precursory thread can be, for example, an acrylic, a pitch (oil or coal residue) or a cellulosic. The manufacture of a carbon fiber requires four steps: reticulation or controlled oxydation, carbonization, graphitization, and a surface processing. Many studies have been conducted on ligament implants made of carbon fiber (Modmor, Morganite; Graphil, Courtaulds Carbon Fiber), and four models have been commercialized (Integraft, Lafil, Biocarb, Jenkins).

The chemical inertia, fineness, and weak strain of carbon fiber seemed attractive at first. Many clinical trials were conducted and suggested a good result of this material used in extraarticular soft tissue reconstruction but a poor performance in intraarticular reconstruction. In knee surgery the rigidity and fragility of the carbon fiber led to an unacceptably high early failure rate. Report of cases of early rupture and severe synovitis confirmed the tendency of this material to fragment and then to migrate. These facts have considerably limited its use, and to our knowledge no carbon fiber ligament is still commercially available.

Polyester PET

Polyester PET is generally called Dacron but also exists under numerous commercial names, including Lavsan, Terylene, Tersuisse, Dacrilene, and Trevira. It has been produced and marketed in 1947–1948 by Du Pont of Nemours (United States) and by Imperial Chemical Industry (United Kingdom). This industrial polymer is the material of choice for medical textile construction, and it is hence very common within implanted structures, prosthesis, implants, or surgical equipment. The polymer itself is biocompatible and relatively resistant to attacks of the physiological tissue (surface degradation). Polyester was introduced within synthetic ligaments in the early 1960s, and is still used today.

It is interesting to notice that two of the most successful commercially available artificial ligament systems, the Ligament Advanced Reinforcement System (LARS) and Trevira ligaments, are both made of polyester PET although their ligament structures and recommended surgical techniques differ from one another. This confirms that polyester PET is currently the non-biodegradable synthetic material of choice for the realization of permanent intra- and extraarticular soft tissue implants.

Polyethylene and Polypropylene

The synthesis of PE was discovered in 1930 by Imperial Chemical Industry, and it was distributed in 1956. It is generally considered for parts of implants used for joint arthroplasties (e.g., hip, knee). PE came into use in 1975 for the manufacture of a nontextile ligament prosthesis (Polyflex). Marlex PE (Phillips Petroleum) was introduced in 1957–1959 for soft tissue repair [12–14]. For ligament reconstruction Marlex was experimented in 1978–1979 [9, 17]. A textile ultrahigh molecular weight PE appeared recently for ligament substitution (polyethylene high-performance, PHP). Marois et al. [8] evaluated the biocompatibility of PHP fiber and compared it to that of polyester PET. Different types of PE are available in fibrous or nonfibrous form: high density, low density, ultrahigh molecular weight, ultraoriented (fibers). The latest generations of PE present excellent mechanical and physical properties.

PP was developed in 1954, marketed in 1957 (Montecatini, Italy), and generally distributed in 1962 (Imperial, Shell). Polypropylene is a polyolefine, as is PE. Its introduction in biomaterials is limited, and its essential application to ligament is confined to the Kennedy Ligament Augmentation Device (LAD) stent. The mechanical resistance of PP is not as good as PEs, and its in vivo biostability must be evaluated according to its exact chemical composition.

The problem related to these materials lies in the biological reaction to the wear-particle release. PE is currently considered the "weak point" of total hip and total knee implants. Wear particles are thought to cause foreign body reactions, leading to the release of lytic enzymes causing bone resorption and loosening of the implants. The same may be the case with PP, given its chemical composition. In the clinical setting reports of synovitis and cartilage damage with the Kennedy LAD have been published confirming the foreign body reaction to wear particles. The frequency of this is probably underestimated as the surgical technique recommended with this device is a stent attached to the bone at only one end. This technique puts no mechanical stress on the implant, and we have seen many failures of the semitendinosus insertion on the tibia with an intact Kennedy LAD in anterior cruciate ligament (ACL) reconstruction surgery. In the cases in which ligament was fixed at both ends early rupture was commonly related to the braided structure of the ligament, and synovitis reaction was more frequent.

Polytetrafluoroethylene

PTFE, commonly called Teflon, was developed in 1938, produced for the first time in 1950, and introduced in 1954 by Du Pont of Nemours (USA patent 2230654–1941). PTFE is a biocompatible substance and demonstrates an inert character in biological tissues. It has been introduced in the construction of arterial prosthesis and surgical soft tissue implants. Gort and Rostrup [3] studied in vivo a Teflon PTFE ligament implant in dogs. PTFE application is limited for synthetic ligaments. It was nevertheless incorporated (TFE) in the composite Proplast prosthesis commercialized for ACL reconstruction.

PTFE is available under two distinct forms: nonporous and microporous. Microporous PTFE (ePTFE) is better known under the commercial tradenames Gore-Tex (Gore, Flagstaff, Ariz., USA) or Impra (Impra, Tempe, Ariz., USA). ePTFE possesses the same basic properties of the normal PTFE (intrinsic biocompatibility, inertia, chemical resistance) but differs radically in term of microstructure. The microporous aspect is obtained industrially by the chemical and mechanical processing described by Gore (USA patent 4187390–1980). The Gore-Tex microstructure is characterized by the presence of transverse nodules connected between them by microfibrils. The Gore-Tex and the Impra are currently used as alloplastic nontextile arterial prosthesis. Park et al. [10] proposed a ligament prosthesis made of ePTFE (Gore-Tex). Production of this prosthesis was stopped in 1993.

Others

Industrial production of linear polyamide, commonly called Nylon, began in 1939 (Du Pont of Nemours). As with polyester, it is a currently used industrial textile polymer. Experimented with in vascular surgery during the 1950s, it has seen limited in vivo biomedical application. Its insufficient in vivo stability and mechanical failures have prevented it from finding a wider application in prosthesis or implants. Introduced during the 1960s in experimental synthetic ligament surgery [7, 15], its production stopped shortly thereafter, and polyamide ligaments have not been tried again since.

Polyaramides are aromatic polyamides and exist also as aromatic polyamides-imides. They were introduced during the 1960s as fibers of Kevlar. Three categories of polyaramides exist: polyaramides I are related to poly-*p*-phenylene terephthalamide (Kevlar, Du Pont of Nemours), polyaramides II are related to poly-*m*-phenylene isophtalamide (Nomex, Du Pont of Nemours) and aromatic polyamide-imide (Kermel, Rhne Poulenc Textile). Their introduction in the medical area is relatively recent and limited to polyaramides I and II. They were introduced into the composite Proplast ligament (Nomex fibers) and later into the Ligaid ligament. Polyaramide Nomex is more popular than Kevlar for ligament implants due to a greater in vivo stability.

Polyurethanes were discovered in 1937. There are two types of polyurethanes: polyester-urethanes and polyether-urethanes. They are applied in many biomedical systems including breast and arterial prostheses, catheters and pace-makers. The biocompatibility and the biostability of some polyurethanes are still a matter of discussion in breast implants. Their introduction in ligament implants was limited [11], and no commercial prosthesis released since.

Biodegradable polymers are essentially the same polymers used for biodegradable sutures: polylactic acid (PLA), polyglycolic acid (PGA), PLA and PGA copolymers (PLGA, Vicryl), polydioxanone, and polycaprolactone. These polymers, the linear polyesters, are biocompatibles and bioabsorbables, and are commonly used in the biomedical field. However, their initial mechanical performance is too weak and their biodegradability too important for their use as a biodegradable ligament implant. The reconstituted collagen, and sometimes chemically reticulated collagen, is also evaluated as a biodegradable ligament implant. The number of biodegradable polymers is relatively limited, reducing the possible choices for biodegradable ligament experimentation. New polymers made of polycarbonate or polyhydroxybutyrate are currently studied for such applications.

Composites

Composites are defined by the homogeneous association of two totally different materials by their basic form. Generally, synthetic textile fibers represent one of the constituents, usually the uni- or bidirectional reinforcement constituent (anisotropic). The best example is resin structures strengthened by textile fibers. Complex structures are not truly composites but are very often assimilated to them. Complexes, as composites, present two distinct constituents by their basic form. Nevertheless, the association of these two constituents is not homogeneous, and the two constituents are rather joined intimately than overlapped totality. Filmand filmassociations are the best examples. Several quasicomposite structures (or complexes) exist in synthetic ligaments. Proplast (Vitek, Houston, Tex., USA) is an example. It consists of a textile core of Nomex fibers drowned within a coating of fluor polymer (fluoroethylene propylene) and a strengthened PTFE (Teflon)/carbon fiber component. Richards also proposed a rather complex composite structure associating a central polyolefine braid within a polymeric nontextile bloc. These complex and composite structures are relatively few and seldom present within artificial ligament prosthesis. There is no evidence at present that these composites can improve the biomechanic behavior and the clinical performance of the reconstructed knee ligament.

Textile Structures

Types of Filaments

The filament form is the main aspect under which appears the synthetic polymers used as ligament substitutes. It is the synthetic basic equipment of

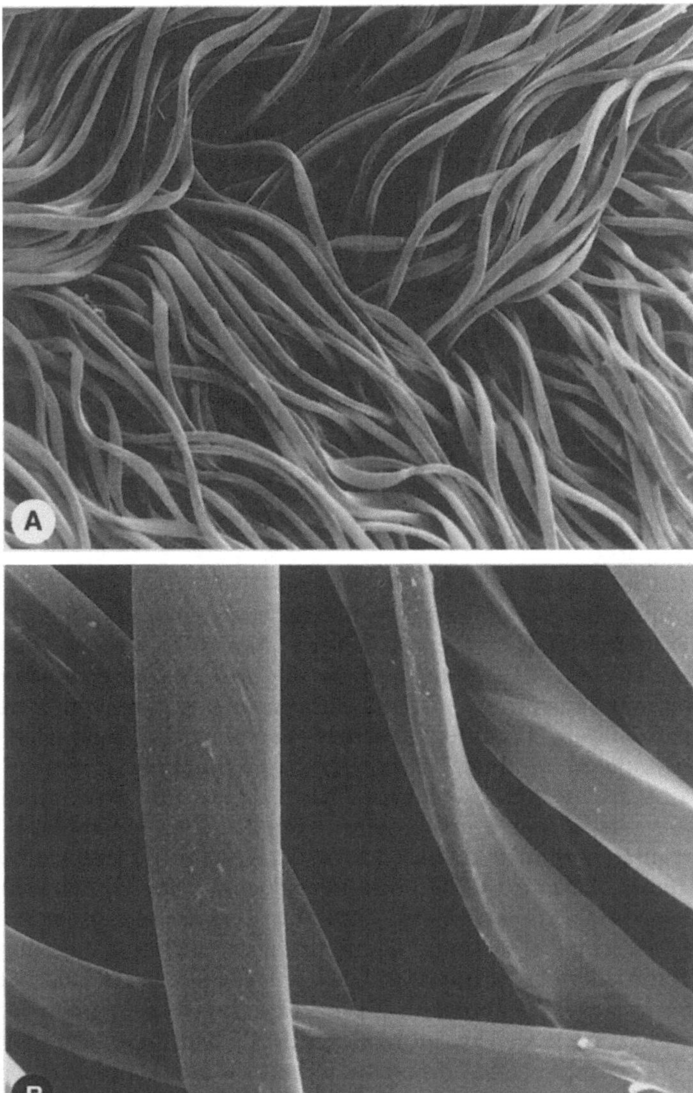

Fig. 1A,B. Microscopic view of polyester PET fibers in an artificial prosthetic structure. A General view of the multifilament polyester threads within the structure. Scanning electron microscopy, ×37.5; *bar*, 10 mm. B Polyester PET filaments within a multifilament strand. Scanning electron microscopy, ×375; *bar*, 100 mm

structure and textile technology. It is defined by the relationship binding the maximal diameter (D) and the length (L): D = L/1000. Notorious differences between synthetic filaments exist however, for example differences in form, in section, in regularity and in general aspect.

Polyester PET is common in the form of filaments or fibers. Ligament substitutes made of PET are generally constructed of quasicircular filaments. The section of these filaments is therefore almost circular, sometimes leaving microridges, especially postprocessing. The range of PET filament diameter varies from 10 to 40 m (Fig. 1). The filament of PP is quasicircular and possesses a diameter of 50 m. PE in ligament substitutes possesses a changing filament aspect. Knit Marlex consists of quasicircular filaments of strong diameter (50 m), while PHP filaments have no real circular section (Fig. 2). They present, rather, an irregular ribbon form with a very particular section. Their aspect is unique and specific compared to that of others synthetic filaments.

The strongest filament is the Gore-Tex macrofilament. Its diameter is approximately 500 m, and this aspect is typical of microporous Gore-Tex structure. It possesses a circular section and remains flexible. The finest filament is in carbon. Carbon fiber is circular, very regular and very small (8–10 m; Fig. 3). The surface can present a slightly streaked aspect. The carbon fiber is flexible due to its small size but is fragile and brittle especially during manipulation and transformation process. Filaments of polyamides, PTFE, and biodegradable polymers generally have a quasicircular section, and their diameters are similar to those of polyester filaments or larger.

An important parameter for the comparison of the various filaments is fineness. The fineness of a textile fiber is defined according to the section (surface) and the specific density of the polymer. It is measured by the title that corresponds to a weight for a certain length of filament. It is expressed by different units (e.g., Tex, Denier) according to the type of fiber. dTex corre-

Fig. 2. Microscopic view of PHP fibers within a PHP prosthetic ligament. Scanning electron microscopy, ×375; *bar,* 10 mm

Fig. 3. Microscopic view of a carbon fiber tow. Scanning electron microscopy, ×750; *bar*, 10 mm

sponds to the weight measured for 10 000 m of filament, while the Denier corresponds to the weight measured for 9 000 m of filament. The title can be determined directly by weighing a determined fiber length or indirectly by measuring the average diameter and the specific density. It is an important parameter, for example, in determinating relative tensile performance. Indeed, to compare the resistance of various textile fibers under tension, measurementss are often made of tenacity (or specific resistance). The tenacity is the tensile resistance of a fiber to rupture divided by its fineness, and is expressed, for example, in N/dTex. It is calculated experimentally by measuring the tension force to rupture and the title of the fiber tested.

The optical aspect of synthetic filaments also change according to the synthetic polymer studied. The polyester fiber is often characterized by its "sheen," an aptitude to reflect luminous rays. This sheen can vary according to the addition of carbon oxide within the polymer and the fiber. Generally, textile fiber appears as white or translucent. Carbon fiber possesses a very specific black color, and the aramides have a rather yellow color.

Types of Thread

As with the filament, the thread can be defined by its fineness and by its title. It is sufficient to relate the title of the fiber to the number of fibers of the thread (nominal title): title (thread) = n title (filament), where n is the number of filaments of the thread.

For the same type of fiber, a thread with more fibers has a higher title and a lower fineness. The nominal title can be determined experimentally by means of the title of the filament or by direct method. The fineness of ligament prosthesis and implants vary enormously. Braided multifilament structures generally have threads of very large title (strands). Woven and knitted structures possess finer thread. The mechanical function and the protection of filaments govern the choice of the thread within the textile structure of the synthetic ligament. Threads or textile strands are often straight linear structures. The physical fiber maintenance by twist or false twist is relatively limited for threads used in textile ligament structures. Thermal and mechanical processings give the thread an inflated and disorganized aspect. Texturization of polyester threads increases the apparent volume and the porosity of the thread. Textured fibers are nonlinearly oriented and take a curled and tortuous aspect. This type of thread has been introduced essentially within polyester arterial prosthesis and to a lesser degree within polyester ligament substitutes. The "velour" effect is usually obtained by the addition of textured threads. Texturization must be used carefully as it can induce a lesser mechanical fiber resistance, leading to fatigue failure.

Textile Structures

Textile ligament prostheses and implants are produced from four distinct textile structures: multifilament strand, woven, braided, and knitted. We discuss each of these synthetic textile structures briefly.

Multifilament Strands

The multifilament strand is a thick thread consisting of a high number of continuous filaments (several hundred to several thousands) depending on their fineness. It is therefore not a textile structure but rather a basic simple structure (Fig. 4). One or several associated multifilament strands can constitute a ligament implant without more textile assemblage. Filaments are then free to move and are not restricted within a textile construction. This structure minimizes the release of wear particles under repetitive loading condition. The

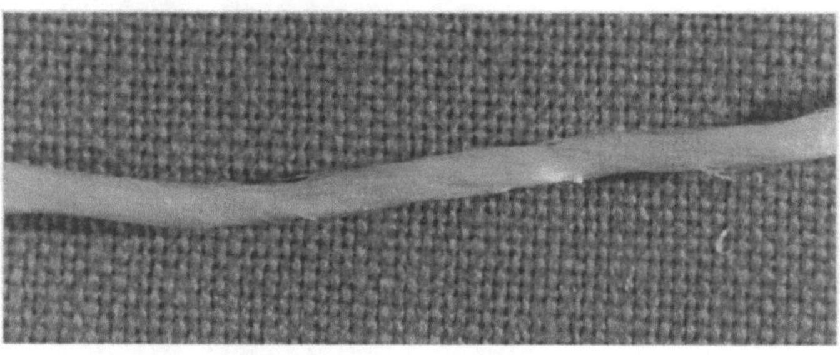

Fig. 4. Macroscopic view of a multifilament strand of high coarseness entering in the manufacture of prosthetic ligaments

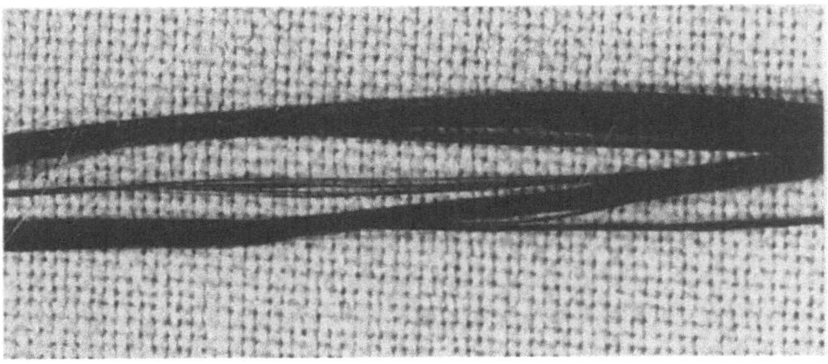

Fig. 5. Macroscopic view of a carbon fiber tow of a Integraft artificial ligament. Note the typical aspect of this material and the fine diameter (7–8 mm)

stress-strain properties of the implant rely on the intrinsic material properties under such loading conditions. Ligament prostheses have been built following this principle, especially for carbon (Fig. 5) and polyester: Jenkins, Integraft (carbon), Apex (polyester), and the intraarticular portion of the LARS.

Multifilament strands can also be used to form more organized textile constructions as braids. One or several multifilament strand can reinforce an independent and organized textile structure.

The Woven

Simple woven constructions of the linen type are obtained by perpendicular intertwining of two groups of threads, the threads of each group being parallel between them. For simple weavings the intertwining is alternated: one over and one under. The strain of the woven structure is limited by the autoblocking effect of the 90° thread orientation. The strain of the weaving is close to the

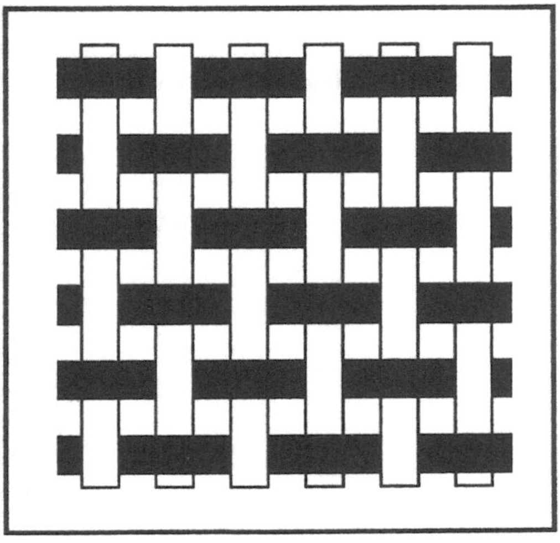

Fig. 6. Principle of construction of a plain woven textile structure. The threads (*black and white*) cross and intertwine alternately in the two main directions (warp and weft)

composing material (fiber or filament) strain, and its variability strain is low (Fig. 6).

Woven structures of ligament implants or prostheses are generally simple structures of plain weave or satin type. The weaving allows to intertwine at 90° two distinct thread groups (weft and warp) and gives resistant and stable textile constructions. The woven have been introduced within polyester arterial prosthesis, but their application to synthetic ligament implants has been limited. Woven vascular prostheses are tubes, but ligament implants are mainly narrow ribbons and sometimes cylindrical envelopes. Examples of woven ligaments are the Stryker, Lygeron, and Trevira, all three artificial polyester ligaments. The core of the Stryker prosthesis is made of several woven narrow ribbons, the Lygeron ligament is made of several internal narrow ribbons with an external cylindrical envelope (Fig. 7), and the Trevira ligament is a simple ribbon.

In woven artificial ligaments the density of the weft threads is often stronger than the density of the warp threads. In fact, the density of weft threads and

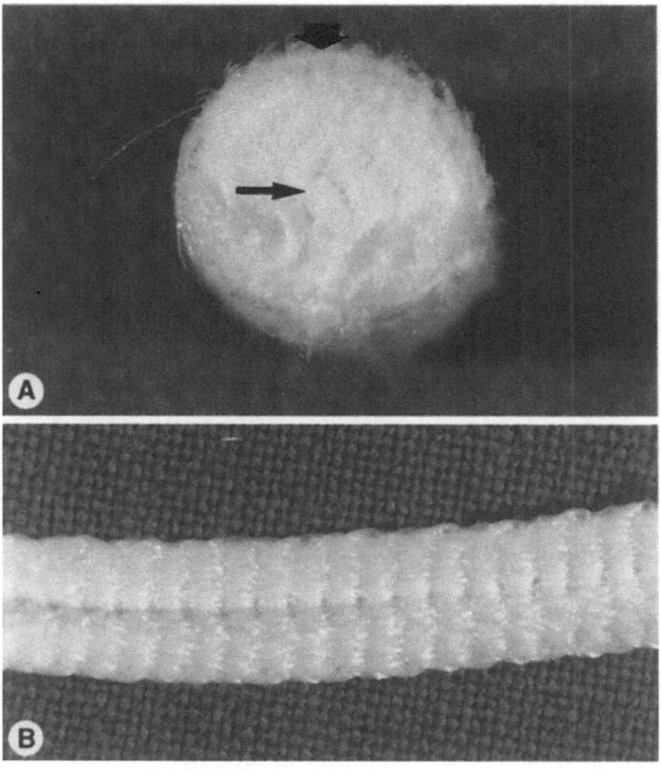

Fig. 7A,B. Macroscopic view of the cross-section of a polyester woven artificial ligament (Lygeron). **A** Aspect of different structural components of the ligament: woven envelope and narrow flat woven ribbons. **B** General view of the internal structure in narrow flat woven ribbons

fibers is increased because it is the axis of the biomechanic work. Warp threads and filaments are not directly in work but ensure rather the general maintenance of the woven textile structure. Under repetitive strain conditions the warp threads move perpendicular to their axis, and the edges fray out, leading to the release of wear particles and eventually failure of the ligament. To minimize this a locking thread system was recently added to the Trevira ligament (Fig. 8). This addition requires further investigation to determine its improvement to the basic woven structure.

The Braided

The braided structure is obtained by means of two groups of threads following symmetrical distinct directions, but not necessarily perpendicular. The braid is generally characterized by the axis of braiding around which the two groups of threads intertwine at symmetrical and therefore identical angles. The angle between the direction of the groups of threads and the axis of braiding is called braiding angle, or half-braiding angle, depending on the definitions. It varies from 20°–170° and determines part of the strain of the braided structure (Fig. 9). The other factor to consider is the strain of the constitutive material.

Fig. 8. Textile structure of the locked woven Trevira polyester ligament

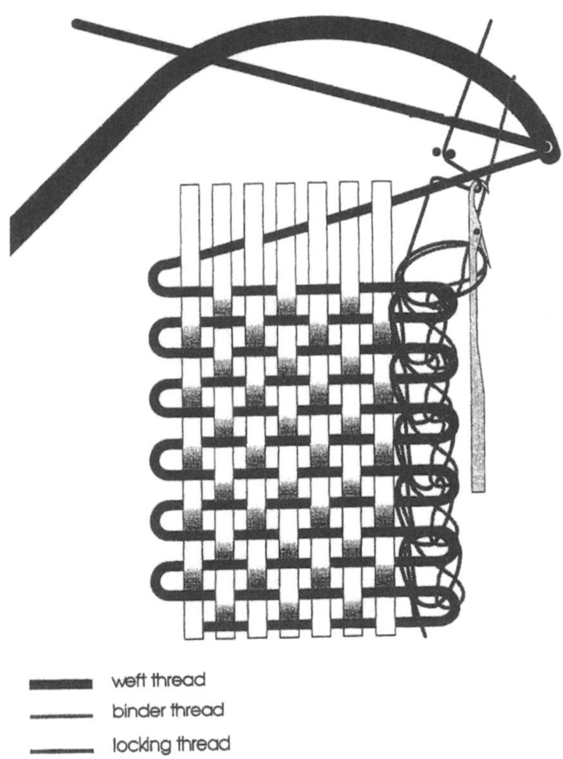

Fig. 9. Principle of construction of a braided textile structure. The threads cross and intertwine at an angle of 90°. Note the course of one thread (*black*) within the braided structure

Several studies have favored braided structures as the construction of choice for synthetic ligaments. These textile structures are recognized and used for their strong density in fibers or filaments, for their mechanical resistance, and for their general unidirectional orientation. Two types of braided synthetic ligaments have been manufactured: a flat ribbon and an elliptic or cylindrical similicord. Examples of braided cylindrical ligaments are Gore-Tex, PHP 1 and 3, Proflex, Ligapro, MBL, and SEM (Fig. 10). The Kennedy LAD, Biocarb, and Lafil are braided flat narrow ribbons (Fig. 11).

On our current knowledge of this structure from laboratory and clinical results it is the worst for ACL reconstruction. The 15% strain to which the reconstructed ACL is submitted during normal knee flexion and extension, and the even greater strain imposed by athletic activity, causes friction between the groups of threads and favors the release of wear particles. The clinical interpretation of this behavior is knee foreign body synovitis generated by ligament debris and early rupture of the implant with recurrence of knee instability.

The Knitted

The stitch is the geometrical basic unit of the knitted structure. It consists of a thread forming the head of the stitch, the stems, and the foot of stitch.

Weft-knits (simple jersey) are obtained from one thread that forms successively, in parallel and linear rows, open and symmetrical simple stitches (Fig. 12). The structure is built according to two basic directions: the direction of construction of the knit (vertical direction) and the direction of the weft

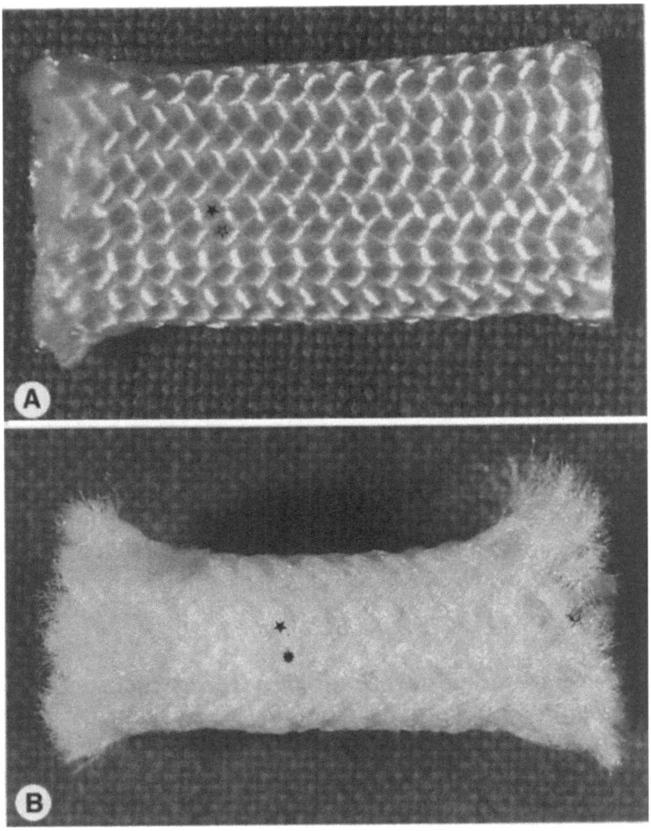

Fig. 10A,B. Macroscopic view of cylindrical braided polyester artificial ligaments. **A** A portion of the cylindrical polyester braid of an artificial ligament. **B** A cylindrical braided structure of a polyester artificial ligament. Note the differences between these two structures although their common braided construction

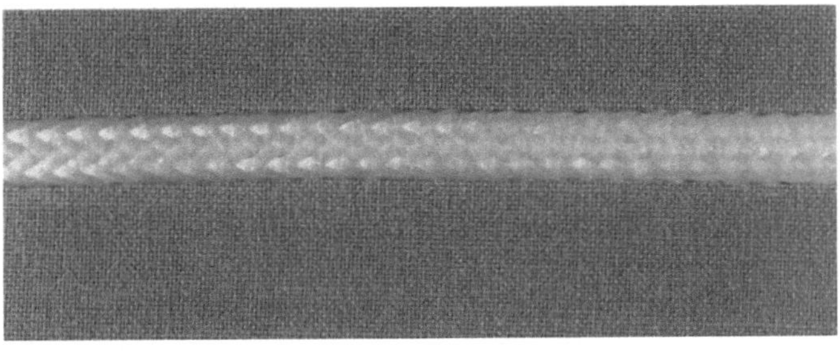

Fig. 11. Detail of a portion of the polypropylene Kennedy LAD. The construction of the implant is a flat braided narrow ribbon

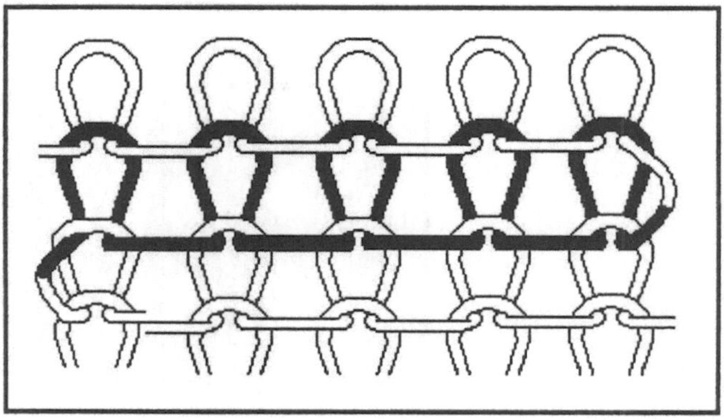

Fig. 12. Principle of construction of a weft-knitted textile structure (simple jersey). A unique constitutive thread forms successively, in parallel and linear rows, open and symmetrical simple stitches (a row is *colored black*). Note the horizontal bonds between the stitches

(horizontal direction). The strain and strength of these structures vary according to these two directions. There is no geometrical or directional homogeneity of mechanical properties and deformation.

Warp-knits are made of several independent threads forming identical and parallel stitches. Stitches are in the direction of knitting; however, these stitches do not form linear columns but a zig-zag, on two neighboring columns. Each column or row of stitches is therefore composed of independent stitches of different thread linked between them by more oblique bonds (rather at 45°). There exists a great diversity of warp-knits according to the possible stitches arrangement. Simple warp-knits are common (Fig. 13, open stitches), but there are also doubled knits (two stitches by site, bond double at 45°; Fig. 14). This diversity in possible structures is more important than for weft-knit structures.

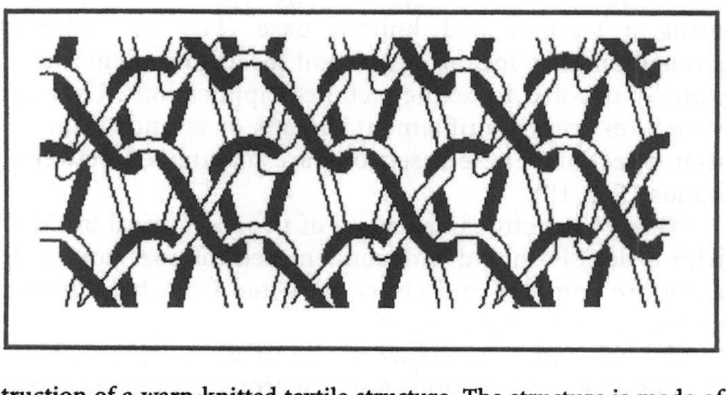

Fig. 13. Principle of construction of a warp-knitted textile structure. The structure is made of several independant threads (*alternately in black and white*). Stitches formed by a unique thread are disposed on different rows and columns. Note the 45° bonds between the independant stitches

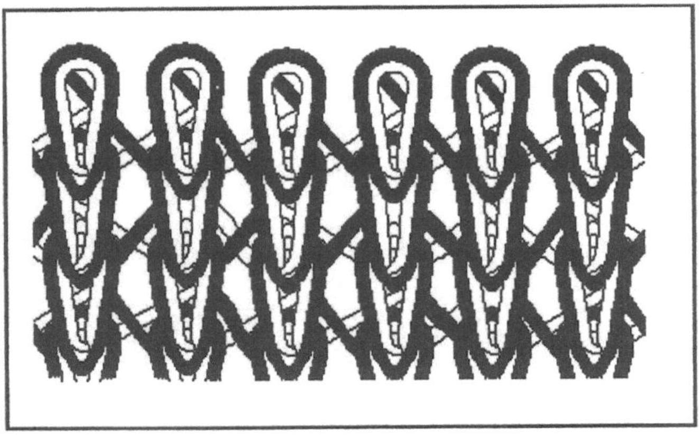

Fig. 14. Principle of construction of a double warp-knitted textile structure. Each stitch site is occupied by two stitches

The mechanical and geometrical stability of warp-knits is more important than for weft-knits. The strain and variability (extent) of weft-knits are reduced especially in the transverse direction. The general rule is: the more the knit is dense in stitches and in oblique bonds, the more it is resistant and strain limited. For synthetic ligaments, a great diversity of knits has been used for the external protective envelope of some models, as for the core of some artificial ligament. The weft-knits "simple jersey" type, the light warp-knits type "marquisette," and warp-knits "Raschel" are examples of such ligaments. The envelope of the Stryker ligament prosthesis consists of assembled weft-knitted bands (Fig. 15). The PHP 2 prosthesis is comparable to a typical Raschel-type knitted net (Fig. 16), while the Ligastic and the intraosseous ends of the LARS are made of light warp-knits strengthened by means of multifilament threads of high densities. These knitted structures, originally flat, are given a form and represent the core and the main structure of the knitted synthetic ligament. The Ligastic and LARS ligaments are typical of the new artificial ligaments built using a strengthened knitted base (Fig. 17). Warp-knitted constructions strengthened by means of filament insertions are more recent textile structures initially devoted to textile technical applications. They consist of warp-knitted structures with multifilament threads or strands inserted in the transverse or weft direction. These insertions are maintained between stitches and oblique bonds (Fig. 18).

"Gimb" structures are made of two cylindrical braided end-zone structures with a double-thread independent central core zone (Fig. 19). These threads are more complex and are constructed from two distinct threads of different constitution brought together.

The value of these structures must be considered on an individual basis as the possible combinations are numerous. More laboratory and clinical data are necessary to define the usefulness of these structures for knee ligament surgery.

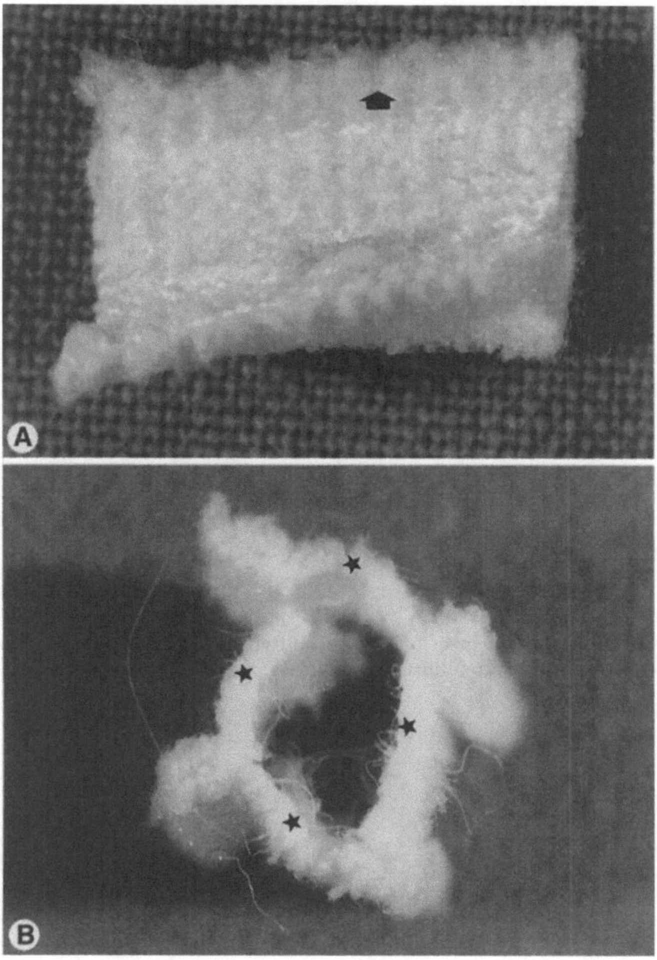

Fig. 15A,B. Macroscopic views of the external knited structure of a polyester artificial ligament (Stryker). **A** The envelope of the ligament is composed of four sutured knitted ribbons ("velour" effect). **B** The aspect in section of the four-sutured ribbons giving the external envelope of the ligament (internal ribbons are not present on this figure)

Impregnated and Coated Structures

The impregnation or the coating of textile ligament prosthesis exists, but only to a limited degree. These impregnations allow fiber and structure protection from fiber-fiber rubbing, manipulation, and contact with external surfaces, for example, bony surfaces. Carbon fiber is an example. It is protected by a coating of biodegradable polymers: polylactic and polyglycolic acid and copolymer relatives. PET fibers were also available with a polyurethane coating (Ligastic).

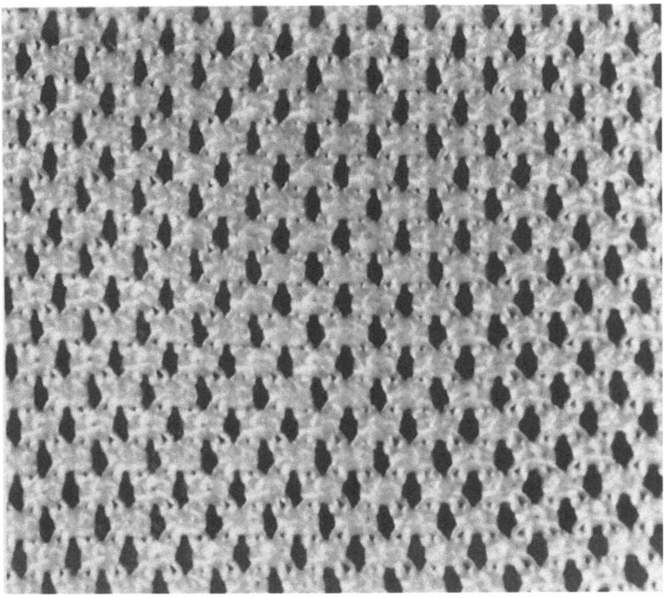

Fig. 16. General aspect of a warp-knitted flat structure: the Raschel ligament. Note the open construction of this knitted structure

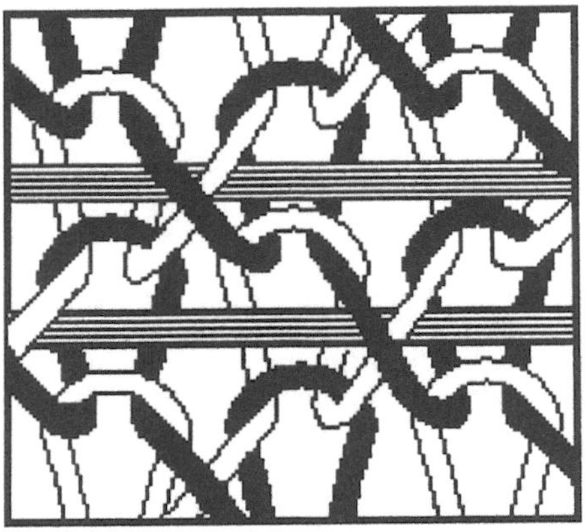

Fig. 17. Principle of construction of a warp-knitted textile structure reinforced by multifilamentous insertions. Insertions of multifilamentous threads or strands are present in a transverse or weft direction. These insertions are maintained within the structure between the stitches and oblique bonds

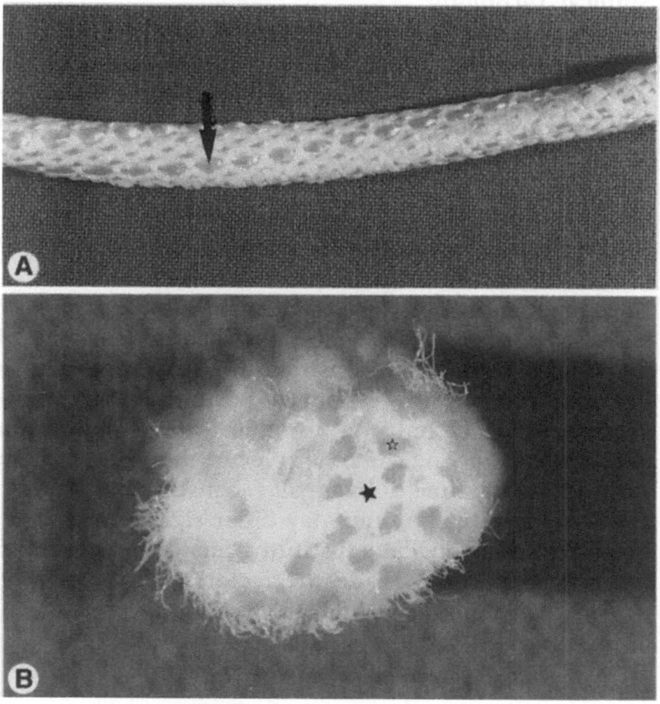

Fig. 18A,B. Macroscopic view of knitted polyester structures with multifilamentous insertions. A General aspect of the knitted structure of the artificial ligament model LARS. Note the presence of the main polyester knitted structure and polyester filamentous insertions. B Section of the Ligastic artificial ligament. Note the presence of the main polyester knitted structure and the polyester filament reinforcements

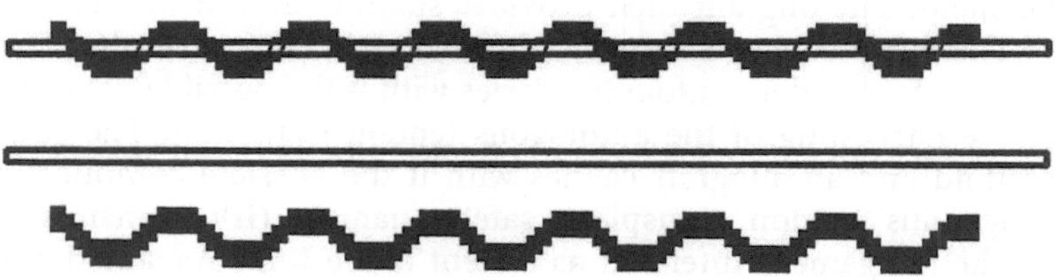

Fig. 19. Construction of a complex multicomponent thread by the association of two constitutive threads with distinct aspects and strains

Surgical Technique

This section deals principally with clinical observations at the time of primary revision of ACL and posterior cruciate ligament surgery. It is a framework for our present understanding of the surgical techniques, leading to the hypothesis of our current studies.

The Problematic Reproduction of a Surgical Technique

Knee ligament surgery is very demanding. Teaching orthopedic residents and surgeons confirms this. One part of the problem comes from the conception of the "ideal" technical reconstruction. This implies an ideal substitute for the ligament and an ideal location to position it. Too often these concepts are not verified and rely only on common sense, and common sense varies from surgeon to surgeon. The other part of the problem is the technical skill of the surgeon. At the time of ligament reconstruction, what the surgeon wants to do and what is actually done may differ significantly. The end result may create excessive stress on the substitutive material causing premature failure of the reconstruction. We divide the factors in this section in two groups: those related to the surgical philosophy and those related to the surgical anatomic technique.

Surgical Philosophy

Two different surgical philosophies are related to artificial ligaments. The first considers the ligament as a stent or a scaffold to augment a soft tissue implant (autograft or allograft). The implant is then regarded as a temporary structure giving stability while the soft tissue matures as a new biological ligament. The second considers the ligament as a true prosthesis to replace the ruptured ligament. The implant then provides stability until it mechanically fails.

These two concepts show no differences in clinical results. The combination of a scaffold and an autograft carries with it the possible complications related to the harvesting of the autologous tendon transplant. The combination of a scaffold and an allograft carries with it the possible complications related to allogenous tendon transplant safety, namely HIV transmission. The only synthetic ligament intended as a stent is the LAD. As stated above, attaching the stent at only one end does not stress the stent mechanically and is by no means different than a regular autograft or allograft ligament substitution. A true prosthesis presents the problem of soft tissue coverage inside the knee joint and the limited life span associated with repetitive loading.

Surgical Anatomic Technique

Many publications describe the anatomy of the cruciate ligaments. Surgical techniques have tried in several ways to reproduce this anatomy. The problem of cruciate ligament reconstruction is the problem of working with bony tunnels. While there is agreement among most surgeons that the intraarticular reconstruction cannot truly be isometric, there is still no agreement on the ideal tunnel entrance inside the joint, the ideal bony tunnels orientation, or the ideal tension under which the graft should be positioned. Some important technical points have been communicated in the literature to explain some early implant failure. These can be divided into three groups: the femoral tunnel, the intraarticular ligament, and the tibial tunnel.

We briefly discuss some points of each group and their importance to ACL reconstruction. A more detailed discussion on all the aspects of these groups is beyond the scope of this chapter. For the sake of facilitating the discussion posterior cruciate ligament reconstruction is not included in the following discussion.

The Femoral Tunnel

The most common mistake about the femoral tunnel in ACL reconstruction is to place it too far anteriorly on the medial aspect of the lateral femoral condyle. Many works report an increase in the tensile stress inside the implant with limitation of knee flexion. As more surgeons become increasingly aware of this fact, one sees this problem somewhat less than previously.

The second point is related to the direction of the tunnel itself. As the device is subjected to stress in three axis (bending, twisting, tensile), it is not known at the present what position would allow a reconstructed ligament to receive stress corresponding to its mechanical properties. It is likely that a better knowledge of these stresses in vivo would lead to a better survival of a synthetic implant.

The last point concerns ligament fixation. As the cortical bone on the lateral aspect of the femur is hard, one must make sure the fixation is anchoring the device strongly enough to prevent slippage of the device in the tunnel.

The Intraarticular Ligament

An important point must be realized as the reconstructed ligament travels from the lateral femoral condyle to the tibial plateau: there is usually some soft tissues in this area. Should these soft tissues be removed or retained? Do these tissues still have some proprioceptive function? Two possibilities are offered to the surgeon. One is to clean these tissues to obtain the best possible view inside the joint and leave the synthetic ligament without soft tissue coverage. The other is to keep all these tissues to protect as much as possible the synthetic ligament from the synovial fluid and perhaps preserve or promote some proprioceptive function. To do this an imaging technique must be added to the arthroscopy as the view inside the notch is limited. Usually fluoroscopy has been chosen for that purpose. So far there is no proof that one choice is superior to the other.

Any friction between the artificial implant and the roof or walls of the notch must be prevented. Notchplasty should be carried down as soon as one suspect a friction point. A special attention to the roof as the knee approaches full extension is important. Notchplasty is probably not necessary in the acute reconstruction and is more likely to be necessary in the chronic situation when the notch is invaded by osteophytes.

The last point is the tension under which the ligament should be positioned. As the normal ACL strain during a full range of motion of the knee is 15%, and most of the artificial implants fail at less than 10% strain, it is mandatory that an artificial ligament has zero tension at the time of fixation. Noncompliance to this observation should highly predispose the implant to premature fatigue failure.

The Tibial Tunnel

The most common mistake about ACL reconstruction is a bad position of the tibial tunnel. To place it too far anteriorly on the tibial plateau causes impingement of the ligament under the roof of the notch. This leads to a limitation of knee extension and early rupture of the device. This situation is common at the time of revising a failed artificial ligament.

The unknown ideal direction of the tunnel and the ligament fixation problem are similar to those of the femoral tunnel. The only difference is related to the soft cancellous bone of the upper tibia, making the fixation problem of the implant more obvious.

Discussion

Problems of the Ligament Prostheses

DiGiovine and Shields [1] have described the common clinical history of the artificial ligament: excellent results during biomechanic testing followed by enthusiastic preliminary clinical results with an overimplantation period, then complications and recognized clinical failures, and finally abandonment of the ligament and return to a conventional ACL reconstruction technique. Our current knowledge is that synthetic ligaments do not seem to be able to offer an adequate long-term biofonctionnality and a biodurability. The material and structure selection, physiological function, surgical procedure, and ligament fixation are items that have not given full satisfaction so far. The artificial ligament has not shown a capacity to replace the ACL on the long-term basis. Its integrity and functionality are remaining problems. Finally, the presence within the articular environment of polymeric materials of various forms can be harmful for the patient. There is therefore a need for less aggressive materials and structures allowing a reconstruction of ligament tissue while limiting the presence of foreign artificial equipment into the articulation.

Synthetic Ligaments: A Lack of Materials or Adequate Structures?

The diversity of synthetic polymers and textile structures observed within ligament prostheses demonstrates that few new materials or structures could be introduced. The concept of synthetic prostheses must be questioned. The ideal resides in a viable human ligament tissue allowing the reconstruction and the regeneration of a ligament tissue. The promotion of new synthetic implants will have to answer the questions raised by the high failure rate in the past. Unfortunately, no current material with miracle performances for the ligament implant exist. Experiments with new materials and implants are underway, but excessively realistic and rigorous basic sciences and clinical documentations are necessary before these ligaments could be considered for general clinical use.

Biodegradable Ligaments: Dream or Reality?

There is currently a lack of knowledge on biodegradable materials and on the exact healing process and results of reconstructed tissue reinforced with a biodegradable support. Numerous studies are needed to provide sufficient informations on biodegradable polymers: their optimal properties, the efficient structures for ligament reconstruction, and the ideal conditions to favor, control, and promote – in vitro and in vivo – collagen growth, regeneration, and orientation.

The Ideal Prosthesis

The knee joint is a very demanding biological environment for artificial textile ligaments. It is submitted to approximately 2 million flexions per year and to tension and twisting motion associated with these flexions. Prostheses are abraded by bony surfaces and ridges, and fiber-fiber friction occurs within elaborate textile structures. The ideal artificial ligament should be safe and biomechanically efficient. It should maintain its properties over time. The surgical implantation procedure should be simple and easy to reproduce. It can be summarized in three points: biocompatibility, functionality, and durability. One of the most difficult aspect for the development of a ligament prosthesis resides in the lack of knowledge on applied real constraints during ligament work. Cadaver studies and biomechanical evaluations performed on older patients do not allow a realistic evaluation of the exact biomechanic needs. The wear of the ligament and the release of particles related to the stress generated on the ligament are still lacking a reliable method of evaluation.

Studies of textile structures are trying to duplicate the physiological structures that the prosthesis will have to replace. The biocompatibility, preparation, and washing methods of textile industrial structures must be developed and systematized. The interaction cell-fiber and tissue-textile needs a better characterization. The method of fixation of ligament prosthesis requires improvement. Many systems of fixation have failed in the past. New ideas are nevertheless proposed with the biofixation by anchorage and the bony blockade on the textile structure. These methods seem promising but they require further evaluation.

Conclusion

An artificial ligament should be considered as a temporary solution to replace a ruptured ligament. The material, the structure, and the surgical technique are important outcome predictors. Artificial ligaments can be useful. An attempt to optimize their survival time in the knee joint requires a definition of the factors leading to failure. More research is needed to identify these factors and to modify them in order to improve today's results. This is leading the way to tomorrow's surgery.

References

1. DiGiovine NM, Shields CL (1991) Synthetic ligaments in ACL reconstruction. A review. Am J Knee Surg 4(1):42–48
2. Emery MA, Rostrup O (1960) Repair of the anterior cruciate ligament with 8 mm tube Teflon in dogs. Can J Surg 4:111–115
3. Gort J, Rostrup O (1959) Teflon fabric for ligament reconstruction: an experimental study. Can J Surg 3:75–78
4. Harrison JH (1957) Ivalon sponge (polyvinyl alcohol) as a blood vessel substitute: failure in experimental animals. Surgery 41:729
5. Harrison JH (1958) Synthetic materials as vascular prostheses. I. A comparative study in small vessels of Nylon, Dacron, Orlon, Ivalon sponge and Teflon. Am J Surg 95:3–15
6. Harrison JH (1958) Synthetic materials as vascular prostheses. II. A comparative study of Nylon, Dacron, Orlon, Ivalon sponge and Teflon in large blood vessels with tensile strength studies. Am J Surg 95:16–24
7. Johnson FL (1960) Use of braided Nylon as a prosthetic anterior cruciate ligament of the dog. J Am Vet Med Assoc 137:646–647
8. Marois Y, Roy R, Vidovsky T, King MW, Bélanger AY, Chaput C, Guidoin R (1993) Histopathological and immunological investigations of synthetic fibres and structures used in three prosthetic anterior cruciate ligaments: in vivo studies in rats. Biomaterials 14:255–262
9. Meyers JF, Grana WA, Lesker PA (1979) Reconstruction of the anterior cruciate ligament in the dog. Am J Sports Med 7(2):85–90
10. Park JP, Grana WA, Chitwood JS (1985) A high-strength Dacron augmentation cruciate ligament reconstruction. Clin Orthop 196:175–185
11. Peterson CJ, Donachy JH, Kanelak A (1985) A segmented polyurethane composite prosthetic anterior cruciate ligament. In vivo study. J Biomed Mater Res 19:589–594
12. Usher FC, Wallace SA (1958) Tissue reaction to plastics. AMA Arch Surg 76:997–999
13. Usher FC, Fries JG, Ochsner JL, Tuttle LLD (1959) Marlex mesh, a new plastic mesh for replacing tissue defects. II. Clinical studies. AMA Arch Surg 78:138–145

14. Usher FC, Gannon JP (1959) Markex mesh, a new plastic mesh for replacing tissue defects. I. Experimental studies. AMA Arch Surg 78:131–137
15. Vaughn LC (1963) A study of the replacement of the anterior cruciate ligament in the dog by Fascia, skin and Nylon. Vet Rec 75:537–541
16. Voorhees Jr AB, Jaretzki A, Blakemore AH (1952) The uses of tubes constructed from Vinyon "N" cloth in bridging arterial defects. Ann Surg 135:332–336
17. Winston LA, Tercy AF, Jardine JH, Parrish FF (1978) The result of replacement of partial or total collateral ligaments with Marlex mesh in the knees of dogs. Clin Orthop 137:287–290

II.3 Computer-Assisted Planning of Prosthetic Anterior Cruciate Ligament Insertion

M. Sati, J.A. de Guise, S. Larouche, and G. Drouin

Introduction

The knee's articulation is hidden beneath layers of skin and soft tissue involved in its function. Unseen beneath these structures is a complex and sensitive mechanism whose injury can lead to loss of mobility and premature degeneration. Movement of the joint is guided by a complex interaction between ligament stiffness and muscle actions. The presence of soft tissue surrounding the bone renders diagnosis and correction of pathologies difficult. One example of these difficulties is the surgical replacement of a ruptured anterior cruciate ligament (ACL) via a prosthesis (see Fig. 1). Optical fiber technology (arthroscopy) has been developed to help the surgeon see internal structures hidden beneath the skin and between the femoral condyles. Even with this technology it is difficult to know where and how to properly position the ACL prosthesis replacement [12].

Prosthetic ligaments have experienced limited success because of premature failure. The main causes of failure are overload, fatigue wear [13, 14], and inflammatory reaction [4]. To understand the mechanisms of rupture, the concept of prosthetic ligament loading has evolved over recent decades. Since natural ligaments fail due to impacts or excessive loads, it is not surprising that the earliest prostheses were designed to possess a high ultimate breaking strength. After the failure of prostheses based only on the high ultimate breaking strength concept, detailed biomechanical studies led to designs which more closely reproduce the elastic nonlinear force deformation curve of the natural ligament [7]. However, the plastic deformations in the material led to permanent elongation [13] and rupture by fatigue. To minimize fatigue due to tension, the concept of "isometric" attachment of the ligament, i.e., finding two points on the femur and tibia, respectively, between which no elongation occurs during knee flexion, was introduced into the surgical procedure [11]. Although this concept has been, and still is, the subject of active debate [2, 5, 16, 18], it has highlighted the importance of adequate surgical procedure for good long-term clinical results. To help avoid the catastrophic failures experienced in earlier replacements in vitro fatigue testing by cyclic elongation was introduced to predict in vivo performance. Cyclic loading on a not perfectly elastic material can cause premature elongation resulting in loss of its prescribed function. Unlike the natural ligament, the prosthesis cannot re-

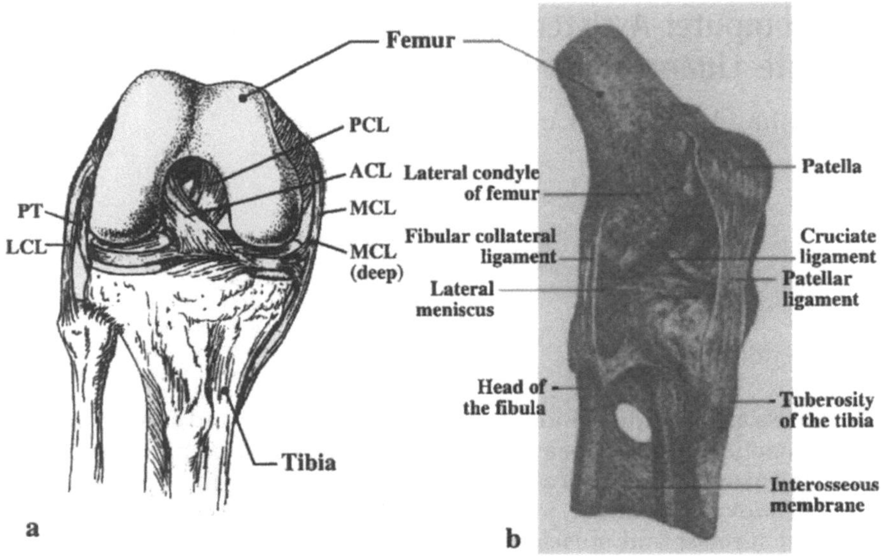

Fig. 1a,b. Knee geometry of bones (femur, tibia, fibula) and main ligamentous structures. **a** Anteroposterior view. *ACL*, Anterior cruciate ligament; *PCL*, posterior cruciate ligament; *MCL*, medial collateral ligament; *LCL*, lateral collateral ligament; *PT*, popliteus tendon [15]. **b** Anterolateral view [24]

generate its material. Even low rates of fatigue wear can over time result in catastrophic failure since the ligament is exposed to at least 1×10^6 cycles per year in vivo [8, 25]. This makes surgical placement which minimizes fatigue wear essential. Even failures due to foreign body inflammatory response can be related to the fatigue of the ligament under in vivo conditions. Fatigue wear on a prosthetic ligament, due to both interfiber and ligament-bone abrasion, produces small particles which can evoke an inflammatory response even if the material is biocompatible in its bulk form [4].

A biomechanical study by Gely et al. [12] revealed that simple elongation fatigue testing does not tell the whole story about in vivo conditions. In fact, the ACL is subjected to combined elongation, flexion, and torsion deformations. Unlike the natural ACL ligament, prosthetic ACL ligaments are anchored via tunnels in the femur and tibia. The angles formed between the tunnels and the ligament body create combined elongation, flexion, and torsion deformations in the ligament as the knee is flexed. Elongation of the ligament depends on the sites where the ligament body enters the femur and tibia and the knee's kinematics. Keeping these entry sites constant, for a given knee kinematic movement, it is possible to change flexion and torsion in the ligament by varying the angle between the tunnel and the ligament body (Fig. 2a). Ligament torsion is defined as the twisting of the functional part of the ligament about its long axis. Bending of the ligament is defined as the angle between the ligament's functional body with respect to either the femoral insertion tunnel (femoral bending) or the tibial insertion tunnel (tibial bending).

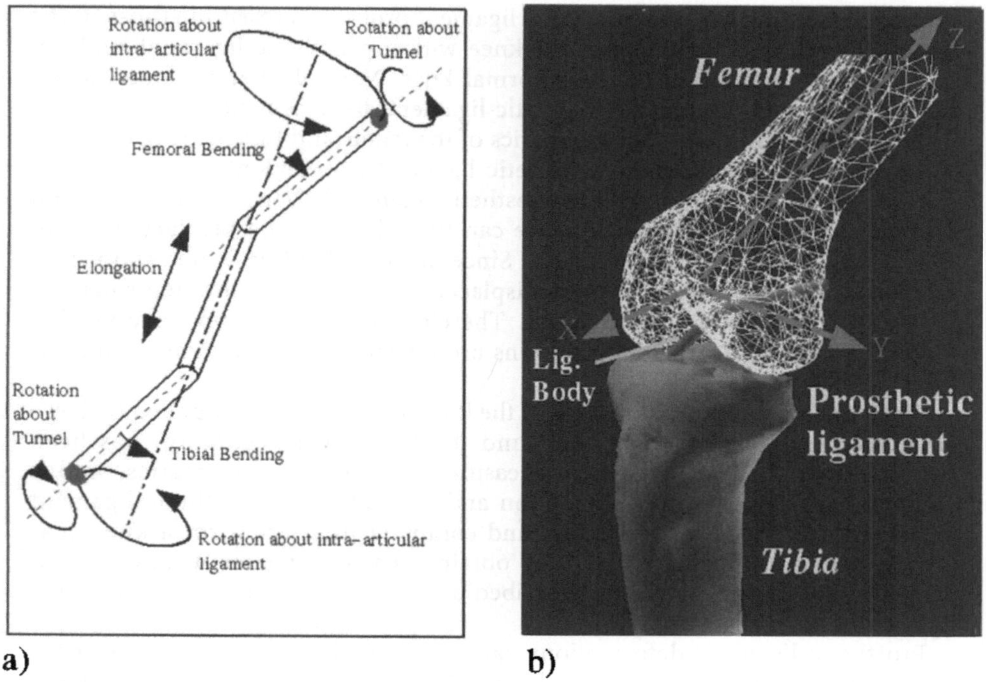

Fig. 2. a Prosthetic ligaments are subjected to combined elongation, bending, and torsion deformations. **b** Prosthetic ligament information generating system (PROLIGS). Prosthetic ligaments are simulated in a personalized three-dimensional computer image of the patient's knee

Most in vitro ligament testing has involved simple elongation or combined flexion-tension cyclic deformations [4]. For example, Dürselen and Claes [10] have performed bending fatigue tests which showed poor fatigue life: 14077 and 1.4×10^6 bending cycles until failure for the Lafil and Gore-Tex prostheses, respectively. The importance of combined elongation, flexion, and torsion has been recently shown through mechanical tests [9]. Combining cyclic 110 N tensile load, 45° bending and 40° torsion resulted in ligament failure after 5×10^6 cycles [9] with significant decrease in fatigue life with increased tensile loads. Although this study has shown decreased fatigue life due to combined loading, the detailed in vivo loading conditions the ligament is subjected to is to date unknown [12, 27].

To investigate a ligament's in vivo loading condition requires accurate measurement of knee movement. Kinematics of dynamic in vivo knee movement are a result of the interaction between active structures (muscles) and passive structures (connective tissue). Tendon loading, for example, depends on forces generated by the associated muscle and wrapping around bony structures. Unlike tendons, ligaments are loaded by bone movement since their function is to maintain the knee's movement within a safe "envelope" of motion. Assuming that the prosthetic ligament properties resemble those of the natural ligament, kinematics of the knee with a prosthetic ligament replacement should resemble those of a normal knee. Normal knee kinematics can therefore be used to analyze prosthetic ligament displacement in the knee.

By accurately tracking the kinematics of the femur and tibia during dynamic movements and defining the prosthetic ligament's placement in the knee we can determine the motion of the prosthetic ligament's bony insertions. From the movement of these insertions we can fully describe the displacements to which the ligament is subjected [12]. Since ligament loading is determined by the displacement of its ends, these displacements determine the deformations to which the ligament is subjected. Therefore in this chapter relative displacements of the ligament's insertions are referred to as ligament "deformations."

Since noninvasive measurement of the knee's movement is impeded by large amounts of soft tissue which surround the knee, instrumentation has been recently developed to accurately measure in vivo knee kinematics, and a computer interface permits simulation and calculation of prosthetic ligament deformations. The instrumentation and computer interface form a knee analysis system used in this chapter to obtain prosthetic ligament deformation data. This technology is briefly described before presentation of the prosthetic ligament deformation analysis.

Prosthetic ligament deformations vary as a function of their three-dimensional surgical placement in the knee [12]. Improving surgical procedures therefore requires a relationship between ligament placement and deformation. It is difficult, however, to interpret and communicate three-dimensional ligament deformations to ligament orientation in space. Finite helical axes [26] and gyroscopic angles [18] are traditionally used in biomechanics to describe three-dimensional joint movement. Helical axes describe knee movement by finite rotation axes which move in space during knee flexion. Euler-based systems such as that of Grood and Suntay [17] describe knee movement as a sequence of rotations about orthogonal axes. Interpretation of ligament deformations with respect to such moving or multiparameter descriptions is complex. Therefore the "average" flexion-extension (FE) axis of the knee is presented in this chapter to serve as a three-dimensional reference direction to help interpret the relationship between ligament insertion orientation and deformation.

Chapters II.4 and II.6 demonstrate that several surgical procedures have been developed for prosthetic ligament placement. These placements have been developed over the years through some quantitative studies, empirical knowledge, and trial and error. This chapter presents a quantitative in vivo analysis of combined elongation, flexion, and torsion deformations as a function of surgical insertion. These data are presented with the objective of improving the understanding of prosthetic ligament deformations and are not intended to criticize or compare previous techniques. It is through the understanding of ligament deformations that surgical placement can be improved to help reduce prosthetic ligament fatigue wear, which is the principal cause of failure of these devices.

Three-Dimensional Knee Analyzer

Movement is often analyzed, for example, in gait analysis, by tracking sensors taped onto the surface of the skin to measure limb movement. However, soft tissue surrounding the knee moves during knee flexion causing a large artifact in the measurement of underlying bone motion [22]. This artifact must be reduced by one order of magnitude to allow in vivo analysis of prosthetic ligament deformations [21]. Although implantation of invasive pins into underlying bone can avoid this artifact, a noninvasive technology was developed to avoid possible complications associated with such invasive techniques. A new mechanical device was designed to reduce movement noninvasively between motion sensors and underlying bone [21] (Fig. 3b). Magnetic sensors are fixed onto this device which allows accurate tracking of the underlying bone movement. Knee kinematics measured in this way were compared to those obtained via a very accurate three-dimensional X-ray technique. This analysis showed that the noninvasive system reduced skin movement artifacts which are between 2 and 17 mm root mean square (rms) to between 3 and 4 mm rms making prosthetic ligament analysis possible.

This mechanical device was equipped with magnetic sensors to noninvasively track underlying bone movement which provided kinematic input to the three-dimensional knee analyzer. Three-dimensional imagery techniques based on computed tomography have been developed to obtain accurate three-dimensional reconstruction of knee geometry. A mathematical calibration method allows real-time correspondence between the spatial position of the patient's femur and tibia and the "virtual" three-dimensional medical image of the corresponding personalized geometry. This provides a "virtual" and personalized representation of the subject's knee in the computer thus revealing movement and geometry normally hidden beneath soft tissue. Software allows recording of three-dimensional kinematics, interactive simulation of prosthetic ligament insertion, and calculation of the resultant ligament elongation, bending, and torsion deformations. Analysis can then be performed in the "virtual model" to predict in vivo ligament deformations.

The system integrates several technologies which are combined together and can be seen in Fig. 3. This technology provides detailed measurements of in vivo knee movement which is the basis of the prosthetic ligament deformation analyses which follow.

Methodology

The three-dimensional knee analyzer is used to obtain prosthetic ligament deformation data for different simulated ligament placements. These data are to be used to help understand the relationship between surgical insertion orientation of the prosthetic ACL and its deformations which may lead to its premature fatigue.

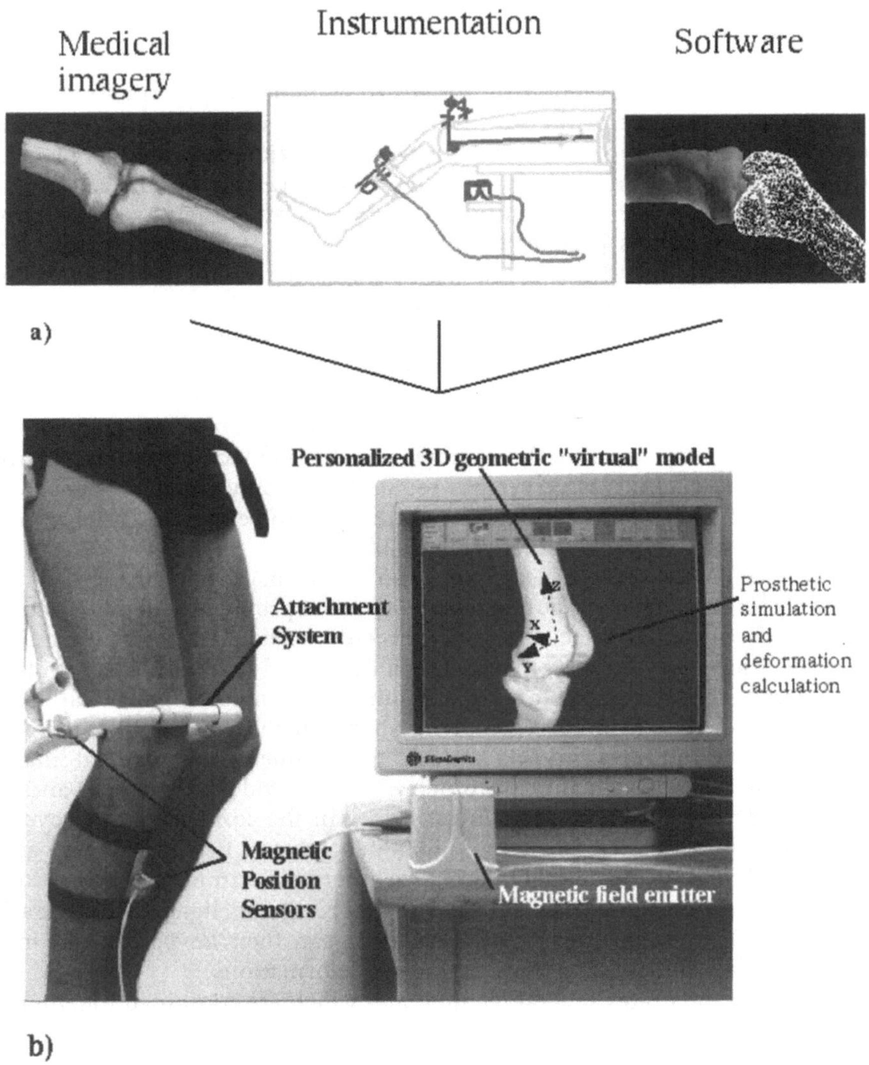

Fig. 3a,b. The three-dimensional knee analysis system. The combination of three components in **a** a mathematical calibration method allows real-time correspondence between the subject's knee (*left* in **b**) and the personalized "virtual knee" (*right* in **b**)

The underlying assumption of the present analysis is that the simulated prosthesis would not alter the knee kinematics. This implies that the prosthetic ligament properties are somewhat similar to those of the natural ligament. It is assumed that the ligament is attached using tibial and femoral tunnels and that the internal ends of the tunnels are near the natural ligament attachment sites (Fig. 2). Attaching the ligament in a different region than the natural ligament could throw off the normal rotation axes of the knee. If the ligament is stiffer than the natural ligament, hence altering the natural kinematics, its deformations could be less than those predicted by our model and vice versa.

Ligament elongation, torsion, and bending deformations in the ligament vary with surgical placement [12]. A previously developed interactive computer program, the Prosthetic Ligament Information Generating System (PROLIGS) [23], was used to calculate prosthetic ligament deformations based on the "inverse deformation principle" [12]. Calculations of prosthetic ligament deformations using PROLIGS and kinematic data from the three-dimensional knee analyzer have been previously evaluated as sufficiently accurate for this type of analysis [21].

To calculate prosthetic deformations, the ligament is first simulated within the three-dimensional kinematic model of the knee. The ligament insertion tunnels are simulated via cylindrical objects, one in the femur and one in the tibia, through which a simulated artificial ligament is passed (see Fig. 2b). The intraarticular attachment points of both femoral and tibial tunnels often correspond to the attachment sites of the natural ligament. In this study these intraarticular points were chosen according to a suggested placement in the literature [3]. The exterior entry point of both the femoral and tibial tunnel therefore determine the orientation of the ligament's femoral and tibial insertion, respectively. Once the ligament was placed, the customized kinematics were applied to the bone-ligament geometry and deformations on the ligament calculated. Data analysis was performed on three subjects to demonstrate the feasibility of the approach. The subjects were asked to perform knee squats and a heel raise off the ground towards the buttock, over approximately 0°–90° knee flexion.

Since it is very difficult to relate three-dimensional ligament deformations to traditional interpretations of knee movement by Helical axes [26] and gyroscopic angles [17], a simpler parameter is proposed. The "average" FE axis of the knee was calculated to serve as a three-dimensional reference direction to help interpret the results. A least-squares algorithm is employed to find the axis which "best fit" the knee's rotation. To calculate this axis, the coordinates of a point on the FE axis is described in the femoral reference frame by $[a]_F$ and the tibial reference frame by $[b]_T$. For this situation, the coordinates of $[a]_F$ and $[b]_T$ remain constant with respect to the femoral and tibial reference frames respectively for all knee positions "i" and are related by:

$$[a]_F = R_i x [b]_T x [t_i] \qquad (1)$$

where R_i and $[t_i]$ are respectively a 3×3 rotation matrix and a translation vector relating the two reference frames at some kinematic knee position "i." We cannot hope to attain exact equality in an experimental situation, but a good solution of $[a]_F$ (or $[b]_T$) can be obtained using a least-squares optimization on the data [20] (P. Genoud, personal communication, 1994).

Ligament deformation is first investigated as a function of different simulated surgical placements. With this analysis a relationship between surgical insertion orientation and deformation is established.

The purpose of prosthetic ligament deformation calculations is to determine insertion orientations which augment the fatigue life of prosthetic ACLs.

Therefore several insertion orientations were investigated in the interactive computer graphics environment to determine relationships between ligament tunnel orientation and ligament deformations.

To study the range of possible deformations, prosthetic ligament tunnels were oriented either parallel or perpendicular to the knee's FE axis. It is shown below that the use of the FE axis as a reference orientation helps to interpret

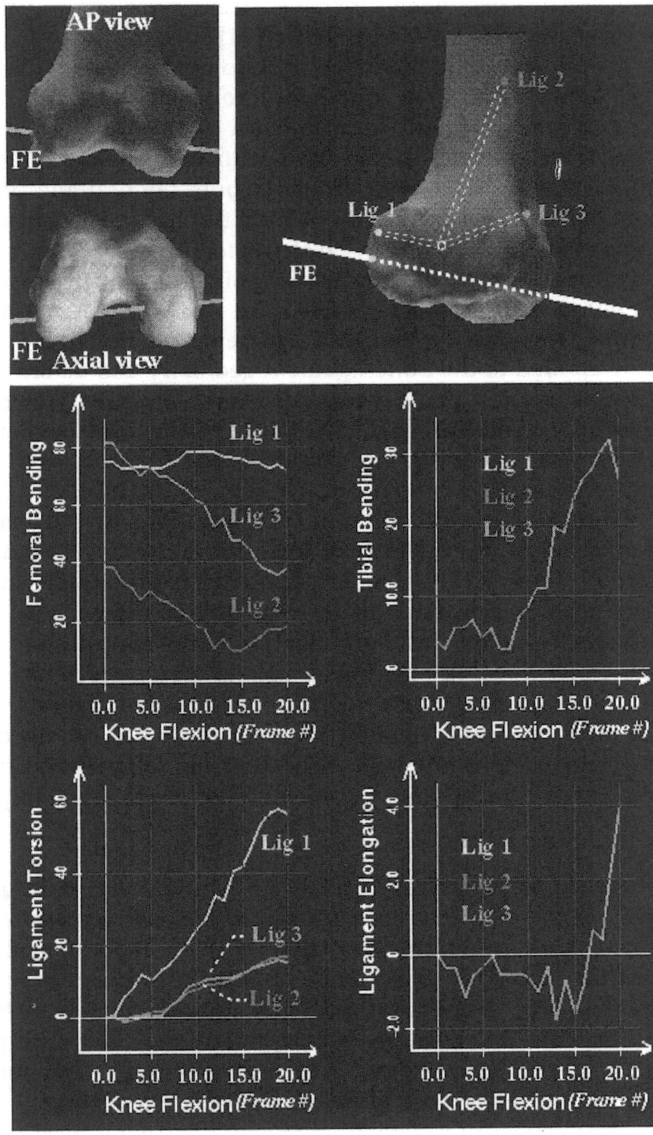

Fig. 4. Deformations of ligaments placed in extreme orientations with respect to the FE axis

the results (see Fig. 4). In Fig. 4, tunnel "lig1" is oriented parallel to the FE axis, tunnel "lig2" is oriented perpendicular to the FE axis and points towards the proximal femur, and tunnel "lig3" is also oriented perpendicular to the FE axis but points towards the anterior of the femur. Variation of tibial tunnel orientation was not investigated since it has been found to have much less influence on ligament deformations than the femoral insertion [12, 23].

In this study the intraarticular entry points of the ligament into the femur and tibia remain constant so that only the effect of orientation on bending and torsion could be studied. Since the elongation of the ligament depends only on their intraarticular entry points, ligament elongation did not vary between lig1, lig2, and lig3 placements.

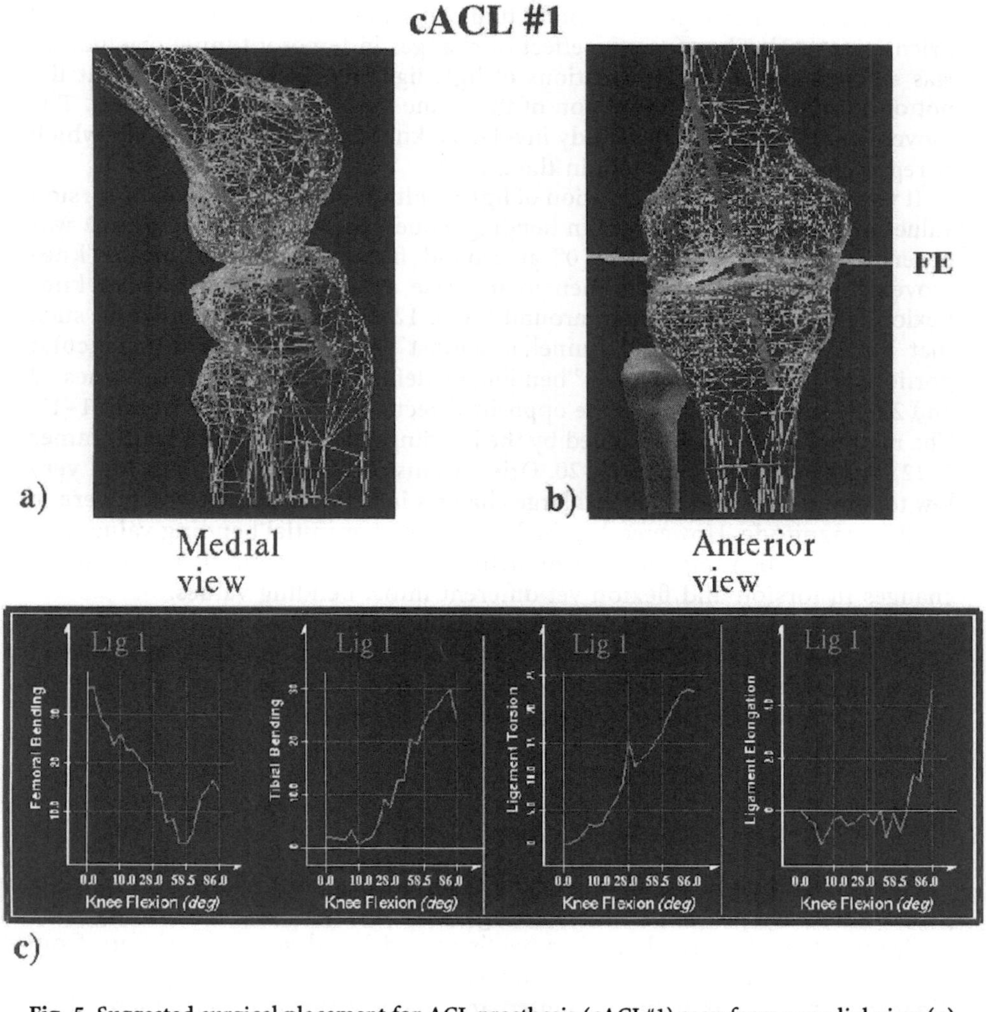

Fig. 5. Suggested surgical placement for ACL prosthesis (cACL#1) seen from a medial view (**a**) and anterior view (**b**). **c** Deformations corresponding to this placement

A suggested orientation for a commercially available ACL prosthesis [19] (Ligament Advanced Reinforcement System, France, company manual), described in Chap. II.4 and referred here to as cACL#1 (see Fig. 5), was also investigated using the knee analyzer system. This shows how ligament deformations can be explained and analyzed to help improve their insertion. The surgical orientation of these ligaments with respect to the FE axis was used to help describe the deformations to which they are subjected.

Results

For the real knee movement data, prosthetic ligament deformation values varied more with changes in femoral tunnel orientation than with tibial tunnel orientation [23]. Therefore the effect of changes in femoral tunnel orientation was studied in detail. Deformations of lig1, lig2, and lig3 can be seen at the bottom of the Fig. 4 as a function of the frame number in the animation. The movement recorded for this study involved a knee flexion from 0° to 90° which is represented by frames 0–20 in the graphs.

It was found that the orientation of lig1 results in very large ligament torsion values and very small changes in bending values. Femoral bending of lig2 was observed to decrease to near 0° at around frame number 12 of the knee movement investigated and then to increase at the highest degree of knee flexion. This occurred since at around frame 12 the knee is in a position such that the ligament's femoral tunnel is almost aligned with its intraarticular portion, giving approximately 0° bending by definition [12]. Between frames 12 and 20 the ligament bends in the opposite direction to that during frames 1–12. The material is therefore fatigued by the bending change observed from frames 0–12 plus that from frames 12–20. Orientations of lig2 and lig3 both had very low torsion values, but both had large changes in bending values which were of similar magnitude. However, lig3 had a much higher initial bending value than lig2. Therefore bending and torsion deformations on lig2 and lig3 had similar changes in torsion and flexion yet different initial bending values.

Deformations on the commercially available ligament cACL#1 can be seen in Fig. 5. It was noted that the femoral tunnel orientation of cACL#1 was close to that of lig2 which is perpendicular to the FE axis. Therefore, as expected, its femoral deformation patterns resembled those of lig2.

Discussion

The understanding of prosthetic ligament failure has been a long and winding path with many pitfalls. Our knowledge has, however, evolved through the realization that the prosthetic ligament's material and geometry differs from that of the natural ligament, which explains why prosthetic ligaments fail differently than their natural counterparts, namely through permanent elongation, fatigue wear, and inflammatory response. The study of ligament fatigue itself has evolved from cyclic tension to combined elongation, bending, and torsion analysis presented in this chapter.

From the ligament deformation analysis we see that prosthetic ligaments can be subjected to large bending and torsion deformations in vivo if improperly placed. Unlike the natural ligament, the prosthesis does not possess the capacity to regenerate its material over time. Furthermore, also unlike the natural ligament, the prosthesis is usually anchored via a bone tunnel, whose movement with respect to the ligament is responsible for changes in bending and torsion. It is these changes in bending and torsion, coupled with tension and fraying (from rubbing against bone) that may cause premature failure [4, 9, 12]. Therefore prosthetic ligaments fail due to fatigue wear failure whereas natural ligaments fail due to sudden impacts or overload. This is precisely why the minimization of bending and torsion deformations is such an important consideration for ligament prostheses.

The results of ligament deformations presented in this chapter indicate a tradeoff between minimizing flexion and torsion in the ligament. That is, an insertion orientation which decreases torsion augments bending in the ligament. Interpretation of these findings may be simplified by relating the ligament tunnel orientation to the average FE axis of the knee. For ligament lig1 the femoral tunnel is approximately parallel to the average flexion axis of the knee. Therefore flexion of the knee rotates the femoral tunnel about its axis, creating torsion in the ligament, but the bending angle between the femoral tunnel and the ligament is high yet approximately constant throughout the flexion. For lig2 the femoral tunnel is approximately perpendicular to the average flexion axis. Therefore, flexion of the knee rotates the femoral tunnel about an axis perpendicular to its long axis, causing much change in bending yet little torsion in the ligament. Femoral bending of lig2 appears on the graph to change less than that of lig3 although its change in bending is in reality almost identical. As mentioned in the observations of the results, this is due to the definition of femoral bending. Bending of this ligament changes direction thus passing through a minimum near 0° during the prescribed movement at about frame 12. Femoral bending of lig2 between frames 12–20 should therefore be added to that between frames 0 and 12 to reflect the change in bending that lig2 is subjected to. A more accurate reflection of the fatigue in the material could therefore eventually be the area under the derivative of the ligament bending curve.

The placement of lig3 is also perpendicular to the FE axis but is oriented differently than lig2; therefore it has the same change in torsion and bending as lig2 but starts out with a higher initial flexion angle, which explains the offset in the bending calculations.

It can be appreciated from these tunnel placements resulting in maximal deformations that the ligament insertion tunnel orientations can be varied to provide a broad range of possible ligament deformations. Studying these extreme placements helps our understanding of the relationship between tunnel orientation and ligament deformation and allows us to situate presently proposed ligament placements within the spectrum of possibilities.

The cACL#1 prosthesis showed relatively little change in torsion deformation and a large change in bending deformation. This can be explained by the fact that the femoral tunnel is oriented perpendicularly to the FE axis giving similar deformation patterns as lig2 in Fig. 4. Therefore this ligament is placed in a region where there is a maximal change in femoral bending. However, the ligament is specified to form a straight line when the knee is flexed at 40°. The principle behind this suggested surgical placement is that 40° is the most common position of the knee during daily activities, therefore the ligament should be as straight as possible with the knee in this position [19].

To minimize mechanical wear from ligament torsion, change in torsion from the ligament's "natural" state should be minimal, and fibers should rub against one another as little as possible throughout an entire knee flexion. The cACL#1 ligament is designed to have pretwisted fibers in the intraarticular portion of the knee which un-twist during knee extension. The pretwist is set at approximately 90° clockwise or counterclockwise, depending on whether it is to be used for a right or left knee. The design of the twist of this ligament prosthesis is based on the observed rotation of the natural ligament's attachment site. Since the natural ligament is attached on the medial aspect of the lateral condyle, its femoral attachment rotates in the sagittal plane. This is true for the natural ligament since its "attachment orientation" is approximately perpendicular to the articular surface where it enters the femur. This is similar to having a prosthetic ligament tunnel oriented mediolaterally (or parallel to the FE axis) which would also result in a large change in torsion. However, it was observed from the calculations that ligament torsion corresponding to the suggested cACL#1 placement remained below 25° over 90° knee flexion. This relatively low torsion value for cACL#1 is due to the fact that the suggested femoral tunnel insertion is perpendicular to the FE axis. Therefore, the pretwist designed into this ligament should be much smaller for the suggested femoral insertion orientation. Results of this study could eventually help improve the fatigue life of this ligament by matching the pretwist in the design with the in vivo torsion to which it will be subjected.

As mentioned above, the present approach assumes that the prosthesis would not significantly alter the knee's kinematics. Comparison of kinematics before and after a ligament replacement could verify change in knee kinematics following prosthetic ligament replacement.

Deformation data generated from this study can help to determine optimal placement of prosthetic ligaments. Fatigue properties of a prosthesis are a function of its material and its design. Data on the relative displacements of tunnel insertion for some surgical insertion could be reproduced by an in vitro tension-torsion-flexion fatigue testing machine previously developed [9] to closely simulate in vivo conditions. Information on the number of cyclic loads to failure could help understand the mechanisms of particle wear and failure and better predict the in vivo survivorship. Since it has been observed that changing ligament insertion orientation can result in a wide range of deformation values, ligament placement can be customized as a function of the specific ligament's fatigue properties. For example, a prosthetic ligament whose fatigue life is low when subjected to bending deformations should be placed such that bending is minimized. A choice between bending and torsion deformations must be made since it has been found, and explained with respect to the FE axis, that insertions which reduce torsion increase bending and vice versa. It should eventually be possible to design prosthetic ligaments so that their design is a function of the deformations to which they will be subjected in vivo. For example, a ligament could be designed to have little sensitivity to torsion deformations and be inserted in a region of high torsion and high yet constant bending. Of course the choice of optimal insertion is a function of many factors. The placement must (a) be surgically accessible, (b) not disturb or interfere with anatomical structures involved in the knee's function, (c) have sufficient amounts of cortical bone to allow proper fixation, and (d) not alter the joint's normal kinematics.

A similar analysis of insertion versus deformation may be useful for nonprosthetic ligament replacement. Presently the replacement of a ligament by an autograft or allograft is quite popular. The future surely holds in store the development of "bioprostheses" [6] consisting of biodegradable scaffolds which promote natural tissue regeneration of the ligament. Living tissue ligament replacements are generally inserted into the knee via insertion tunnels in a similar fashion to a prosthesis. They therefore undergo similar deformations as an artificial prosthesis while they "heal" in the knee. It has been observed that cells which produce the natural ligament matrix do so as a function of mechanical strain or stress [6, 28]. Therefore it may eventually be possible to insert the autograft, allograft, or bioprosthetic ligament such that strains or stresses are "optimal" to favor cell proliferation and matrix production.

It has been shown how the knee analyzer system can be used to help plan prosthetic ligament insertions. There must, however, be continuity between diagnostics, planning, and intervention for the system to be efficient. Improper diagnosis of the pathology makes exact planning useless. Moreover, planning of the perfect procedure could lead to catastrophic results if it is not properly implemented. For example, it is now possible with the knee analyzer exactly to match the designed "pretorsion" in a prosthetic ligament such as cACL#1 to the torsion to which it is subjected in vivo. However, it is surgically difficult to exactly place the ligament in the prescribed position for which deformation values are known. It is also difficult to predict what ligament placement or tension would best restore the three-dimensional kinematics of the joint. Continuity between the aspects of diagnostics, planning, and implementation would ensure quality and efficiency.

The technology presented in this chapter could provide a basis to support the process of ligament instability correction. Knee kinematics and laxity are the basis for detection of ligamentous instability in the knee. To date, detection of pathologies is performed principally through one-dimensional analyses such as the AP drawer or Lachman test. A common and increasingly popular three-dimensional laxity test is the pivot shift, which it is presently difficult to quantify numerically. Such three-dimensional quantitative analysis could possibly be performed by the knee analyzer to aid in diagnosis of ligament insufficiency. The possible use of the knee analyzer as a three-dimensional diagnostic tool is presently under study. The technology used by this system can also be applied to surgical intervention by placing a motion sensor on a surgical pointer or tool and performing the same mathematical correlation between the image and navigation system reference frames. This allows the system to track surgical tool position during an intervention. The surgeon controlling the surgical instrument is therefore provided with visual feedback of his instruments and patient bone position throughout the procedure. Similarly to the knee movement analysis application, sensors can be directly placed on the bones of the subject allowing the surgeon to perform complex manipulations, typical of orthopedics, without losing valuable image-generated data. A working surgical assistance prototype based on some concepts presented in this work has been developed and applied for the placement of pedicle screws in the spine [1].

The future scenario to resolve a ligamentous instability in the knee could therefore involve diagnosis of the pathology through kinematic analysis, preoperative planning of "optimal" surgical placement, and perioperative guidance towards the predetermined trajectory. Perioperative guidance by the virtual interface could be fulfilled via visual, auditory, or tactile cues to improve accuracy of the intervention. Visual control of surgical tool movement, even when working on structures normally hidden from view, gives image-guided orthopedic surgery the potential to diminish trauma by avoiding open surgery.

Conclusion

The relationship between prosthetic ligament insertion and deformation has been established through the development of new instrumentation and analysis software. Software has been developed to simulate prosthetic ligaments and calculate their deformations. The combination of the analysis software and hardware form a three-dimensional knee analyzer capable of predicting in vivo ligament deformations.

It was suggested that prosthetic ligament deformations could be explained in reference to their orientation with respect to the FE axis of the knee. Insertion of the tunnel parallel to the FE axis maximized torsion while minimizing bending and vice versa for tunnels placed perpendicular to this axis. This indicated and explained a tradeoff which exists when choosing a prosthetic ligament insertion orientation.

The analysis in this chapter suggests that the pretorsion designed into the cACL#1 prosthesis is too high and should be reduced to fine-tune its desired function. It was also shown that the ligament was placed in a region where femoral bending deformation is very high and torsion very low, since the femoral tunnel is oriented perpendicularly to the FE axis. The knee analyzer system offers a powerful tool which should help to understand prosthetic ligament fatigue in vivo, augment long-term ligament prosthesis survival and improve their future designs. Further analyses are required before an "optimal" surgical placement can be determined for this and other prosthetic ligaments.

Acknowledgments. The authors thank Dr. N. Duval, Alain Richard, and Stephane Baldo for their help on technical aspects of this work and Nicola Hagemeister for her help on the literature review. This research was funded by the NSERC and FCAR organizations

References

1. Amiot LP, Labelle H, de Guise JA, Sati M, Brodeur P, Rivard CH (1995) Computer assisted pedicular screw fixation: a feasability study. Spine 20(10):1208–1212
2. Blankevoort L, Huiskes R (1991) ACL isometric is not the criterion for ACL reconstruction. 37th Annual Meeting, Orthopedic Research Society 4 March 4, Anaheim, Calif.
3. Cazenave A, Laboureau JP (1990) Reconstruction isométrique du ligament croisé antérieur. Détermination pré et post-opératoire du point fémoral. Rev Chir Orthop Reparatrice Appar Mot 76:288–292
4. Claes LE, Ludwig J, Margevicius KJ, Dürselen L (1995) Biological response to ligament wear particles. J Appl Biomater 6:35–41
5. Covey DC, Sapega AA, Sherman GM, Torg JS (1992) Testing for isometry during PCL reconstruction. 38th Annual Meeting, Orthopedic Research Society, 17 February, Washington
6. DesRosiers EA, Méthot S, Yahia LH, Rivard C-H (1995) Réponse des fibroblastes ligamentaires la stimulation mécanique. Ann Chir 49(8):768–774
7. Dorlot JM, Ait Ba Sidi M, Tremblay GR, Drouin G (1980) Load elongation behavior of the canine anterior cruciate ligament. J Biomech Eng 102:190–193
8. Drouin G (1986) The prosthetic replacement of the cruciate ligament. Clin Orthop 208(7):59–60
9. Drouin G, Masson M, Yahia L (1991) In vitro fatigue testing of prosthetic ligaments: a new concept. Biomed Mater Eng 1:159–165
10. Dürselen L, Claes L (1990) Mechanical properties of artificial ligaments. Clin Implant Mater Adv Biomater 9:439–444
11. Flemming BC, Beynnon BD, Nichols CE, Renstrom P (1992) In vitro comparison between predictive isometry measurement and elongation in the reconstruction of ACL. 38th Annual Meeting, Orthopedic Research Society, 17 February 17, Washington
12. Gely P, Drouin G, Thiry PS, Tremblay GR (1984) Torsion and bending imposed on a new anterior cruciate ligament prosthesis during knee flexion: an evaluation method. J Biomech Eng 106:285–294
13. Gillquist J (1990) Stryker dacron ligament: techniques and results. In: Proceedings of the Seventh Annual International Symposium on Advances in Cruciate Ligament Reconstruction of the Knee: autogenous vs prosthetic. Palm Desert, Calif., pp 106–116
14. Good L, Odensten M, Pettersson L, Gillquist J (1989) Failure of a bovine xenograft for reconstruction of the anterior cruciate ligament. Acta Orthop Scand 60:8–12
15. Griffin LY (1995) Rehabilitation of the injured knee, 2nd edn. Mosby, St. Louis
16. Grood ES, Hefzy MS, Lindenfield TN (1989) Factors affecting the region of most isometric femoral attachments. I. The posterior cruciate ligament. Am J Sports Med 17(2):197–207
17. Grood ES, Suntay WJ (1984) A joint coordinate system for the clinical description of three-dimensional motions: application to the knee. J Biomed Eng 105:136–144
18. Hefzy MS, Grood ES, Noyes FR (1989) Factors affecting the region of most isometric femoral attachments. II. The anterior cruciate ligament. Am J Sports Med 17(2):208–216

19. Laboureau JP, Bercovy M (1993) Indications actuelles des ligaments artificielles. 3rd European Congress of the Orthopaedic Traumatology Society of the East of France. Strasbourg
20. Richard A (1995) Recherche d'un axe fixe de rotation dans le genou. Undergraduate Thesis, cole Polytechnique, Montreal
21. Sati M, de Guise JA, Drouin G (submitted) Improving in vivo knee kinematic measurements: application to prosthetic ligament analysis. The Knee
22. Sati M, de Guise JA, Larouche S, Drouin G (1996) Quantitative assessment of skin-bone movement at the knee. Knee (in press)
23. Sati M, de Guise JA, Martel Y, Masson M, Drouin G (1991) An application of 3D medical imagery to the kinematic modeling of knee motion. International conference on 3-D analysis of human movement. July 1991, Montreal, pp 159–162
24. Schider F (1957) An atlas of anatomy for artists. Dover, New York, p 19
25. Seedhom BB (1992) Reconstruction of the anterior cruciate ligament. Proc Inst Mech Eng [H] 296:15–27
26. Veldpaus F, Woltring HJ, Dortmans L (1988. A least squares algorithm for the equiform transformation from spatial marker coordinates. J Biomech 21:45–54
27. Yahia L, Hagemeister N, Drouin G, Sati M, Rivard CH (1994) Conceptual design of prosthetic anterior cruciate ligaments: the need for a biomimetical approach. Biomimetics 2(4):309–330
28. Yen EHK, Pollit DJ, Whyte WA, Suga DM (1990) Continuous stressing of mouse interparietal suture fibroblast in vitro. J Dent Res 69(1):26–30

II.4 Isometric Reconstruction of the Anterior Cruciate Ligament: Femoral and Tibial Tunnel Placement

J.P. Laboureau and F. Marnat-Perrichet

Introduction

Study of our anterior cruciate ligament (ACL) plasties shows that more than one-fourth of failures are due to technical faults. This is true for all types of repair and especially for those involving the use of artificial ligaments, which are less tolerant than autogenous ligaments. These observations have led us to propose a set of rules, the first of which being the respect of isometry. Isometry has two components: "extrinsic" isometry depends on the choice of the femoral and tibial insertions, and "intrinsic" isometry depends on the ligament architecture.

This chapter presents the development and clinical implementation of isometric surgical insertion for an ACL replacement. Definitions of both extrinsic and intrinsic isometry are first provided. Methodology and results of experimental data used to determine isometric surgical placement are presented, followed by a description of the corresponding clinical implementation of the proposed technique.

"Extrinsic" Isometry

Determination of the Femoral Isometric Point

The results of this anatomic and radiological study have previously been published [5]. As a whole, the ACL is not isometric. It is taut in complete extension and relaxed for a 90° knee flexion. However, some transitional fibers are close to isometry, as has been shown by many authors, notably Daniel et al. [6], Abbink [1], and Bradley et al. [4].

The femoral isometric point F, i.e., femoral attachment of these fibers, is the center of a 140° circular arc formed by the posterior border of the lateral condyle (Fig. 1). The radius of this circle varies between 17 and 26 mm depending on the subject. Saragaglia et al. [17] and more recently Elias et al. [8] arrive at the same conclusion. On average, point F is located at 59% of the anteroposterior length of the condyle, measured on a line parallel to the Blumensaat line and starting from the top of the posterior condyle. It is re markable that this circular arc of 140° has the same numerical value as the knee's normal range of motion. Moreover, the center of the arc is the only

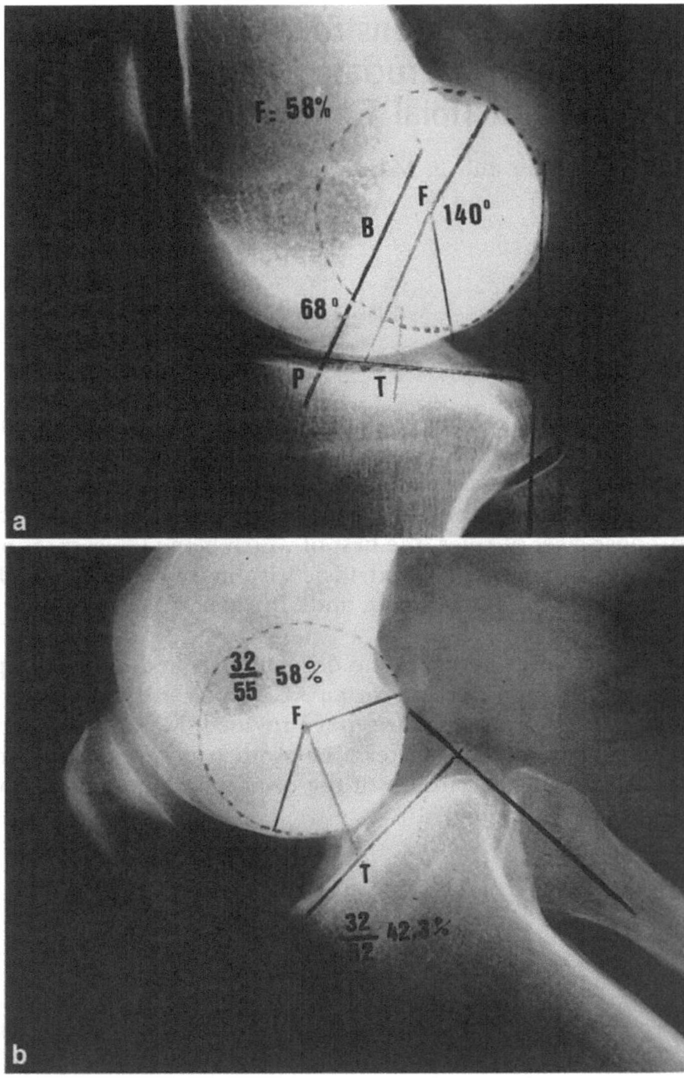

Fig. 1. a The isometric point F is situated at the center of a 140° circular arc and corresponds for this case to 58% of the anteroposterior length of the condyle (i.e., 55 mm for the example shown). The angle formed between the Blumensaat line and the tibial plateau is 68° in maximum extension. Note how the increased condylar radius in **a** (and hence increase in distance FT) explains the loading of the ACL with respect to the knee in flexion seen in **b. b** Same knee in flexion, the isometric tibial point T, center of the tibial plateau along its longest length, is at 42.3% of the length (52 mm) of the superior face of the internal tibial plateau. The FT distance is practically constant between flexion and extension (32–31 mm) since there is little to no hyperextension in this knee

geometrical shape possibly compatible with the definition of isometry during complete flexion-extension of the knee.

Determination of the Tibial Isometric Point

The stabilization by the ACL can be considered as a geometric system. The tibial plateau appears as a tangent to the circle formed by the posterior border of the lateral condyle. The contact point moves from front to back when knee flexion varies from 0° to 140°. Since the medial tibial plateau is relatively immobile when ACL is normal (unlike the lateral tibial plateau), and the ACL inserts near the axial border of the medial tibial plateau, this insertion is also relatively immobile in the anteroposterior direction. Our choice of the center of the tibial tunnel is based on three considerations: (a) anatomic and radiological study of the center of the tibial insertion of the ACL, (b) the absolute necessity of the absence of conflict between the plasty and the roof of the intercondylar notch in complete extension, and (c) perioperative clinical measurements of length variations in the plasty.

"Intrinsic" Isometry

One should in fact speak of "limited anisometry." The anterior fibers relax during extension. The posterior fibers relax during the flexion. The ligament, as a whole, is taut during complete extension. During the movement from extension to complete flexion, the most posterior and proximal femoral insertions become anterior and distal, and roll around the anterior fibers. The main axis of the ACL femoral insertion rotates 140°. This movement is clockwise for the right knee and counterclockwise for the left knee. This movement partially compensates for the decrease in the distance between the tibial and femoral insertion points so that the individual fibers remain relatively isometric (Fig. 2). The natural elasticity of the ligament or the autogenous transplant compensates for these movements without fiber stress.

The drawback of artificial fibers, which is their lack of elasticity, must be compensated by the ligament's architecture. This architecture must have two principal anterior and posterior bundles. These two bundles are made up of longitudinal fibers without transversal structures because transversal fibers would limit the winding around of longitudinal fibers during torsion in the intraarticular part of the ligament (Fig. 3).

With the ligament being placed for practical reasons in the flexed knee, the two bundles must be pretwisted around one another in a clockwise direction for the right knee and a counterclockwise direction for the left knee. This implies different ligaments for left and right knees. When the knee moves from complete flexion to extension, the twist disappears. In complete extension all fibers are parallel and longitudinal, thus at their maximum length. The reproduction of this architecture allows the physiological mechanism of progressive fiber recruitment.

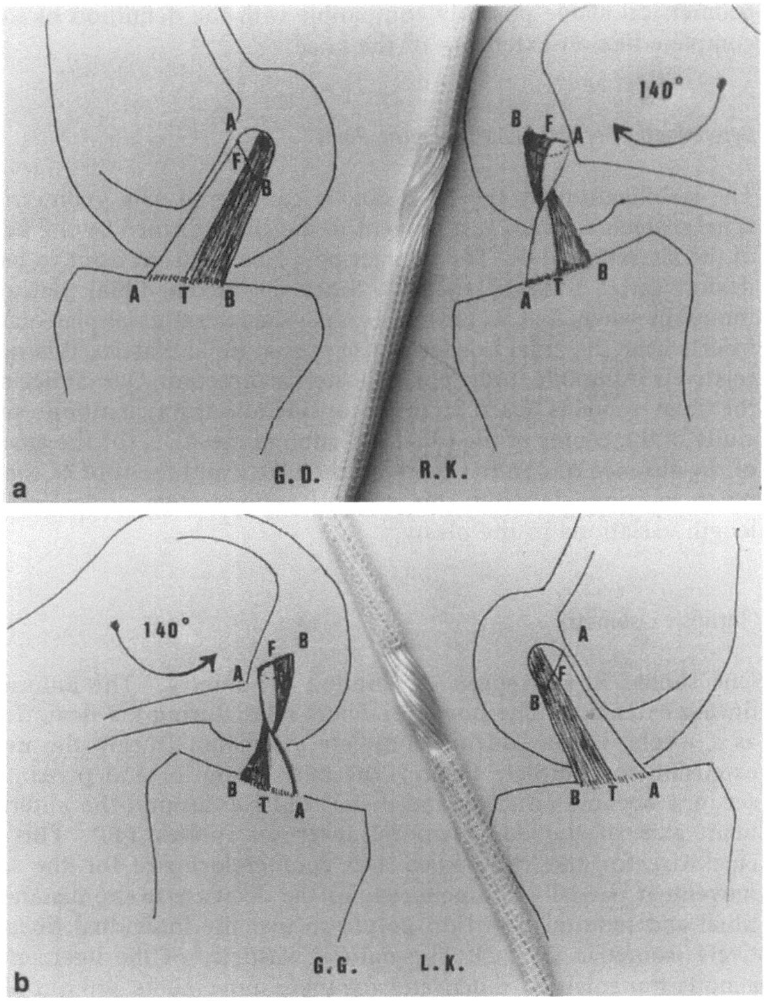

Fig. 2. a Right knee. In extension the ACL fibers are taught, longitudinal, and parallel. In flexion at 140° the line AB has turned by the same number of degrees, which invokes clockwise torsion in the ligament fibers. The artificial ligament must during manufacturing be pretwisted in the same direction as the torsion to sustain it without damage. **b** Left knee. The torsion here is counterclockwise and requires a specific ligament

Methodology and Results

Anatomic and Radiological Study of Tibial ACL Insertion

Materials and Method

This study was performed on 25 knees operated by open surgery for total or unicompartmental prostheses, and having a normal ACL (14 men, 11 women).

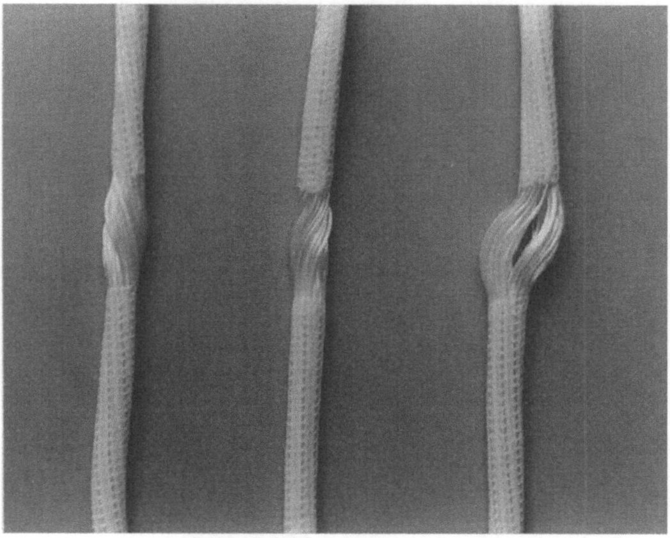

Fig. 3. *From left to right:* LARS ligament for left ACL (counterclockwise) as inserted at 90° knee flexion; right ligament (clockwise) the two anterior and posterior bands which form the ligament. The free central fibers correspond to the intraarticular zone. In the intraosseus zone the longitudinal fibers are united by a special knitting procedure

Metallic landmarks were fixed in the cartilage at the anterior and posterior boundaries of the ACL insertion. The spacing between the landmarks was measured and marked with an electrical bistoury on the adjacent cartilage of the medial tibial plateau. A lateral X-ray was taken. The center between anterior and posterior marks was determined on the X-ray. The resected bone pieces and the ACL insertion were then measured (Fig. 4).

Results

The ACL insertion is large: from 16 to 21 mm (mean 19 mm). Its center is located near the middle of the projection of the largest anteroposterior length of the tibial plateau. The length ranged from 45 to 65 mm (mean 54.4 mm; Figs. 1, 5). The distance between the center of the tibial ACL insertion and the most anterior border of the plateau was 22–33 mm (mean 26.92 mm), that is, 49.6% of the total length measured from the anterior border. Measurements were also performed relative to the medial tibial plateau (MTP). Its longest length was 39–56 mm (mean 46.6 mm). The center of the ACL insertion relative to the anterior border of the MTP was 17–23 mm (mean 19.8 mm), that is 42.5% of the total length measured from the anterior border.

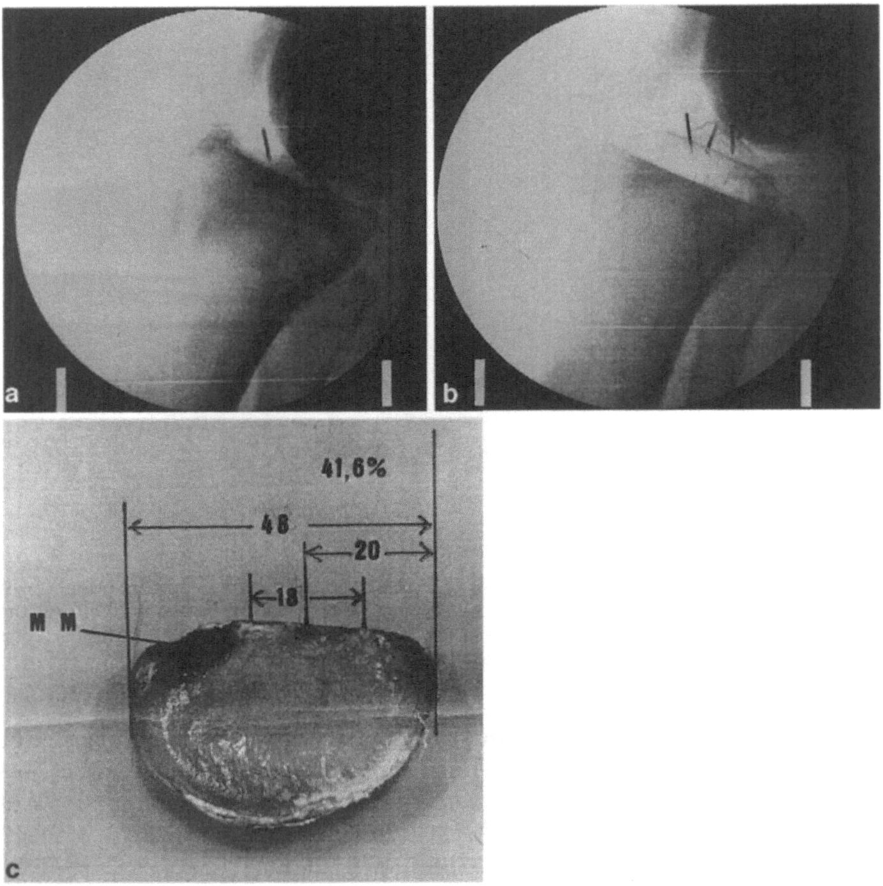

Fig. 4a–c. Locating the ACL insertion for a unicompartimental prosthesis. **a** View before resection. **b** View after resection of the internal tibial plateau. The median landmark is dead center of the plateau and corresponds to the apex of the spine. **c** The measurements are taken on a resected plateau. The ACL insertion is 18 mm, and its center is at 20 mm, or 41.6% of the total length 46 mm. Note that the posterior insertion of the medial meniscus is flush with the most posterior fibers of the ACL

Absence of Conflict with the Intercondylar Roof

The conflict which seems to be the cause of technical faults has led us to study the relationship between the intercondylar roof and the tibial plateau

Materials and Method

Fifty patients without pathology or previous traumatism on at least one knee, were recruited: 29 men and 21 women aged 18–36 years. A lateral X-ray of the

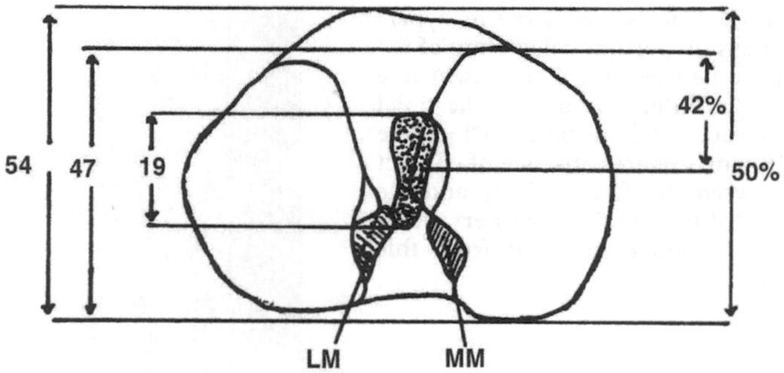

Fig. 5. Schematic view of tibial plateau and average measurements reported in millimeters. Note the close relationship between the ACL and the posterior horn of the lateral meniscus (*LM*)

knee of a standing subject at maximum extension was taken while ensuring superposition of the condyles. The source-plate distance was 1.50 m to prevent any magnification and control penetration to permit proper visualization of the Blumensaat line. On this X-ray we measured: (a) the posterior angle formed between the Blumensaat line and the tibial plateau, and (b) the space between the intersection of these lines and the middle of the tibial plateau (i.e., between the most anterior projection and the posterior border of the internal tibial plateau; the presumed center of the ACL insertion).

Results

The B-TP angle (Blumensaat line, tibial plateau) varied between 82° and 45°, with a mean of 65.0° (64.9°±8.3° in men and 65.1°±7° in women). Of the 50 patients 35 (70%) had an angle in between 60° and 70° (65°±5°) The value of this B-TP angle depends on two factors: (a) the angle between the intercondylar roof and the femoral axis; (b) the degree of added hyperextension. The distance between the Blumensaat line and the center of the tibial plateau was 14 –3 mm, with a mean of 9.5 mm (9.8±1.9 mm in men and 9.3±2 mm in women; Fig. 6). This intersection appears always anterior to the tibial plateau center for all knees studied. The larger the B-TP angle, the closer the intersection was to this center. In 2 of 50 knees (4%) the distance was 3 and 4 mm.

Perioperative Clinical Study

Materials and Method

We measured the movement of the artificial ligament at the tibial tunnel for 50 operated patients according to the following procedure. Only the femoral extremity of the ligament was fixed with an interference screw. The ligament was

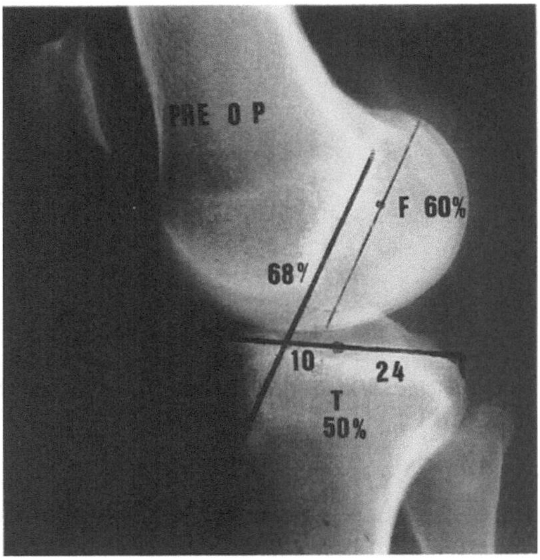

Fig. 6. The lateral X-ray in hyperextension permits calculation of the angle between the Blumensaat line and its intersection with the tibial plateau. The distance PT (here 10 mm) ensures absence of conflict between the future plasty and the roof of the notch. The intervention can be totally planned from this view

held manually in tension, marked by a thin needle at the tibial tunnel exit first at 100° knee flexion and then at the complete extension. Then the distance between the landmarks was measured.

Results

In 10 cases the two landmarks were superimposed, and in 16 they were in contact with one another. We accepted an error of 0.5 mm in these cases. In 16 cases the distance was 1 mm, in 6 it was 2 mm, and in 2 it was 3 mm. The mean movement was 0.84 mm.

Clinical Application

Principles and Techniques

The isometric femoral point can be determined from an exact preoperative lateral X-ray projection, with superposition of the two condyles and the plate-source distance at 1.5 m to avoid any magnification. We used a transparent template to determine the circle and its center more easily.

The position of the center F can then be determined with respect to anatomical reference points. These reference points can be the circumference of the circle, the anterior border of the intercondylar notch roof or its posterior border which corresponds to the "over-the-top" point, that is at the posterior edge of the Blumensaat line on the X-ray (point O). The distance between points F and O can be measured on the X-ray. It varies from 6 to 12 mm depending on the subject. (Fig. 7).

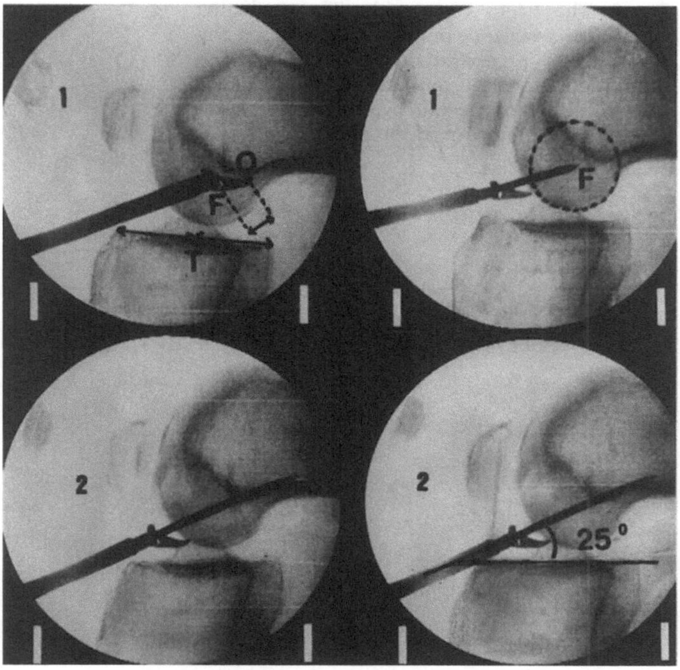

Fig. 7. Perioperative view under fluoroscopy. *1*, The isometric guide is opened to the FO spacing which allows placement of the K-wire guide at the center of the circle. The extremity of the guide arches around the posterior border of the roof of the notch and is oriented at 10 o'clock or 2 o'clock depending on whether it is a right or left knee. *2*, The K-wire is penetrated into the condyle with an approximate angle of 25° with respect to the tibial plateau

The femoral guide FAC (Ligament Advanced Reinforcement System, LARS), the end hook of which attaches onto the point O, is adjustable. The length between the tip of the hook and the drill guide, which passes through the handle of the instrument, represents the distance between the points F and O and can be adjusted thanks to a graduated wheel (Fig. 8).

The body of the instrument represents the future plasty and must be oriented in the notch to avoid contact with its walls, therefore guaranteeing the absence of friction between the neo-ACL and the surrounding bone. This body of the instrument crosses the anterior face of the posterior cruciate, which allows visualization of the future relationship between the two ligaments after the ACL reconstruction, a relationship which guarantees the stability of the knee.

Use of the guide permits accurate placement of the K-wire which serves as a guide for a cannulated drill bit.

In the absence of a preoperative X-ray, the center of the circle, and adjustment of the guide can be determined using fluoroscopy. Open or arthroscopic surgery can be performed. In the latter case the guide with the hook in closed position is introduced by a microincision at the intersection of the

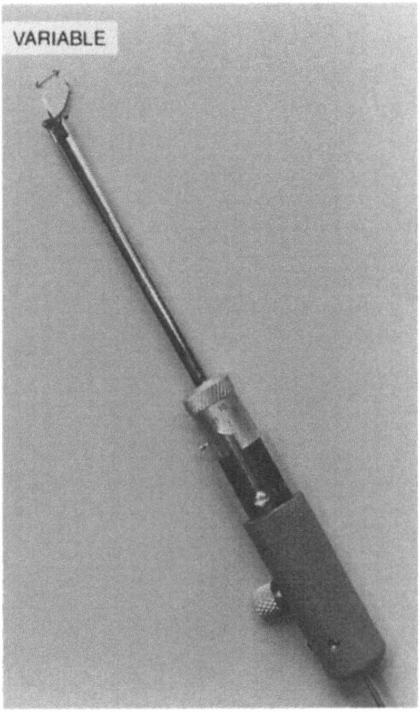

Fig. 8. The LARS FAC guide. The graduated wheel allows variance in the K-wirehook spacing according to preoperative measurements. The intraarticular portion must remain at a distance from the walls of the notch, which ensures absence of later abrasion of the plasty

medial border of the patellar tendon and the anterior border of the tibial plateau. The knee is flexed at 90°–100°. The guide penetrates into the middle of the notch, somewhat closer to the anterior border of the medial condyle than to the lateral condyle, yet keeping a certain distance. The hook is then opened to the desired spacing and fixed at the "over-the-top" position, at 10 o'clock or at 2 o'clock, depending on whether it is a right or left knee. The body of the instrument is directed obliquely upwards and out with about a 25° angle with respect to the tibial plateau and a 20° outwards angle with respect to the sagital plane. This orientation actually represents that of the normal ACL with a knee flexed at 90°.

The 3 mm rigid K-wire is motor-driven through the guide. When penetrated by a few centimeters, the guide is retracted. The K-wire can then penetrate totally. It passes by, at a distance, the posterior methaphysal cortical bone and exits quite high on the anterolateral femur and out through the skin. The choice of the center of the tibial plateau as the center of the tibial tunnel corresponds to the anatomy and guarantees the absence of conflict between the plastie and the roof of the notch in the majority of cases. Nonetheless, in the rare cases where there is coexistence of a vertical roof and a hyperextension, a conflict between the plasty and the roof is not eliminated if the roof-center distance is inferior to the radius of the tunnel.

We cannot move the insertion back since this would risk making the plasty vertical, which would deprive it of its mechanical efficiency. In these rare cases a plasty of the roof of the notch must be performed. These cases must be

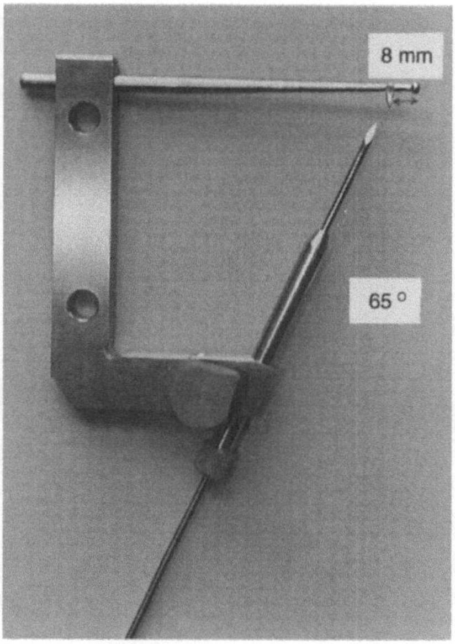

Fig. 9. The LARS TAC guide. The rounded extremity of the intraarticular rod touches the anterior aspect of the PCL. The tibial K-wire, which must be oriented at 65°, penetrates into the articulation, 8 mm in front of the PCL, i.e., at the center of the plateau

detected clinically, by looking for hyperextension, and radiologically. The lateral X-ray in maximal extension must be part of a preoperative assessment and permit planification of the surgical intervention (Fig. 6).

The difficulty in arthroscopy is determining the field of depth. The LARS tibial guide has a ball on its extremity which one applies on the anterior face of the PCL and it provides a 65° angle with respect to the tibial plateau (Fig. 9).

In this situation the point pierced into the cartilage of the tibial spine is situated 8 mm more forwards, which is on average 42% of the MTP length. This corresponds to the intraarticular emergence of the K-wire put into place from the middle of the medial face of the metaphysis, just above the tendons of the Pes Anserinus. Another method to position the K-wire is based on a geometrical relationship (Fig. 10). The femoral drill guide was inserted in the external condyle from the center of the posterior circle as indicated above. This K-wire exits at the skin. With the driller placed at the femoral side the K-wire is retracted upwards and outwards from the notch until it is flush with the axial aspect of the condyle. The knee until now flexed at 90° is flexed at 50°. The femoral K-wire then forms a 65° incline with respect to the tibial plateau. The K-wire is then moved downwards into the tibia, which it enters at the center of the anatomical insertion and exits through the skin on the medial face of the tibial metaphysis. This K-wire serves as a guide for a cannulated bit.

In addition to its simplicity, this method has the advantage of assuring a perfectly rectilinear alignment of the tibial and femoral tunnel for a knee near its functional position. This absence of angulation during loading is important for all plasties, as suggested by Djian et al. [7]. It is fundamental for artificial ligaments to prevent ovalisation of the osseous tunnels or the rupture of the implant.

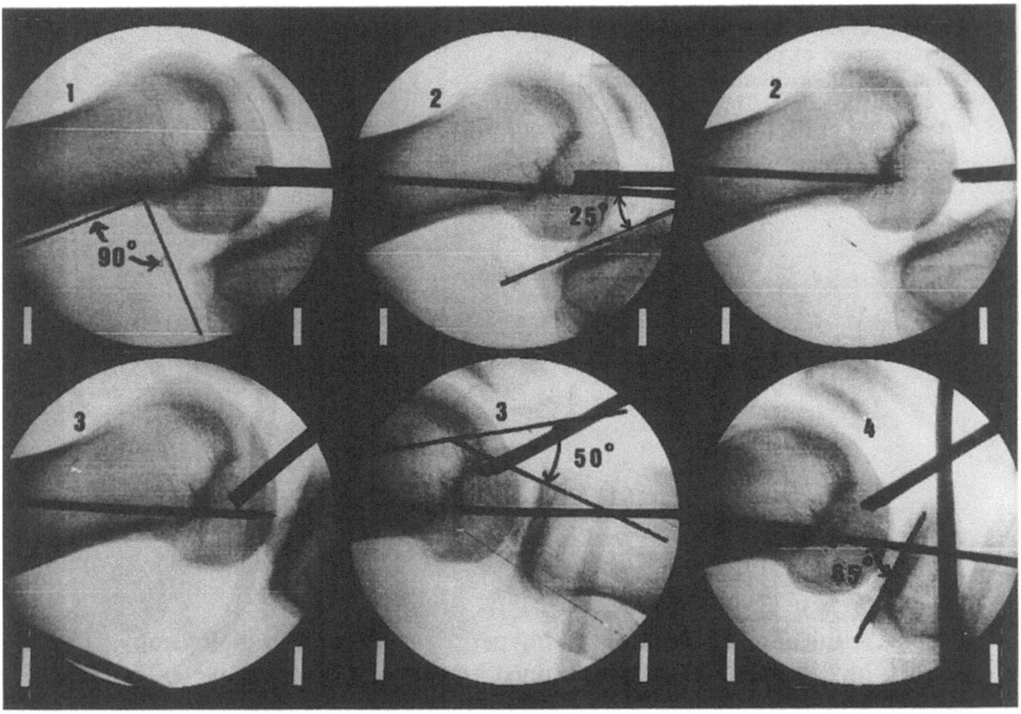

Fig. 10. The four phases in an arthroscopic intervention. *1*, The drill guide, which is placed with the FAC guide or under direct visual control on the fluoroscopic screen, at the center of the circle, penetrates the lateral condyle, the knee at 90°. *2*, The K-wire is penetrated with an upwards sloping angle of 25° and is retracted by its femoral extremity exiting at the skin until its distal point is flush with the wall of the notch. *3*, The knee is placed in 50° flexion. The K-wire then forms a 65° incline with respect to the tibial plateau (25° ± 40°). *4*, The K-wire is penetrated from the top to the bottom of the tibia and serves as a guide for the femoral and tibial drill bits. The two tunnels are drilled one after another to avoid damage of the remaining ACL

Results

Mechanical

The above technique has been used for the past 5 years. Among 126 patients who exhibited a acute ACL lesion reconstructed with sutures reinforced by artificial ligament, 107 have been seen again, with an average follow-up of 30 months (12–54 months). The mechanical tests were carried out according to the criteria described by Bercovy [3] with radiolaximetry using the TELOS under 100, 150, and 250 N (Fig. 11)

Under 250 N, 92 (86%) had a differential Lachman equal to or less than 3 mm, 6 (5.6%) between 3 and 5 mm, and 9 (8.4%) greater than 5 mm. On the Jerk test 91 (85%) had a negative result, 10 an uncertain result, and 6 a positive result.

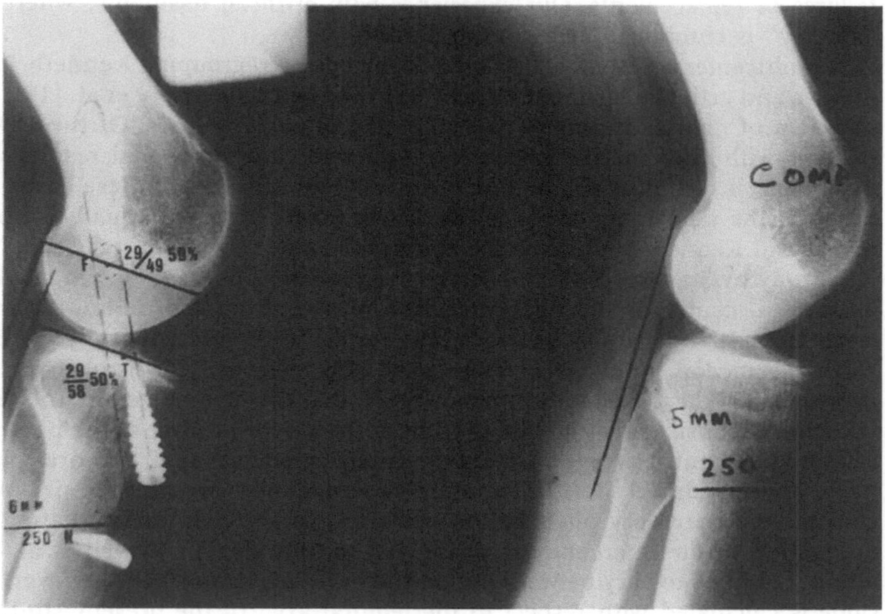

Fig. 11. Comparative radiolaximetry at 32 months postoperatively under 250 N. Differential Lachman 1 mm. The good mechanical result corresponds to a proper technique: F at 59%, T at 50% of the total length of the plateau, tunnels perfectly aligned in the functional position

Functional (for Mobility)

No patient exhibited a deficit in extension. Of the 107 patients 11 (10%) had a flexion deficit always less than 15°. These results are more better than those in a similar series of 100 patients reported previously [13, 14]. In this study, although the femoral tunnel was the same, its placement and orientation in the tibia and the relation with the intercondylar roof was not precisely defined. In this series, the differential Lachman effect was higher than 5 mm for 30% of cases and an extension deficit for 8% of cases.

Discussion

The results of our measurements on the placement of the ACL and the slope of the intercondylar notch [5], lead to a more posterior tibial tunnel and more anterior femoral tunnel than classic techniques. Almekinders et al. [2], in a study on the postoperative problems encountered on a series of 70 ACL reconstructions by patellar tendon, ideally placed the center of their tibial tunnel on a cadaveric knee at 34% of the anteroposterior length of the tibial plateau. They found this to be 29% on their operated patients. The authors consider biological factors to be more important than mechanical factors in explaining failures of replacements. Our experience with artificial ligaments, where the "biology" is eliminated, led us to think differently.

A multicenter series of 575 ACL reconstructions, regrouping Kenneth-Jones plasties and artificial ligament plasties was reviewed by Bercovy et al. [19]. The objective of this study was to determine the cause of failures. Of the 18% of patients with poor mechanical results, even when the functional results were acceptable, a technical positioning error was found in half of these cases as a factor in the slackening of the plasty. These errors occurred equally with artificial and Kenneth-Jones plasties, and gave identical consequences (Fig. 12).

A study by Stauebli et al. [18] on the magnetic resonance images located the anatomical center of the fibers insertion at 43% from the anterior border, which corresponds almost exactly to the results in the present study (42.5%), when the medial tibial plateau was used as a reference. Morgan et al. [15] presented similar results and further illustrated that the center of the tibial insertion is consistently 7–8 mm in front of the anterior aspect of the posterior cruciate ligament. However, we only partially agree with their femoral placement. Their landmarks for the femoral K-wire does not vary with the size of the condyle and may very often be too posterior. In a recent study relating the orientation of the intercondylar notch roof to knee flexion Howell et al. [10] found this slope to vary between 26° and 46° with a mean of 35°. However, this slope is calculated with respect to the femoral axis. In the present study we investigated the real maximal slope, i.e., that which considers the knee in hyperextension. Taking into account this hyperextension, the authors found a variance in the center of the tunnel to be between 26% and 55% of the length

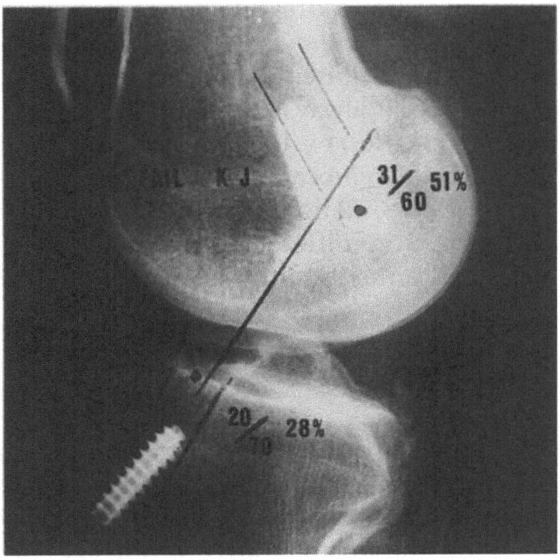

Fig. 12. Failure by error in KJ technique; reproduction of the laxity after an initial period with deficit in extension. The tibial tunnel is partially in front of the Blumensaat line (at 28%). The femoral tunnel is too anterior (at 51%). The revision required obstruction of the previous tunnels with bone grafts and drilling of new tunnels

along the tibial plateau. The safest position we located at 42% of the medial tibial plateau or 50% of the largest length coincides with their results.

The slope of the intercondylar roof, which we found to be on average 65°, establishes not only the central position of the tibial tunnel but also its sagittal plane angle with respect to the tibial plateau. This angle must also approach 65° to avoid bending in the plasty at the intraarticular tunnel exit. Morgan et al. [15] advocate an ideal angle of 68°, which is very similar to the present study.

Regarding the femoral tunnel, our choice is more anterior than classical descriptions. Good et al. [9] place the tunnel at 66% of the anteroposterior length of the femoral condyle, along a line parallel to the Blumensaat line, starting from the edge of the posterior condyle. In our study this percentage was 59%. On an average condyle, this corresponds to 3 or 4 mm, i.e., at the most anterior fibers of the natural ACL which are the most isometric. Furthermore, Good et al. [9] found a mean length of 2.9 mm in their plasty (2.7–3.3 mm) between 0° and 90° knee flexion. We hope to be more strict in the use of artificial ligaments. They concluded that the transplant must be fixed in a knee position which maximizes its length. This is in accordance with our opinion when perfect isometry is not obtained. Following the condition of isometry and avoidance of contact with the intercondylar roof, the absence of tension should be the golden rule when using an artificial ligament.

Dijan et al. [7] in a retrospective study of 36 ACL reconstructions reinforced with artificial ligaments placed their tibial tunnel on average at 34% of the anteroposterior length of the internal tibial plateau with an angle of 74° and a femoral tunnel at 72% of the condylar length. With these landmarks they obtained a 6/36 ratio, that is, a 16% deficit in extension, 8/36 (22%) deficit in flexion, and 5/36 (13%) of differential Lachman effect greater than 5 mm at 89 N, force for which the test has little sensitivity. For Bercovy et al. [3] only the test results at higher than 180 N are sufficiently reliable and eliminate the false negatives.

The deficit in extension appeared to us to be significantly correlated with an overly posterior femoral tunnel and with a tibial tunnel located in front of the Blumensaat line. Contrary to our deductions and those of Howell et al. [11, 12], we did not find a correlation between residual laxity and the position of the transplant. It is logical to think that an overly anterior placed tunnel, leading to a mechanical deterioration of the transplant, would result in a poor mechanical result. Romano et al. [16] also insisted on the influence of the choice of insertions on mobility and laxity of the knee.

It is certain that our choice of more posteriorly placed tunnels on the tibia and more anteriorly placed tunnels on the femur theoretically make the plasty less efficient for controlling anterior drawer. However, the anterior aspect of the plasty consequently buts up against the roof of the notch, along its entire length, in extension which opposes backwards movement of the femur. The ligament's action is no longer restricted to the tibial insertion. As well illustrated by Bercovy [3], respecting the rigidity curve of the knee is more important to consider than its laxity at low applied force values. We prefer the risk of a slight residual laxity to a potentially more damaging risk of a deficit in extension or flexion which in any case would imply a more or less premature degeneration of the ligament by repetitive microtrauma and thus a poor mechanical result.

Conclusion

The use of the artificial ligaments, which are more sensitive than autogenous plasties, led us to a strict surgical technique which allows better definition of the compromise to which all ACL reconstructions are subjected. This compromise is necessary until the exact reproduction of the natural ACL is obtained. The ligament must ensure mechanical efficiency while limiting stresses, permitting normal function, and minimizing social and economical consequences of the accident.

The respect of isometry, absence of conflict with bony surfaces, absence of tension, and alignment of the tunnel at 50° of flexion are the fundamental rules. To these four rules we must add the proprioceptivity aspect, which requires a nontraumatic arthroscopic surgery and the preservation of the recently ruptured cruciate ligament, which possesses indispensable mechanoreceptors. It is certainly the strict adherence to these rules, and use of the most studied ligaments, which presently allows us to obtain competitive results more simply and quickly for the patient.

References

1. Abbink EP (1988) Clinical experience in correction of chronic anterior cruciate ligament defiency with bovine xenograft: a 5 years study. In: Friedman M, Ferkel RD (eds) Prothetic ligament reconstruction of the knee. Saunders, Philadelphia, pp 101–111
2. Almekinders LC, Moore T, Freedman D, Taft TN (1995) Post-operative problem following anterior cruciate reconstruction. Knee Surg Sports Traumatol Arthroscopy 3:78–82
3. Bercovy M, Weber E (1995) Evaluation de la laxite, de la rigidite et de la compliance du genou normal et pathologique. application a la courbe de survie des ligamentoplasties. Rev Chir Orthop 81:114–127
4. Bradley J, Fitzpatrick D, Daniel D, Schercliff T, O'Connor J (1988) Orientation of the cruciate ligament in the sagital plane. J Bone Joint Surg Br 70B:94–99
5. Cazenave A, Laboureau JP (1990) Reconstruction isometrique du ligament croise anterieur. Determination pre et post-operatoire du point femoral. Rev Chir Orthop 76:288–292
6. Daniel DM, Penner DA, Burks RT (1988) Anterior cruciate ligament graft isometry and tensioning. In: Friedman MJ, Ferkel RD (eds) Prosthetic ligament reconstruction. Saunders, Philadelphia, pp 17–21
7. Djian P, Christel P, Roger B, Witvoet J (1994) Evaluation radiologique et IRM des ligamentoplasties intra-articulaires utilisant le tendon rotulien. Rev Chir Orthop 80:403–412
8. Elias SG, Frieman MAR, Gokcay EL (1990) A correlation study of the geometry and anatomy of the distal femur. Clin Orthop 260:98–103
9. Good L, Odensten M, Gillquist J (1994) Sagittal knee stability after anterior cruciate ligament reconstruction with a patellar tendon stip. A two year follow-up study. Am J Sports Med 22(4):518–523

10. Howell SM, Barad SJ (1995) Knee extension and its relationship to the slope of the intercondylar roof. Am J Sports Med 23(3):288–294
11. Howell SM, Berns G, Farley TE (1991) Unimpinged and impinged anterior cruciate ligament grafts: MR signal intensity measurement. Radiology 179:639–643
12. Howell SM, Taylor MA (1993) Failure of reconstruction of the anterior cruciate ligament due to impingement by the intercondylar roof. J Bone Joint Surg Am 75:1044–1055
13. Laboureau JP, Cazenave A (1991) Rupture fraiche du ligament croise anterieur. Technique de renfort par ligament artificiel. Resultats apres 5 ans d'experience. Rev Chir Orthop 77:92–102
14. Laboureau JP, Cazenave A, Dericks GH (1991) The acute rupture of the anterior cruciate ligament. Reconstruction by suture on a synthetic reinforcement evaluated by arthroscopy. Arthroscopy 1:13–22
15. Morgan CD, Kalman VR, Grawl DM (1995) Definitive landmarks for reproductible tibial tunnel placement in anterior cruciate ligament reconstruction. Arthroscopy 11:275–288
16. Romano VM, Grof BK, Keene JS, Lange RH (1993) Anterior cruciate ligament reconstruction. The effect of tibia tunnel placement on range of motion. Am J Sports Med 21:415–418
17. Saragaglia D, Dayez J, Borot E, Butel J (1988) De l'utilisation de Kennedy LAD dans le renforcement des plasties du ligament croise anterieur selon MacIntosh-Imbert. A propos de 100 observations. J Traumatol Sport 5:83–91
18. Stauebli HU, Rauschming W (1994) Tibial attachment area of the anterior cruciate ligament in the extended knee position. Anatomy and cryosections in vitro completed by magnetic resonance arthrography in vivo. Knee Surg Sport Traumatol Arthroscopy 2:138–146
19. Bercovy M, Laboureau JP, Dericks GH (1993) Reasons for failure in the surgical treatment of anterior cruciate instability. European Congress of Orthopaedics, European Federation of National Associations of Orthopaedics and Traumatology, Paris, 21–23 April

II.5 Considerations on the Biomechanics of the Knee Joint with Regard to Ligament Reconstruction, Especially with a Polyethylene-Terephthalate Alloplastic Ligament (Trevira Ligament)

U. Witzel

Introduction

An insufficiency of the cruciate ligaments causes instabilities, displacements of the joint, and changes in the joint's hydraulics. The sliding parts of the cartilage, the subchondral osseous structures, and the menisci are then subjected to high risks of destruction. This phenomenon not only appears with untreated active patients but also in reconstructions in which the physiological situation is not restored. Biomechanical studies can explain early and later appearing joint damage by relating the different effects. From this we can assess factors which influence the diagnostic and surgical procedures.

In ligamentoplasties the primary stability during physiological positioning of the knee is of prime importance. The standard reference for manual and operative investigations is the healthy knee. Functional stimulation of the treated ligament during healing or functional stimulation of the tendinous graft during ligamentization is the condition for long-term stability. Under no circumstances must overloading or load peaks exceeding the maximum load be applied to the replacing structure. These load peaks occur, for example, when short load impacts are applied on the knee and when the reflex-arc did not have the time to produce an effective muscular compensation. They lead to ongoing microtrauma of the collagen structure. In active patients these load peaks cannot be avoided. We therefore propose augmentation or protection of the autogenous structure by the use of a synthetic Trevira prosthesis with matching rigidity. The rigidity of the prosthesis depends on its width. The Trevira ligament can be used when the surgeon fears the use of autogenous structures for the reconstruction of the anterior or the posterior cruciate ligament because of operative consequences, or when transplants have already failed. Our investigations show that it resists loading peaks and prevents the chondral and the subchondral structures from overloading. Normal load bearing and full extension are obtained as early as 3 weeks postoperatively.

Shape of the Joint Surfaces and Their Loading

The load vector in the knee joint changes in magnitude and direction with knee flexion (Fig. 1). Since biological systems are functionally optimized, the tibial

Fig. 1. Changes in magnitude and direction of the load vetor in the knee joint with knee flexion

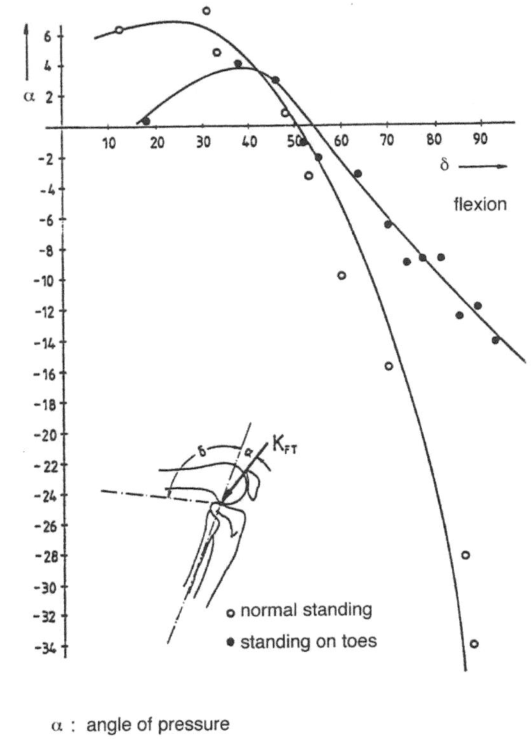

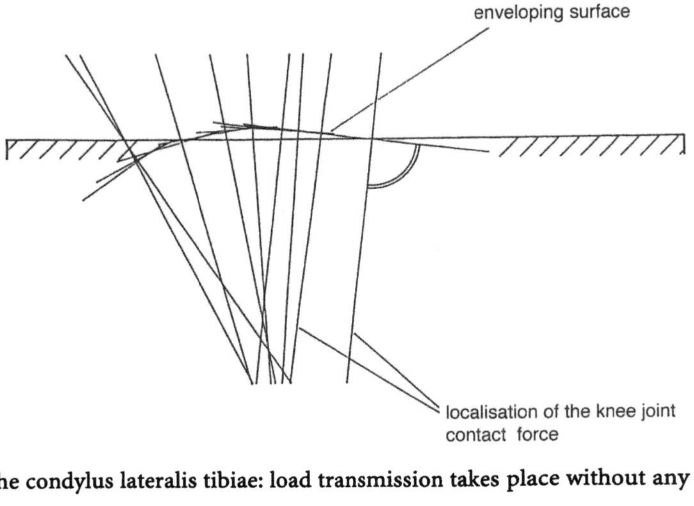

Fig. 2. Construction of the condylus lateralis tibiae: load transmission takes place without any shear stress

Fig. 3. Lateral side of the knee joint

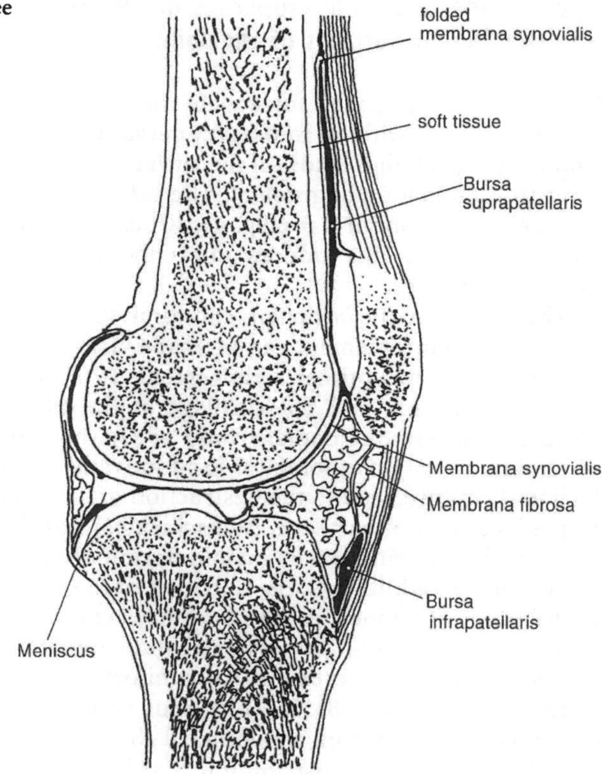

surface is medially concave and laterally convex. This leads, on the one hand, to a tibial rotation and, on the other, to a perpendicular load which guarantees a load transmission between the femur and the tibia without any shear stress (Figs. 2, 3) [1].

The joint capsule and the ligamentous apparatus, especially the cruciate ligaments, control this kinematic process with very low forces. Only in the case of full extension or flexion do the stresses within the ligaments increase, producing a damping effect on the movement.

Deviations from the physiological chondral positions due to a cruciate ligament insufficiency or rupture leads to higher loading of the remaining ligament and capsular structures. Muscular compensation can be obtained only if the reflex-arc is activated, and only if the sensors provided are still responsive. A fundamental rule of thumb is that intact or reconstructed cruciate ligaments carry out the kinematic control function, while reflex muscular compensation takes place only if the impact time is longer than the reflex-arc. This is the case when the movement is slow and planned. Otherwise, cruciate ligaments are in danger of a rapid overloading, which must be taken into account by an augmentation.

Effect of the Structural Properties of the Menisci and the Sliding Surfaces of the Joint

The contact between the surfaces in the lateral joint or between a convex and a concave surface with an important curvature on the medial side would lead to unphysiological high loads on the surface if the menisci were not present in the joint to establish a congruency. The ratio between medial and lateral contact surface area is about 1.5 (Fig. 4), which explains why the resultant force is shifted medially. This induces an evenly distributed compression force on the chondral surfaces.

The force on the articular chondral surfaces is received by the collagen fibers. This reinforcement can be made visible by means of split lines (Figs. 5, 6) and shows an optimal arrangement. It is oriented equally in the femoral condylar cartilage and the cartilage of the tibial plateau and therefore can sustain load without damage only when the articulation is positioned in a physiological way. Pathological joint movements, for example, due to cruciate ligament rupture, lead to the destruction of the tangential fibrous structure and to the local loss of cartilage. Menschick [2] has described the transport of the synovial fluid during extension and flexion of the knee joint. This persuaded us to examine, in addition to the nutrition of the cartilage and the improvement of its tribological properties, the hydraulic effect of the synovial fluid. Our calculations showed that the synovial fluid initially contributes, along with the menisci, to a short decrease in and uniformization of the physiological surface compression (Figs. 7, 8). If a meniscus is destroyed, it can remain unrepaired or can be removed from the joint. In the latter case neither the hydraulic nor the mechanical stress distribution is still present. The resultant point loading leads then to an unphysiological overloading of the cartilage especially of the subchondral structures.

Fig. 4. Contact surface analysis

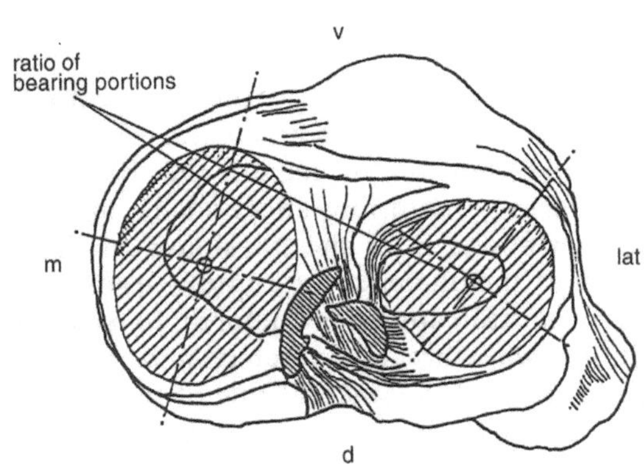

$$\frac{\text{medial area } A_{med}}{\text{lateral area } A_{lat}} = 1{,}5$$

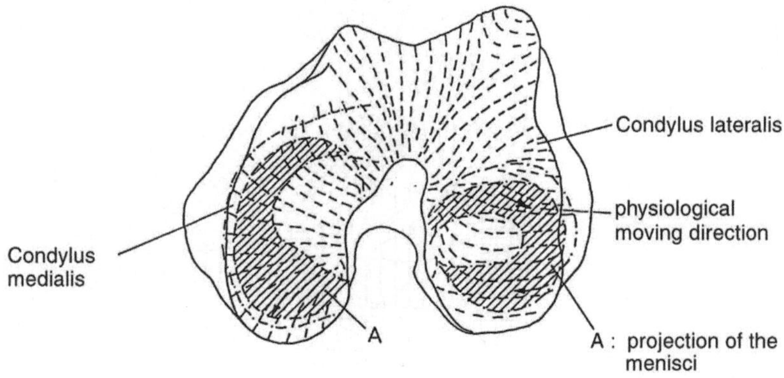

Fig. 5. Split lines on the left femur

Three-dimensional calculations under point loading have been carried out on a sectional joint model and have predicted both chondral expulsions and a higher equivalent stress (according to von Mises; Fig. 9). It is surprising, however, that the maximum equivalent stress lies in the subchondral area at the frontier between cortical and cancellous bone. From the above, we can expect important sclerotic changes or osseous damages. In fact, magnetic resonance imaging shows bony defects which cannot be detected by arthroscopy because they are covered by cartilage (Fig. 10). This leads us to the conclusion that it is important to treat meniscal defects because of their biomechanical importance.

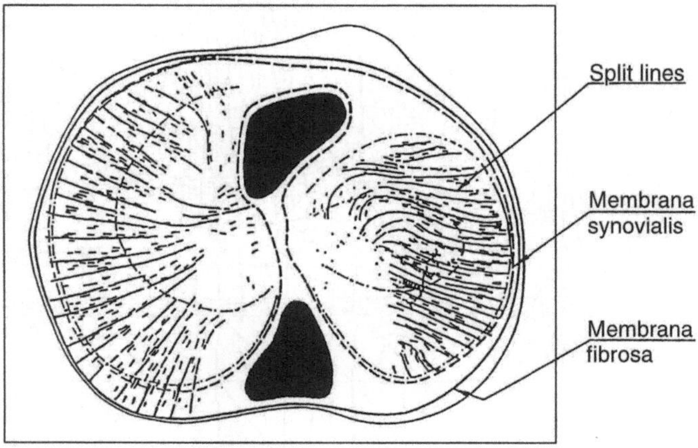

Fig. 6. Analysis of split lines on the proximal surface of the right tibia

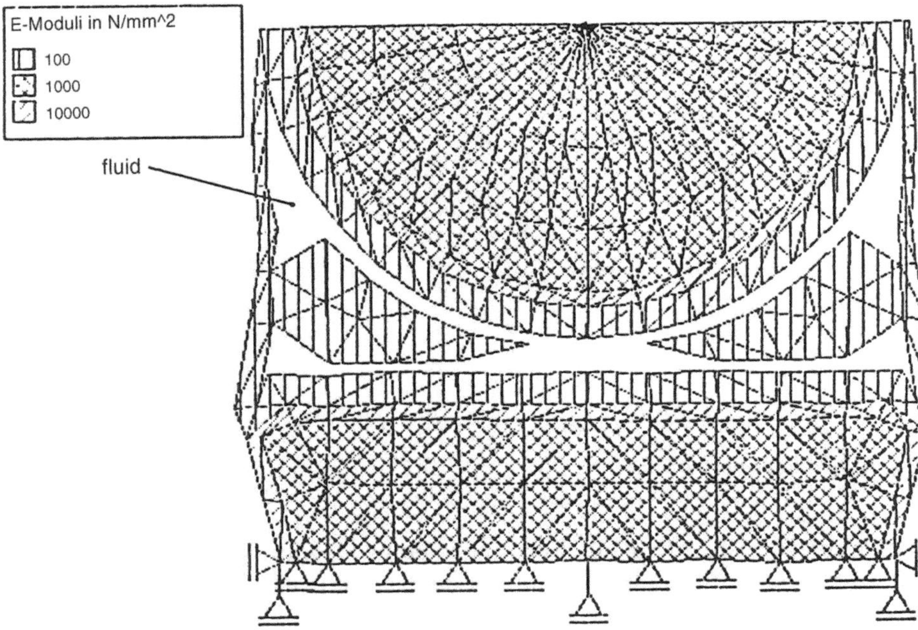

Fig. 7. Young's modulus distribution (finite element method)

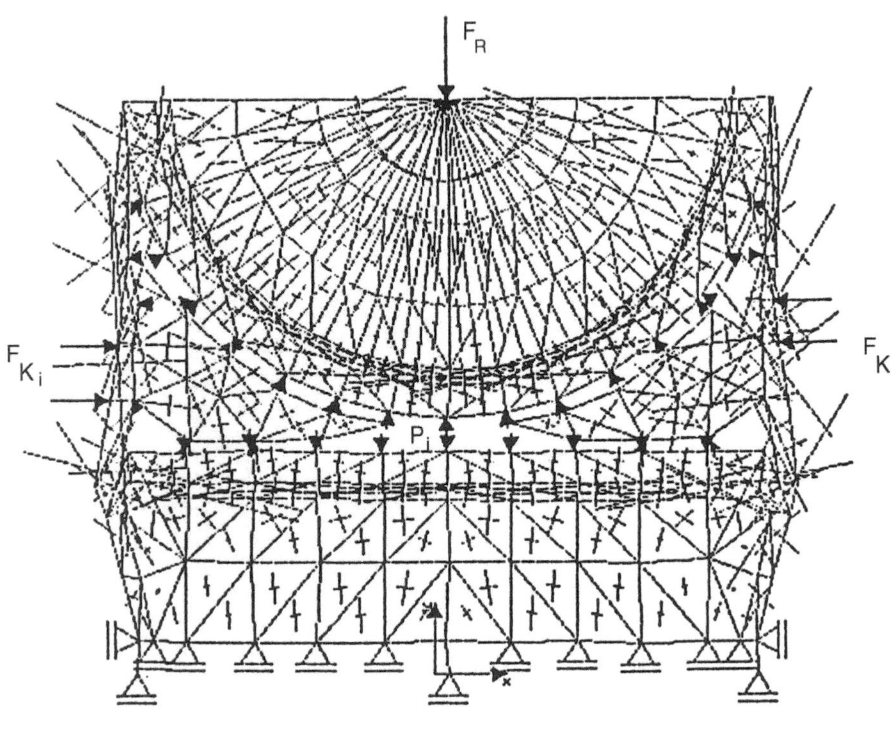

Fig. 8. Stress distribution under hydraulic pressure conditions. F_{Ki}, Forces applied by the joint capsule; F_R, resultant force on the joint; p_i, hydraulic pressure

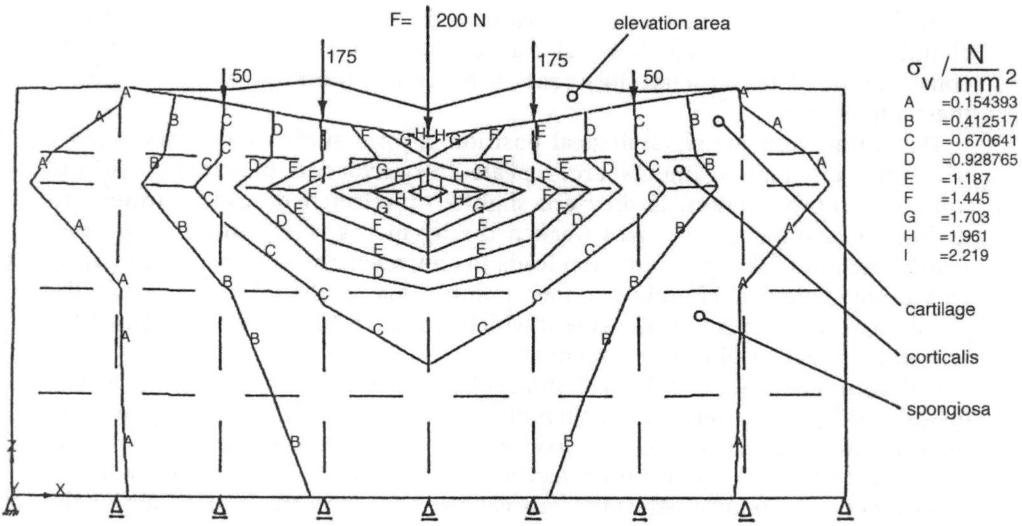

Fig. 9. Subchondral stress concentration after a defect of the hydraulic system

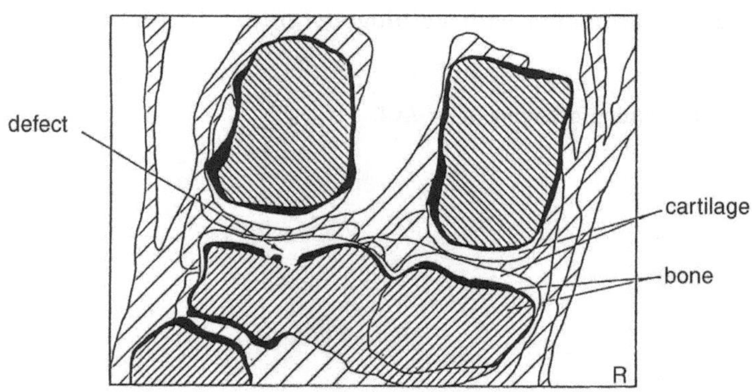

Fig. 10. Defect in the subchondral bone of the lateral tibiaplateau

Ligament Ruptures and Chondral Damages

Pathological deviations of the joint surfaces can be the result of insufficient knee ligaments (for example, because of the failure of a ligamentous suture) and lead to long-term basis to meniscal damage. The short-term effect is the destruction of the chondral structure. If a substantial dynamic load is applied rapidly to a knee with ruptured ligaments, the muscular compensation cannot take place, and joint wear occurs. This results in enormous deviations of the articular surfaces from the physiological position. This can be observed arthroscopically, especially in competitive sportsmen who previously ruptured their ligament. It is increasingly frequent to hear of total detachment of chondral structures, deep chondral cracks, and extractions. Only an early reconstruction of the cruciate ligaments can prevent the joint structures from these phenomena.

Deviation from the physiological position of joint surfaces also takes place when the cruciate ligaments are repeatedly overloaded plastically without muscular compensation. If they are slightly overstretched several times, individual microfibrous ruptures weaken the ligaments which undergo permanent elongation. This phenomenon leads to an irreversible ligament elongation with unphysiological chondral contact points. This results in deviations in the physiological sequence of movement which is related to the orientation of the tangential collagen fiber reinforcement.

Collagen fibers have a helical molecular structure and consist of waved helical fibrils. Moreover, the collagen fibers also show a helical arrangement in the cruciate ligament [3]. Under load such a system has a progressive characteristic curve with a natural protection from overloading before the rupture is observed. The natural straining process takes place in the linear part of the curve (Fig. 11) thanks to a sensory control. To avoid the failure of the suture or the rupture of the collagen fibrils with the resultant elongation it is important to protect these structures against overloading (for example, with Trevira ligaments) after reconstruction of fresh ligament ruptures, after reinsertions, or after replacements with autologous grafts. To provide the autologous struc-

Fig. 11. Loading behavior of an ACL

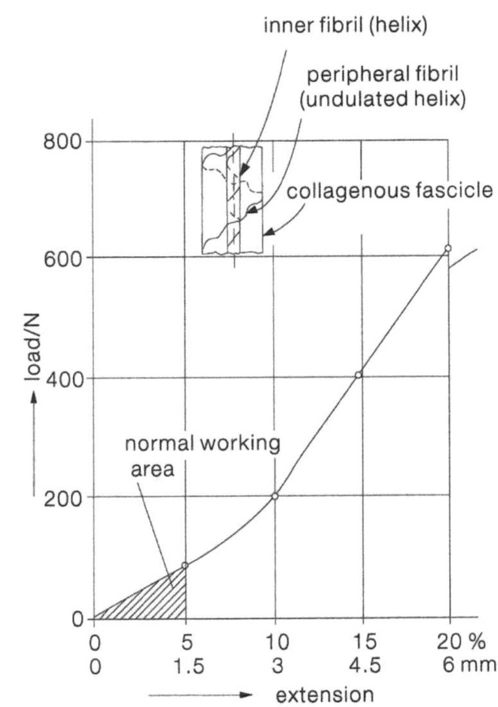

tures a physiological stimulation and permit healing, that is, revascularization of the graft, however, it is important to note that the alloplastic augmentation should not dominate the kinematic process.

Reconstruction of the Cruciate Ligaments

Previous studies have emphasized the necessity of reconstructing the ligamentous apparatus to achieve the physiological positioning of the joint and confirm the important role of the menisci. A reconstruction can possibly avoid or stabilize an arthrosis, or it can stabilize a situation where chondral damage has already taken place. Early postoperative mobilization plays an important role here because it is the only way to involve the numerous mechanoreceptors involved in muscular balance. A disbalance leads to an increase in the loading of the capsule-ligament complex with the above consequences.

Various techniques are available for reconstruction of the cruciate ligament: primary suture, bony reinsertion after a fresh rupture, autologous and alloplastic materials [4]. Today it is well known that sutures and tendinous transplants can be augmented with alloplastic materials which protect the reconstructed graft temporarily from overloading. Early mobilization avoids loss of muscular tone. The augmentation can be obtained by a coaxial or an immediately parallel (in-line) or an "over-the-top" (off-line) reconstruction.

The aim of the augmenting material is to protect the sutured structure from overloading resulting in micro traumas which demonstrate a permanent elongation as a consequence. In autologous transplants the augmentation acts as a stress-sharing structure for the graft during ligamentization. Mendenhall (personal communication) reports that the graft's strength regains 24% of its original value after 7 weeks of implantation. Even after 52 weeks the strength of the transplant reaches only 57% of its original value (Fig. 12).

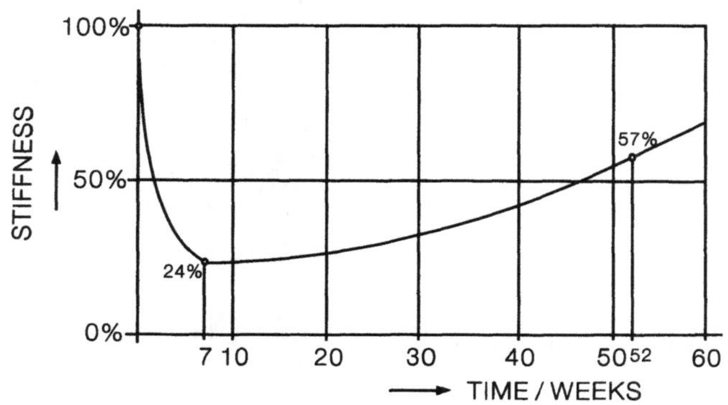

Fig. 12. Time of ligamentization of a tendon graft

The One-Canal Technique

To obtain stress sharing and not stress shearing it is important to match the rigidity of the augmentation material to that of the autogenous structure. This is the reason why Trevira ligaments of 3, 5, and 8 mm have been developed. The 8-mm grafts are designed for reconstruction of the cruciate ligaments by a one-canal technique, whereas the anterior cruciate ligament is led over-the-top and the posterior one over-the-bottom (Fig. 13) [5]. These techniques avoid leading the ligament around sharp edges at the opening of the bony tunnels. Especially in the anterior cruciate ligament the positioning around the lateral femoral condyle can produce substantial wear, biological reactions, and finally rupture of the prosthesis.

The Trevira graft lies flat on the femoral condyle and is thus protected from abrasion. On the other side the graft is led around the dorsal tibial edge without going back through a femoral tunnel, which also protects the graft from the bony structures.

Using the over-the-top technique by preserving the tibial insertion point of the native ligament leads to instability. Mathematical calculations on the changes in the length pattern of an isometrically positioned thread have been carried out. If the thread is positioned during full extension (length change 0° = 0° flexion) it can be shown that every deviation from the isometric point results in a length change of the thread during flexion (Fig. 14): anterior deviations put the thread (or the prosthesis) under excessive tension (danger of destruction of all structures) and distal and dorsal deviations from the isometric point shorten the distance and loosen the thread (loss of stability).

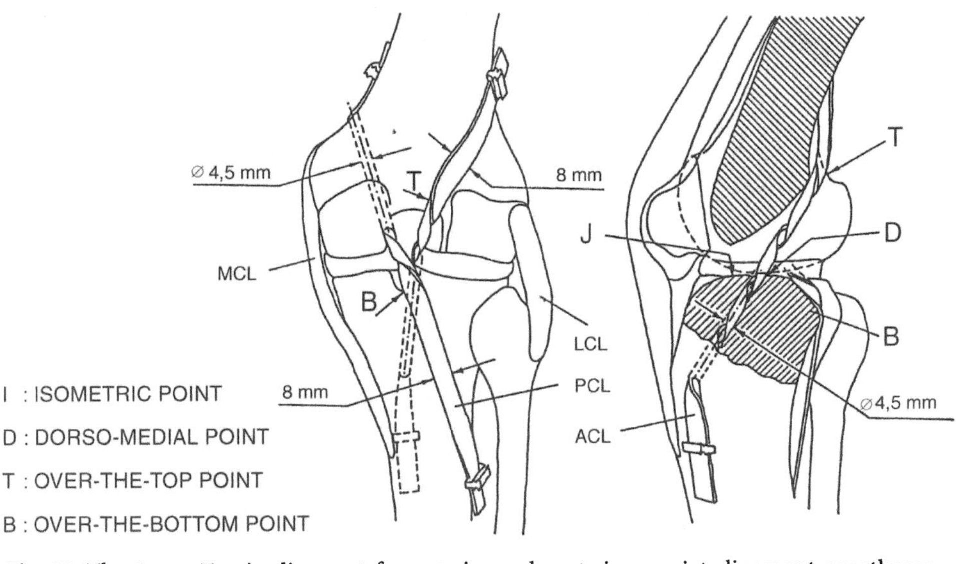

Fig. 13. The 8-mm Trevira ligament for anterior and posterior cruciate ligament prostheses. Diameter of the tunnel: 4.5 mm

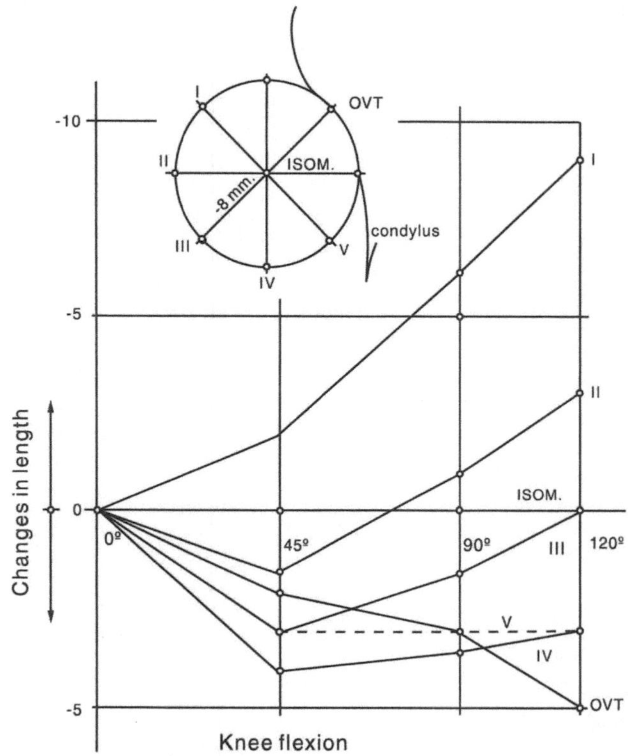

Fig. 14. Changes in the length pattern of a thread during flexion positioned in five different ways. *I/II*, anterior deviation of 8 mm from the isometric point; *III/IV/V*, distal deviation; *OVT*, over-the-top point

Measurements on nine cadaver knees were carried out in which the over-the-top technique was used and the tibial physiological insertion point preserved (Fig. 15). The mean loosening was found to be 7 mm, which is unsatisfactory for the knee's stability.

An Optimization of the Over-the-TopTechnique

The same thread experiment was carried out for the optimized over-the-top technique [5], in which the tibial tunnel was moved dorsomedially compared to the native insertion point. Results show that for 120° flexion, the displacement was reduced to 3.5 mm (Fig. 16). A cruciate ligament implant can therefore be placed with a pretension of 70 N. For small knee flexions (most important loading phase) this results in a pretensioned situation, and for larger knee flexions (swing phase during walking and muscular stabilized bending) a tolerable loosening is obtained.

Fig. 15

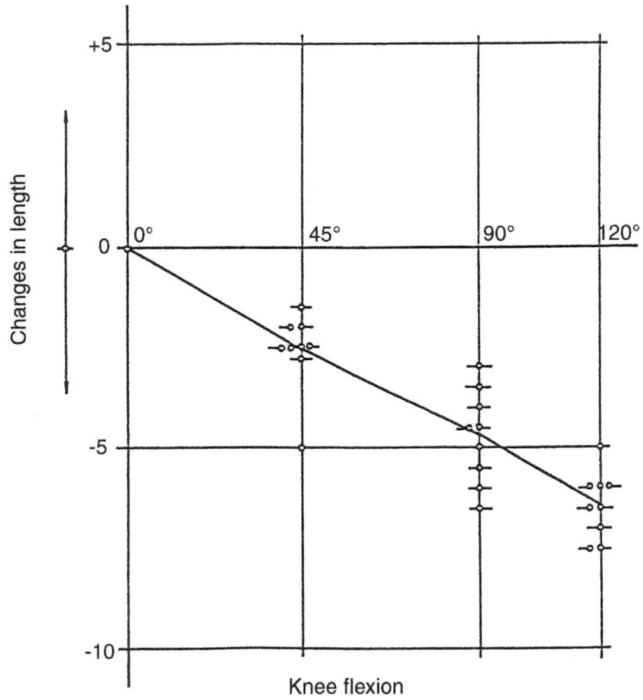
+5
0
-5
-10
Changes in length
0°
45°
90°
120°
Knee flexion

Fig. 16

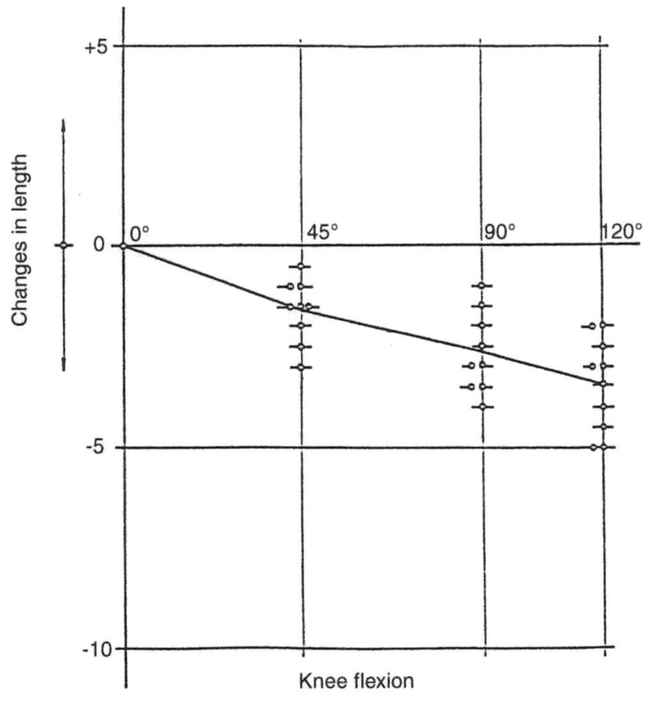
+5
0
-5
-10
Changes in length
0°
45°
90°
120°
Knee flexion

These investigations can be summarized by following kinematic paradox: a kinematic "four-bar system" tolerates, with good precision, various roll-slide couples (important in case of an unilateral joint replacement with intact ligaments), and one anatomic roll-slide configuration allows different "four-bar systems" (important for cruciate ligament reconstruction).

The Stress-Strain Curve for Augmentation, Protection, and Prosthesis

In addition to the above kinematic conditions, a number of other parameters are crucial for a successful cruciate ligament prosthesis, including physical material properties, biocompatibility, and textile construction (Fig. 17). Material and construction define the ultimate load and the elongation behavior of

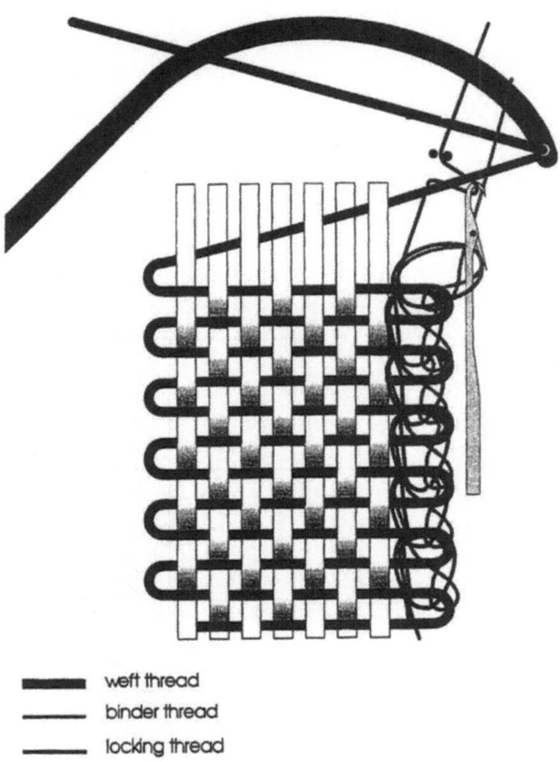

Fig. 17. Textile construction of the Trevira band (Telos, Germany)

Fig. 15. Changes in the length pattern of a thread when the over-the-top technique and the tibial physiological insertion is used

Fig. 16. Changes in the length pattern of a thread for the "optimized over-the-top" technique

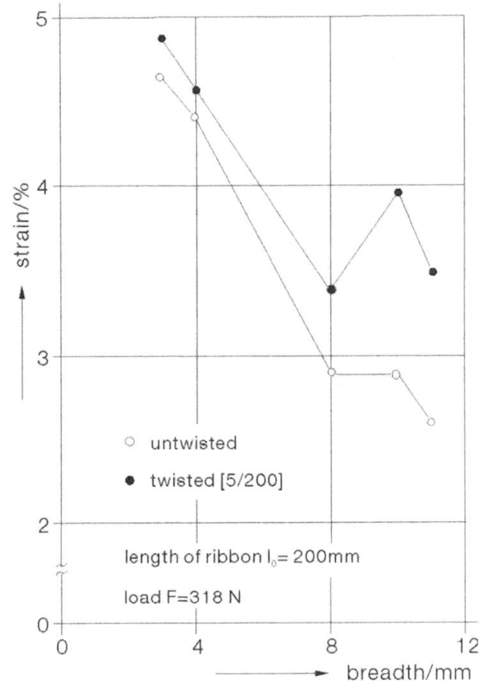

Fig. 18. Strain-breadth function of the Trevira bands

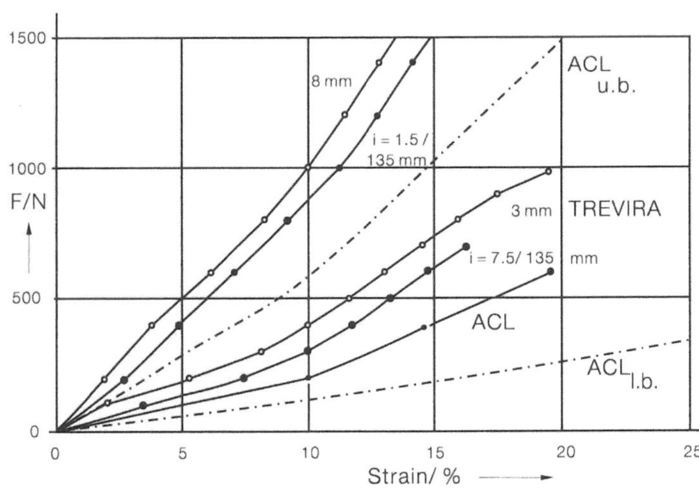

Fig. 19. Load behavior of different Trevira reconstruction structures compared to that of the natural ligament. *i*, Twisting number per length; *u.b.*, upper bound; *l.b.*, lower bound

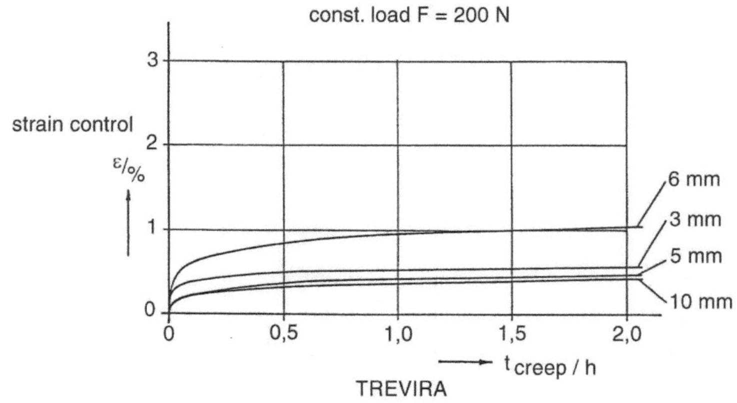

Fig. 20. Creep behavior of Trevira ligaments

the prosthesis. The elongation behavior is a function of the graft width (Fig. 18). The mechanical behavior of the graft must correspond to that of the augmented structure to protect the biological structures. Figure 19 shows that the 8-mm-wide prosthesis has a behavior similar to that of the natural ligament, and that it demonstrates a progressive load behavior. Thanks to a twisting of the synthetic ligament the insertion tunnels have a diameter of only 4.5 mm. The good creep behavior of the graft can be seen in Fig. 20.

The 3-mm-wide prosthesis can be used for augmentation. In this technique the autologous replacing structure and the synthetic graft are inserted at the same point or at two nearby points. This is the reason why the two structures undergo the same strain. Depending on the particular strength of each graft, the loading can then be calculated.

In case I, for example, the load is taken over by the whole reconstruction (Fig. 21): F = 2F1. Therefore the autologous portion must sustain following loads: Fauto = F/2, and for the alloplastic material the value is also: Fallo = F/2. In case II we can assume that an equivalent strain takes place for both structures, which leads to the formula: F = F1+F2 = 3F1. In this case the alloplastic material is twice as stiff. The loading of the autologous part is then: Fauto = F/3. This simplified representation shows that the higher stiffness of a synthetic graft leads to lower loading of the autologous structure and therefore to better protection. This can, however, threaten the biological remodeling of the autologous structure. The strength of the employed synthetic materials and that of the autologous graft must therefore be made consistent with one another.

In the case of protection no simultaneous loading is observed (Fig. 22). Their partial loading depends on knee flexion. During full extension and high flexion the anterior cruciate ligament is more strained than during intermediate values of knee flexion [6]. It is then obvious that the overload-protection must be guaranteed by a synthetic graft of adequate stiffness, which is positioned by the optimized “over-the-top” technique. It should loosen slightly with increasing flexion and permit functional loading.

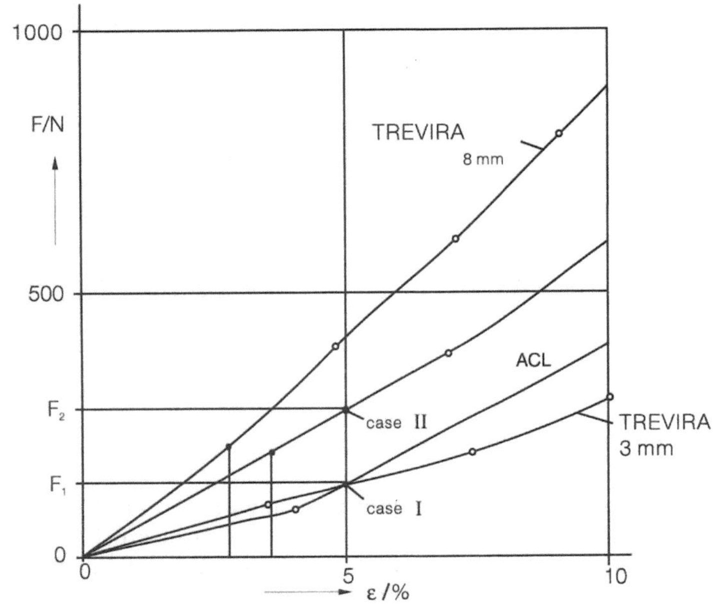

Fig. 21. Load behavior of an augmentation

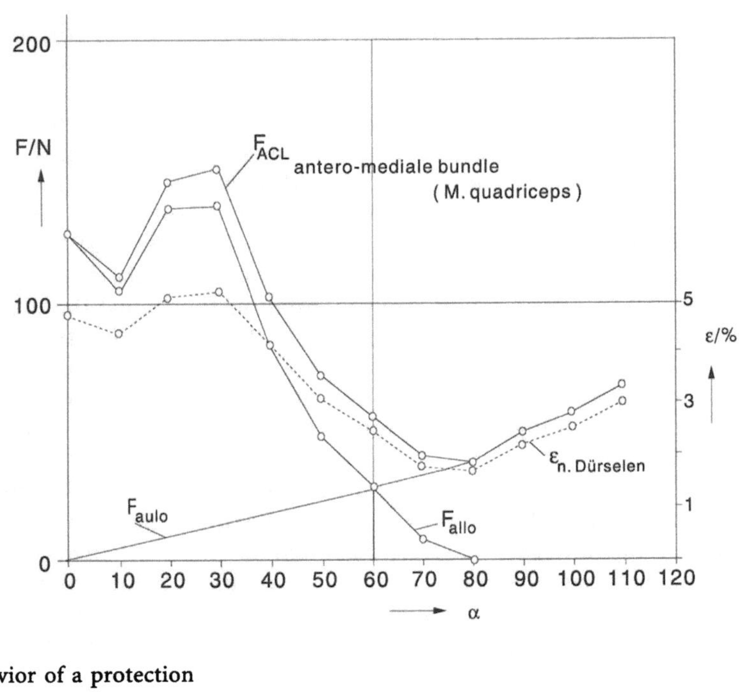

Fig. 22. Load behavior of a protection

Conclusion

During knee flexion the joint surfaces must be in contact without damage only in the physiologically relevant area. Therefore the operative treatment of damaged cruciate ligaments (either anterior or posterior) in active patients is necessary to reestablish the biomechanics of the knee joint. The same can be said for treatment of the menisci, which permits the preservation of the hydraulic and mechanical stress distribution.

Autologous structures, i.e., biological tissues, must be protected from overloading during their ligamentization or healing, thus avoiding partial ruptures or permanent elongation. This protection can be achieved with simple and compact methods such as augmentation by means of 3-mm Trevira bands or protection by means of 5-mm bands. In addition to their protective function, the Trevira bands provide a functional stimulation (mechanical loading necessary for ligamentization) thanks to their optimal stiffness.

In cases postoperative trauma, such as a new rupture or damage to a biological cruciate ligament, the 8-mm Trevira ligament prosthesis can be used for reconstruction of the ligament. This prosthetic ligament should also be used in the case of primary rupture for patients with high sports activity, for whom another injury is a risk or in the case when the surgeon fears postoperative complications due to the excision of autologous structures. In fact, biological structures are not overly abundant but are adapted to the functional loading. Removal of soft tissue leads to temporary or permanent damages.

References

1. Witzel U (1991) Zur Biomechanik des Kniegelenks unter besonderer Berücksichtigung kinematischer, kinetischer und hydraulischer Aspekte. DVM, Berlin
2. Menschik A (1987) Biometrie. Springer, Berlin Heidelberg New York
3. Amiel D, Billings E, Akeson WH (1990) Ligament structure. In: Daniel D, Akeson W, O'Connor J (eds) Chemistry and physiology in knee ligaments, structure, function, injury and repair. Raven, New York
4. Knaepler H, Krudwig W, Witzel U (1994) Überlegungen zur differenzierten Therapie bei Insuffizienz des forderen Kreuzbandes – Grundlagen und klinische Konsequenzen. Aktuel Traumatol 5(24):188–194
5. Witzel U, Paar O, Kerschen J, Krudwig W (1991) Biomechanische Untersuchungen zu isometrischen Zwei- und Einkanaltechniken beim vorderen und hinteren Kreuzband-ersatz. DVM, Berlin
6. Dürselen L (1990) Biomechanische und mechanische Untersuchungen an Kniegelenksbändern und deren alloplastischem Ersatz. Dissertation, University of Ulm

II.6 Reconstruction of Cruciate Ligaments Using a Synthetic Ligament of Polyethylene Terephthalate (Trevira Ligament)

W.K. Krudwig

Introduction

The long-term unsatisfactory results and lack of confidence concerning biological cruciate reconstructions (see Chap. I.5) and especially the goal of reducing operative trauma and physiological disorders have led to the use of an alloplastic ligament in all cases of acute and chronic cruciate ligament tears. The synthetic ligament, implanted since 1980 without any change in its structure, is a 8-mm-wide and 1-mm-thick ligament of polyethylene terephthalate (Trevira-Hochfest 730) and meets all biological and material requirements suitable for cruciate ligament replacements [2, 3, 6-10]. The choice of this ligament followed the considerations and experiences discussed in Chaps. I.5 and II.5. Its biocompatibility has been confirmed by numerous tests and clinical follow-ups [1, 4–6, 10].

Reconstruction of the Anterior Cruciate Ligament Using the Trevira Ligament (8 mm): A Modified Over-the-Top Technique

We prefer miniarthrotomy with a nearly 2.5-cm-long medial incision offering a clear and open view into the notch. Arthroscopically assisted procedure is also possible.

At the over-the-top point the dorsal capsule is opened from inside to outside and is spread a little with a curved clamp. The inside-outside perforation guarantees that the clamp does not slip on the capsule, and that no soft tissue remains between the clamp and the condyle. A possible roughness can be slightly smoothed with a curved rasp, but this must not roughen up the cortex or make a groove.

A nearly 4-cm-long incision follows at the distal lateral side of the thigh a little ventral the palpable corner of the condyle. The iliotibial tract is also cut open. The vastus lateral muscle is ventralized. All soft tissues covering the over-the-top passage must be cut open or removed. Remaining structures atrophy under the pressure of the ligament and result in a functional lengthening of the ligament and an increasing laxity of the knee.

The free over-the-top passage is checked again, and the rounded hook with a long suture-loop is introduced into the notch (Fig. 1). Close attention must be paid to ensure that the hook does not run around or through the posterior

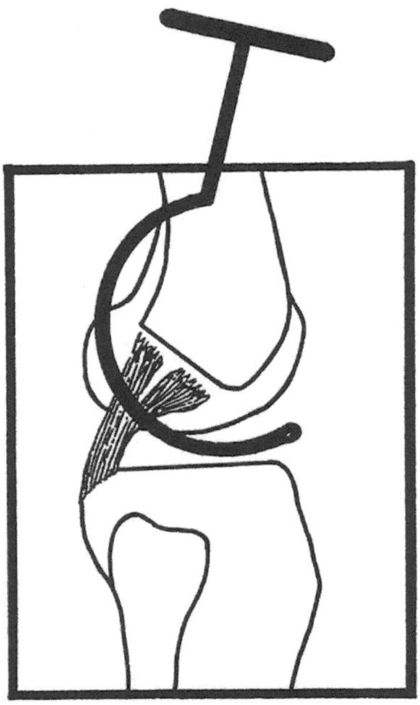

Fig. 1. Rounded hook

cruciate ligament, and that there are no remaining soft tissues between the hook and the femur.

After controllin the correct position of the hook the suture-loop is pulled out, the hook is removed, and the free run of the loop is checked. In the case of a narrow notch the pliable hook can be used from ventral. It is pushed forward over the skin level, and the suture-loop is pulled through. Now the tip of the

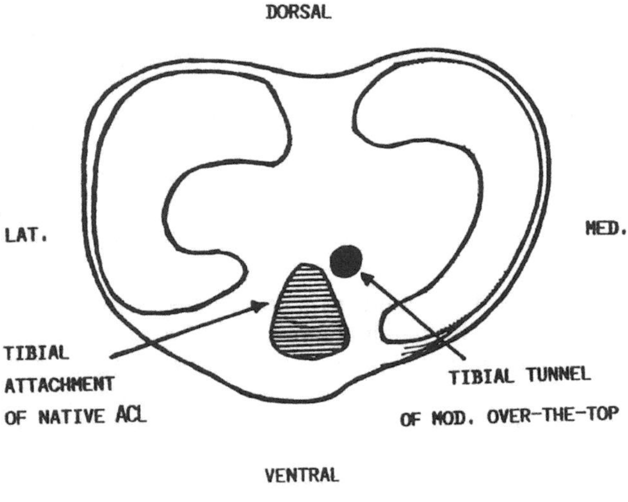

Fig. 2. Dislocation of the tibial tunnel

aiming device is placed on the tibial plateau a little more dorsally and medially behind the native anterior cruciate ligament attachment (Fig. 2). The 4.5-mm shield pointing for the later incision of the tibial tunnel and taking in of the cannulated drill for guiding the K-wire is also placed and fixed.

Under C-arm control in side view the aiming device and 4.5-mm shield is brought in line with the over-the-top point with the help of the externally held drill (Fig. 3). After a 2-cm-long incision and the final check with the C-arm a 1.8-mm K-wire guided by the canulated drill in the shield is drilled through the tibial head still in line to the over-the-top point and crossing the middle third of the tibial plateau.

The drill, shield, and aiming device are removed, and the correct position of the K-wire is checked. The K-wire should run parallel to the roof of the notch in extension; up to 50° of flexion it is always aiming at the over-the-top point. In the case of an incorrect position under movement of the knee the K-wire must be removed and better placed. If the K-wire is correctly placed it is shortened, and the tibial plateau is drilled with the 4.5-mm cannulated drill guided by the K-wire. The intraarticular end of the suture-loop is catched and pulled out through the tibial tunnel (Fig. 4). The suture-loop is replaced by a long thin wire-loop from distal to proximal, a second one is hooked into the first and pulled out distally.

The final C-arm control in side view demonstrating the run of the wire-loop and the later ligament should thereby show an absolutely straight and contactless run of the wire-loop from extension to nearly 50° of flexion. Under greater flexion the wire-loop is slightly kinked at the tibial tunnel. The wire-loops are moved up and down and should not show any curve inside the notch.

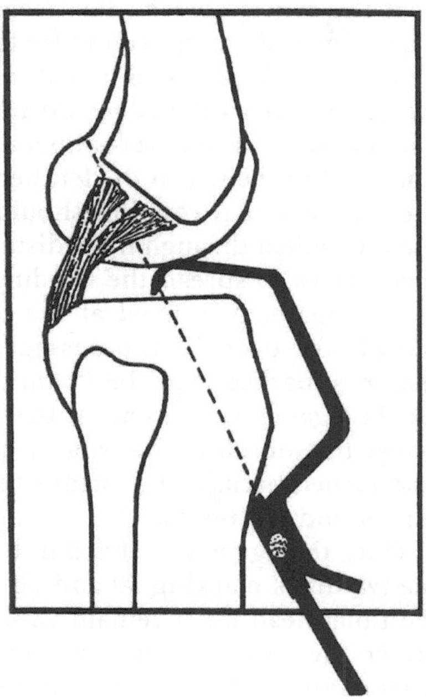

Fig. 3. Aiming device pointing at the "over-the-top" point

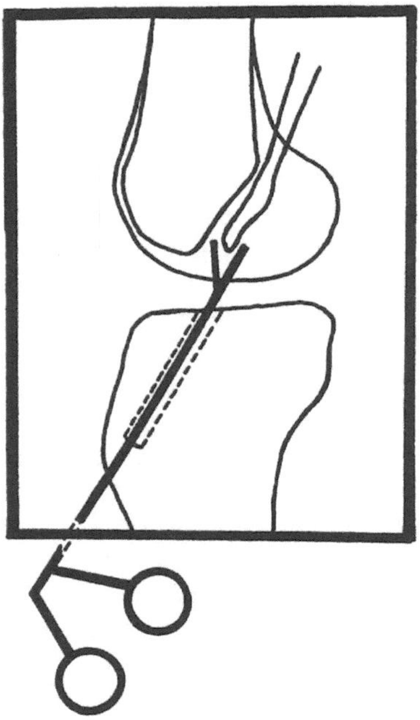

Fig. 4. Catching of the suture loop through the tibial tunnel

In case of a slight curve a contact with the posterior cruciate ligament can be found in the utmost cases. Then the wire-loops must be removed and placed correctly.

If the run is correct, the Trevira ligament is hooked and sutured into the wire-loop and wound round by the stitch. The result is a small cylinder; it can be additionally rounded by flat tongs. The ligament is fixed into the tensor and twisted as far as the edges do not come into contact, even under tension, to avoid abrasions. It is recommended twist the ligament to the right in the right knee and to the left in the left because a slight opening of the windings during the final external rotation should be achieved. Now the twisted Trevira ligament is pulled through from distal to proximal and moved up and down two or three times to spread the windings over the full length.

The ligament is fixed at the femur first. After turning back the windings outside the over-the-top passage, the ligament is laid flat on the femur. The femur is marked with the 10-mm marker, and a staple is driven in strongly to fix the ligament. The end of the ligament is cut until nearly 2 cm and placed under the muscles. The windings of the ligament outside the tibial tunnel are also turned back, which means that the windings remain only inside the tibial tunnel and the notch.

Once the ligament is fixed into the tensor, the pretension is first set between the two lines marking 50 and 70 N. At 90° of flexion a slight anterior draw of the tibial head must remain while the Lachmann test result must be negative under the same pretension. However, if there is a physiologically positive Lachmann result on the healthy side, the pretension should be lowered.

The final fixation of the ligament takes place in full extension of the knee and under individual pretension with a second staple after marking the tibial cortex with the 10-mm marker. The end of the ligament is also cut to almost 2 cm and placed carefully and flatly under the pes anserinus. Figure 5 shows the aimed position of the ligament without windings outside the tibial tunnel and the capsule.

The closure of the wound follows in layers after an intraarticular Redon drainage and a second one at the dorsal capsule. The bandage of the wounds and a compression stocking of the leg complete the operation.

Reconstruction of the Posterior Cruciate Ligament Using the Trevira Ligament (8 mm): The Over-the-Bottom Technique

We prefer a minimized arthrotomy with a nearly 7-cm-long, slightly curved medial incision enabling an open and clear view into the notch and of the medial condyle. In 90° of flexion of the knee the dorsal capsule is opened and spread a little with a curved clamp over the native posterior cruciate ligament attachment. There follows a nearly 4-cm-long incision, approximately 7–8 cm distally to the tibial-fibular joint. The fascia is cut open as well, and the muscles are removed from the lateral tibial cortex first with a bone-rasp and then with the clamp. In 90° of flexion the rounded hook with a long suture-loop is introduced through the dorsal capsule and slipped distally and laterally, always in contact with the dorsal tibial cortex until the tip of the rounded hook

Fig. 5. Aimed position of the ligament

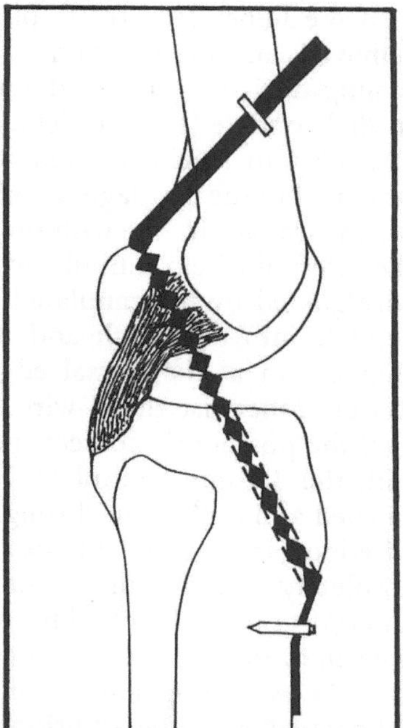

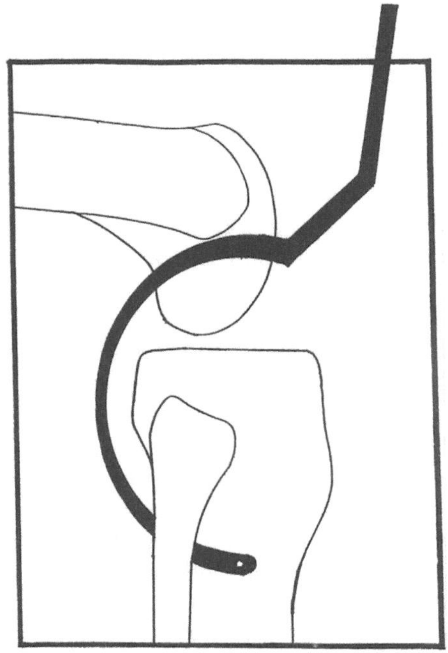

Fig. 6. Rounded hook

appears between the tibia and the muscles (Fig. 6). Close attention must be paid that no parts of the muscles remain between the tibia and the hook. Otherwise the rounded hook must be better placed.

If the hook is correctly placed, the suture-loop is pulled out, the hook is removed, and the free run of the suture-loop is checked. Then the tip of the aiming device is placed at the angle between the roof of the notch and the medial condyle (1 or 11 o'clock position) half-way between ventral and dorsal. The 4.5-mm shield is fixed on the medial condyle nearly 1 cm medial and proximal to the cartilage in only slight flexion of the knee. The aiming device is finally brought in line with the dorsal edge of the tibial plateau with the help of the externally held cannulated drill under C-arm control (Fig. 7). A 1.8-mm K-wire guided by the canulated drill in the shield is drilled through the medial condyle into the notch and its position is checked under C-arm control. It should aim at the dorsal edge of the tibial plateau in extension and slight flexion; otherwise the K-wire must be removed and better placed.

If the position is correct the K-wire is shortened and the condyle is drilled with the 4.5-mm cannulated drill guided by the K-wire. The suture-loop is catched and pulled out through the tunnel (Fig. 8) and replaced by a thin long wire-loop from distal to proximal. Now the C-arm control should show the absolutely straight run of the wire-loop and thereby the later ligament. The anterior draw of the tibial head can also be demonstrated by pulling the wire. If the run of the wire-loop is correct, the Trevira ligament is hooked and sutured into the wire-loop and wound around by the stitch. The result is a small cylinder; this can be additionally rounded by flat tongs. The ligament is fixed

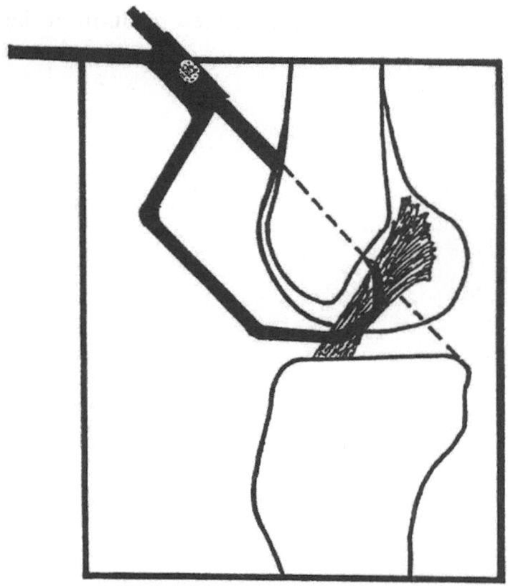

Fig. 7. Aiming device pointing at the "over-the-bottom" point

into the tensor and twisted as far as the edges do not come into contact, even under tension, in order to avoid abrasions.

Now the twisted Trevira ligament is passed through from proximal to distal and moved up and down two or three times to spread the windings over its full length. The ligament is first fixed to the femur. The synovial tissue is cut open

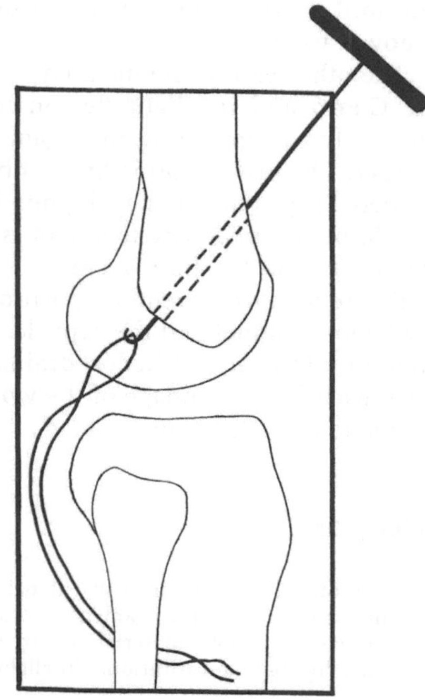

Fig. 8. Catching of the suture loop through the femoral tunnel

Fig. 9. Aimed position of the ligament

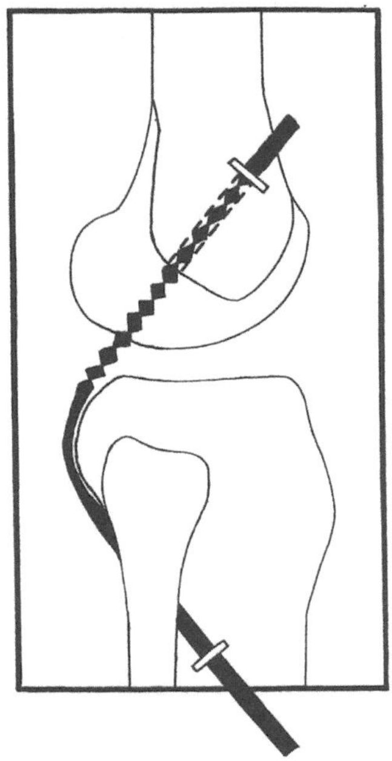

and is used to completely cover the ligament and staple later on. The windings of the ligament outside the condyle are turned back, and the ligament is laid flat on the condyle and strongly fixed with a staple. The end of the ligament is cut until nearly 2 cm and completely covered (including the staples) with the synovial tissue.

The other end of the ligament is fixed into the tensor again. In side view of the C-arm and in slight flexion of the knee the tibial head can be drawn anteriorly by tensioning the ligament. The cortex is marked with the 10-mm marker. The pretension is first set between the two lines marking 50 and 70 N. If there is physiologically higher laxity of the healthy side, the pretension should be lowered. The ligament is laid flat on the tibia, strongly fixed with a staple, and cut to about 2 cm.

Figure 9 shows the aimed position of the ligament without windings outside the femoral tunnel and the capsule. The closure of the wounds follows in layers after an intraarticular Redon drainage and a second one between the tibia and the muscles. The bandage of the wounds and a compression stocking of the leg complete the operation.

References

1. Boszotta H, Helperstorfer W, Pflanzl W (1993) Foreign body synovitis – a limiting factor in use of the Trevira ligament in cruciate ligament surgery? Unfallchirurgie 19:138–143
2. Contzen H (1985) Material requirements and biological conditions for the alloplastic knee ligament substitution. Unfallchirurgie 11(5):242–246

3. Contzen H (1986) Materialtechnische Voraussetzungen und biologische Grundlagen für den alloplastischen Kniebandersatz (am Beispiel des Treviabandes). Hefte Unfallheilkd 181:837–834
4. Kock HJ, Stürmer KM, Letsch R (1991) Biological tissue response to prothetic replacement of the anterior cruciate ligament with PET implants (Trevira Hochfest). Unfallchirurg 94:594–602
5. Letsch R, Stürmer KM, Kock HJ, Schmid-Neuerburg KP (1993) Primary repair of acute tears of the anterior cruciate ligament and protection by synthetic augmentation (Trevira Hochfest). Indication, technique, and results of a five-year-study. Unfallchirurg 96:499–507
6. Letsch R, Stürmer KM, Kock HJ, Wissing H, Schmit-Neuerburg KP (1994) Implantation of a PET-ligament (Trevira Hochfest) as an anterior cruciate ligament prothesis in salvage cases of the knee. Medium to long-term results of a clinical study. Unfallchirurgie 20(6):293–301
7. Mockwitz J, Contzen H (1982) Treatment of inveterate knee ligament lesions with alloplastic material. Unfallchirurgie 8:176–179
8. Mockwitz J, Contzen H (1984) Alloplastische Korrektur chronischer Kniebandinstabilitäten mit einem Band aus Polyaethylenterephthalat In: Rettich H (ed) Biomaterialien und Nahtmaterial. Springer, Berlin Heidelberg New York
9. Mockwitz J, Rau B (1988) Ideal indications for the use of a textile ligament of high strength Trevira in alloplastic replacement of the anterior cruciate ligament in chronic isolated rupture. Unfallchirurgie 14:276–282
10. Neugebauer R, Burri C, Claes L, Piehler J (1983) Vergleichende histologische und biomechanische Untersuchungen zum Bandersatz mit verschiedenen alloplastischen Materialien. In: Schreiber HW (ed) Chirurgisches Forum '83 für experimentelle und klinische Forschung. Springer, Berlin Heidelberg New York, pp 115–119

II.7 Isolated and Combined Posterior Cruciate Ligament Instability

J. C. Fernandes and P. Ranger

Introduction

The posterior cruciate ligament (PCL) stands today as the anterior cruciate ligament (ACL) did some 20 years ago; it has replaced the ACL as the "enigma of the knee" [83]. The issue of surgical versus conservative treatment still remains disputed. Clinical and investigational research on injuries to the PCL lags far behind our understanding of the ACL. Acute injuries to the PCL may go unrecognized, and it is difficult to assign a strict chronological period to acute versus chronic injuries. The belief that injury to the PCL is a rare event is slowly being dispelled due to better imaging techniques and clinical acumen. The incidence of PCL tears is estimated to be 5%–20% of all ligament tears [12, 14, 42]. Most injuries to the PCL occur in either motor vehicle accidents or athletic events [31], and disruption of the PCL as an isolated injury does occur [13, 47, 71].

As the natural history has not been properly delineated, it remains questionable whether surgical treatment is any better than the nonsurgical approach [83]. Keller et al. [43] have shed light on the long-term prognosis of the PCL-deficient knee. After correct diagnosis, surgical treatment of these injuries is warranted only if it alters the natural progression of the PCL-deficient knee.

Anatomy and Biomechanics

Anatomy

The PCL is truncated at its midportion but fans out superiorly toward its femoral attachment and to a lesser extent inferiorly to its insertion into the posterior aspect of the tibia, approximately 1 cm below the joint line [62] (Fig. 1). From the tibial insertion the ligament extends anteriorly as well as slightly medially to insert on the lateral wall of the medial femoral condyle and on the roof of the intercondylar notch. The PCL has a broad femoral origin that forms a semicircle on the medial femoral condyle with an average width of 32 mm [22], then tapers centrally to its narrowest width (11 mm), and widens again to an average width of 13 mm at its tibial insertion [29]. In adults its average length is 38 mm [86]. The ligament is oriented vertically in the frontal

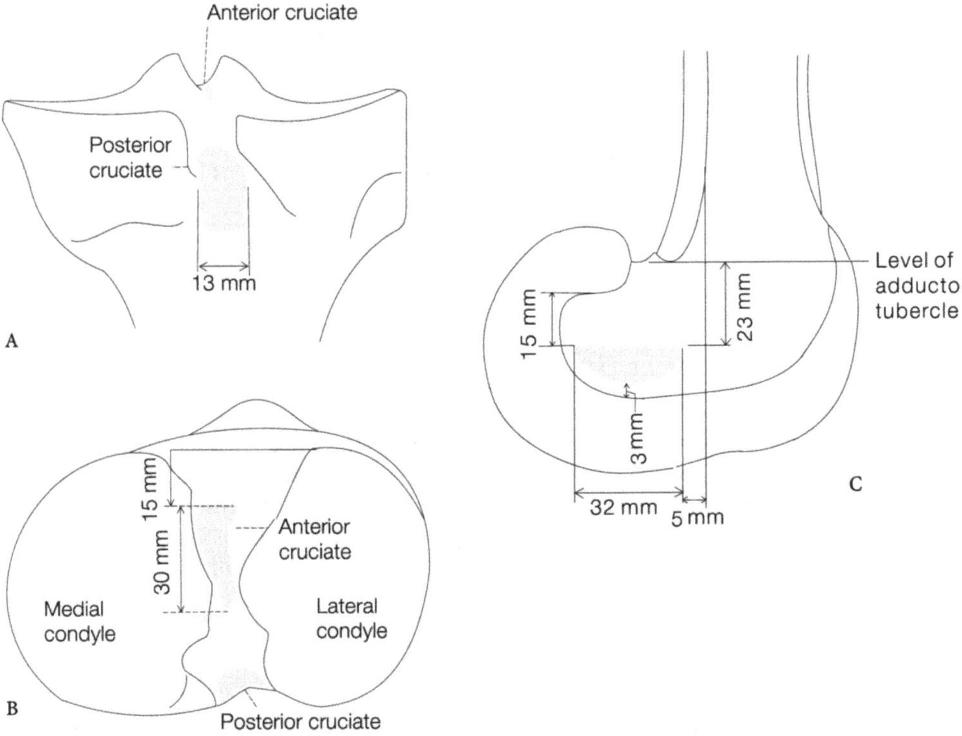

Fig. 1. A,B Tibial insertions of the ACL and PCL and their average dimensions and shapes. **C** The femoral origin of the PCL with its average dimensions and relevant shape. (Reproduced with permission from [22])

plane and angles forward 30°–45° in the sagittal plane (depending on the amount of knee flexion) [61]. The PCL is intraarticular, but it is extrasynovial and is located near the center axis of the knee and is composed of two components: a thick anterolateral portion and a smaller posteromedial portion. The anterolateral component is tight in flexion and relatively lax in extension. The posteromedial component is tight in extension and relatively lax in flexion [22].

Mechanical Properties of the PCL

The PCL serves two major functions: (a) it provides the primary restraint against posterior tibial displacement [8], and (b) in conjunction with the ACL it regulates the "screw home" mechanism of the knee [21]. Biomechanically the PCL represents 95% of the resistance to posterior displacement of the tibia on the femur in 90° of flexion [22]. Secondary restraints to posterior drawer consist of the popliteus tendon, arcuate ligament complex, and the posteromedial and the posterolateral capsule in extension [24]. Information concerning the tensile properties of the PCL is scarce. Its two bundles, the anterolateral and posteromedial, are functionally distinct, although anatomically not easily separated. Prietto et al. [61] tested whole PCLs from young donors (22.5 years of age) and found values for modulus (109 ± 50 MPa) and ultimate tensile strength (26.8±9.1 MPa). The properties of the posteromedial and anterolateral portions of the PCL in older specimens (75 ± 14 years of age) are significantly different than those in younger ones [63]. Although generally accepted that the PCL is approximately twice as strong as the ACL, one study shows ultimate strengths that are similar to those for the ACL [61]. However, studies of ligament strength have been flawed by specimen preparation, age, orientation, and loading and should not be accepted purely at face value [90].

Kinematics of the PCL

The PCL has been shown to be a significant secondary restraint to external tibial rotation with the knee at 90° of flexion, playing a lesser role in resisting external rotation as the knee moves toward full extension [24]. Resistance provided by the PCL to varus-valgus moments applied to the knee is relatively insignificant if the collateral ligaments are intact [24, 27]. However, the PCL appears to play a role in resisting varus stress following sectioning of the lateral structures, again providing its greatest contributions at 90° of knee flexion [24].

Muscles that cross the knee also play a large role in maintaining physiological kinematics for the knee. The activity of muscles can introduce large changes in strains and forces experienced by a ligament or its replacement. Smidt [75] reported that the quadriceps exerts an anterior shear force on the tibia when the knee is flexed between 5° and 60° and a posterior shear force with the knee flexed between 75° and 90°. Both the hamstring and the triceps surae muscles exert a constant posterior-oriented force that keeps a PCL-deficient knee posteriorly displaced especially at flexion angles greater than 40° [18]. This constant force applied even at rest explain why reconstructive procedures using auto- and allografts could elongate before the ligamentization period is completed. To minimize strain on the graft after PCL reconstruction, strong gastrocnemius and triceps surae contractions should be avoided beyond 30° of flexion [18]. Muscular activity can compensate for ligament laxity in some cases [57], but it rarely restores full stability to the joint.

Pathophysiology of PCL Injuries

It has been shown that there are frequent injuries to the lateral side of the knee occurring with a clinically "isolated cruciate ligament" [78]. Articular contact pressures have been studied using pressure sensitive film in a cadaver model, with a 50% increase in pressure over the 15°–90° range of flexion in the intact knees [73]. The average patellofemoral pressures increase 25% regardless of the flexion angle when the PCL is cut. An additional 50% increase in patellofemoral pressure at 15° of flexion results from cutting the posterolateral complex in combination with the PCL. In a static cadaveric knee model with simulated physiological loads before and after transection of the PCL the tibia was observed to sublux posteriorly with loading at 60° of flexion. Significant changes were measured in both absolute pressure and pressure concentration in the medial compartment at 60° of flexion after cutting the PCL. This pressure concentration may relate to the clinical observation of medial femoral condyle degenerative changes in the PCL-deficient knee. Of course a varus-aligned knee increases the medial stress concentration and perhaps expedite medial compartment degeneration [73].

Mechanism of Injury

Three common mechanisms can result in PCL injury: (a) hyperflexion, with or without an anterior tibial force [20]; (b) hyperflexion with a downward force applied to the thigh [36]; and (c) hyperextension, often with an associated varus-valgus force [56]. The mechanism of injury for tearing only the PCL is a posteriorly directed force on the tibia with the knee flexed at 90°. This injury along with an isolated avulsion of the PCL [35] often occurs in motor vehicle-related accidents (the so-called dashboard injury) and contact sports (especially football). A similar mechanism can result from a fall on a flexed knee. Clancy et al. [12] believe that if a person falls on a flexed knee with the foot in dorsiflexion, the force lands on the patella; however, if the foot is in plantar flexion, the tibia is driven posteriorly. We agree with Zarins and Fish that hyperextension would not be a mechanism for tearing the PCL alone [92] because at least a portion of both cruciate ligaments tightens on extension of the knee; it seems unlikely that the ACL or the posterior capsule would be spared in this mechanism of injury

Natural History

In spite of anecdotal evidence of high levels of performance in competitive athletes with PCL-deficient knees, the ultimate fate of the PCL-deficient knee in young, active patients remains nebulous. Nonoperative treatment of isolated PCL disruption in the athlete results in satisfactory function in the short term. Nearly all patients are able to return to full sports activity without surgery [16, 20]. Static posterior stability is not necessarily required for acceptable functional stability and the degree of abnormal laxity is not correlated with the severity of symptoms [15]. Only 20% of patients have a poor functional grade, and only 10% consider their symptoms bad enough to warrant surgical treatment. Of most concern is the reported increase in degenerative arthritis associated with chronic PCL instability [12, 15, 16, 43].

Functional results of PCL injuries clearly depend on the injury mechanism and the magnitude of the damage to other structures about the knee. PCL injuries resulting from athletics give 86% good to excellent results, as compared with 8% resulting from motor vehicle accidents and falls [3]. Knee stability decreases and pain worsens with time when the PCL is deficient. Functional adaptation lasts for 3–18 months when symptoms of pain and instability improve as quadriceps muscle reeducation occurs. Functional tolerance lasts approximately 15 years, allowing athletes to return to the highest level of competition despite slowly progressing deterioration of the patellofemoral and tibiofemoral articular surfaces. Osteoarthritis occurs is less than one-third of patients after 15 years of injury [16].

Factors found to significantly influence prognosis include the degree of instability, a history of meniscectomy, quadriceps muscle insufficiency, and patellofemoral problems. Selective strenghtening of the quadriceps is thought to be the reason for improvement in the ability to run up and down stairs. The subconcious adaptation of these dynamic stabilizers serves to maintain stability of the PCL-deficient knee [3]. Differences in the clinical course of patients with straight unidirectional posterior instability are demonstrated as compared with combined multidirectional instability [83]. Excellent to good results are noted in 86% of the knees with unidirectional instability, as compared with 48% of the knees with multidirectional instability. These should be considered for surgical stabilization because, without surgery, they have a higher incidence of unacceptable function, roentgenographic degenerative changes, and patellofemoral disease.

No well-controlled prospective series with long-term follow-up and sufficient number of patients exist to show that static reconstruction for chronic PCL deficiency can slow down the progression of OA changes. These data raise concerns about the fate of the PCL-deficient knee because the abnormal biomechanics appear to result in progressive articular cartilage deterioration as a function of time [12].

Clinical Evaluation

History

The patient usually presents with a history of trauma that can be related to the above mechanisms: (a) hyperflexion, with or without an anterior tibial force [20], (b) hyperflexion with a downward force applied to the thigh [36], and (c) hyperextension, often with an associated varus-valgus force [56]. Acute pain and swelling is usually minimal. The patient rarely complains of knee instability except with a combined lesion. In chronic cases athletes usually present with patellofemoral pain.

Physical Examination

Reported overall clinical accuracy for PCL injuries diagnosis is 96%, with 90% sensitivity and 99% specificity [65]. Standing alignment of genu varum, tibial external rotation, or genu recurvatum with walking are signs of lateral and posterior tibiofemoral distraction, as well as of posterior subluxation of the

Table 1. Clinical evaluation of the PCL-deficient knee

Posterior instability tests	Posterolateral instability tests
Posterior sag test	External rotation recurvatum test
Thumb test	Prone posterior drawer
Godfrey's test	Posterolateral drawer test
Posterior drawer test	Reverse pivot shift
Quadriceps active test	Tibial external rotation test
Gravity sign near extension test	The voluntarily evoked posterolateral drawer
Active reduction of posterior tibial subluxation	
Passive reduction of posterior tibial subluxation	
Passive posterior drawer near extension (reverse Lachmann test)	
Dynamic posterior shift test	

lateral tibial plateau. Patients may walk with the affected knee in slight flexion to avoid the accentuated external rotation that occurs in terminal extension. Clinical determination of tibial posterior translation, varus rotation, and external rotation is required to detect PCL injury and to distinguish combined PCL and posterolateral corner injuries [17]. Numerous specific tests have been described to diagnose posterior and posterolateral insufficiency (Table 1).

Associated posterolateral instability must be ruled out by physical examination in all patients with a chronic PCL-deficient knee. Varus and external rotation are not significantly altered with an isolated PCL injury. Absence of the PCL alone is associated with abnormal posterior laxity as evidenced by increased posterior translation of the tibia on the femur ("posterior sag"), especially at 90° of knee flexion [17]. Functional disruption of the lateral collateral ligament and the posterolateral capsule results in abnormal varus and external rotation of the tibia on the femur at 30° of knee flexion. The popliteus tendon is an important structure preventing external rotation and posterior translation of the lateral tibial plateau [79]. A combined PCL and posterolateral injury causes large increases in posterior translation, varus, and external rotation at all flexion angles. Marked lateral joint space opening to varus stress testing in full extension without an appreciable end-point may represent disruption of both cruciate ligaments and the posterolateral structures. Examination under anesthesia should be performed during operative procedure on the knee of the patient who is unable to relax during a routine physical examination or when a definitive diagnosis cannot be reached and magnetic resonance imaging (MRI) is either inconclusive or has not been obtained.

Imaging and Instrumented Evaluation

Radiographic Features

A fragment of bone may be seen at the posterior proximal tibia from an avulsion of the PCL tibial attachment. In cases of posterolateral instability an avulsion of the arcuate ligament at the posterolateral tibia ("arcuate sign") or an avulsion of the popliteus tendon at the distal lateral femur should be ruled out [39]. Degenerative changes identified on roentgenograms of the chronic PCL-deficient knee involve predominantly the medial compartment and the patellofemoral joint. Joint space narrowing, subchondral sclerosis, peripheral osteophytes, and alteration in the tibiofemoral axis are observed with post-traumatic degenerative arthritis [12].

Abnormal varus is identified on single-leg, full-length standing anteroposterior roentgenograms. This roentgenographic analysis may provide valuable information for determining the need for a valgus-producing osteotomy. Stress roentgenograms can quantitate posterior subluxation and allow a side-to-side comparison. Lateral roentgenograms are obtained and classification of PCL injuries is based on whether it is isolated or combined and the amount of laxity on posterior drawer test (Table 2) [52].

Scintigraphy

The technetium (^{99}Tc) bone scan is a sensitive test which identifies areas of increased blood flow and bone remodeling. We believe this is one of the most sensitive tests for early detection of patellofemoral degeneration. In fact, a "hot patella" is a strong argument for PCL reconstruction. Some orthopedic surgeons have recommended serial bone scans to monitor the development of degenerative changes and to aid the decision to proceed with PCL reconstruction [11].

Computed Tomography

Computed stress tomography has been used to diagnose posterolateral instability of the knee [72]. The lower extremity is positioned on a support frame with the knee flexed to 25°. External rotation torque (3.4 nm) through a foot plate and a posterior force (45 N) at the level of the tibial tubercle are applied simultaneously. Overlapping transaxial reference lines verify external rotation and posterior subluxation of the lateral tibial plateau.

Table 2. Classification of PCL injuries

Type	Definition	Laxity
I	PCL strecthed	-5 mm
II	PCL torn; MF ligaments intact	5-9 mm
III	PCL and MF ligaments torn	+10 mm
IV A	PCL and LCL/posterolateral injury	+12 mm
IV B	PCL and MCL/posteromedial injury	+12 mm
IV C	PCL and ACL torn	+15 mm

Magnetic Resonance Imaging

MRI of the knee depicts the PCL as a large low-intensity band on both the sagittal and coronal views [7]. MRI boasts 100% accuracy, sensitivity, and specificity in identifying knees with PCL injury [28]. Joint position affects signal intensity [76]. Focal signal changes within the ligament have been attributed to degeneration of the tissue in approximately 60% of cadaver specimens when histological analysis was compared with MRI appearance [32]. MRI is a helpful adjunct especially in determining the location of the tear and in identifying other associated injuries [1] (Fig. 2).

Functional Analysis

PCL strain increases markedly when the knee is flexed greater than 90° during terminal stance phase. Patients with an untreated PCL-deficient knee walk at a fast cadence with significantly less knee extension during midstance. Avoidance of terminal extension is thought to be a compensatory mechanism to avoid secondary stresses on the joint surfaces and posterior capsule. Electromyography shows early activation of the gastrocnemius and soleus muscles during stance phase, probably a dynamic compensatory mechanism.

Instrumented Knee Arthrometers

A variety of instrumented knee arthrometers are now commercially available to measure knee motion, but these measurements are operator dependent and require significant technical expertise [84].

The use of these devices in PCL-deficient knees is much less documented than in ACL-deficient knees [84]. Before neutral alignment, the injured knee might indicate excessive anterior laxity [84]. This is also due to the fact that it is important to have the starting point for measurement precisely identified [9]. We consider arthrometer measurements for PCL laxity or instability neither accurate enough nor reproducible enough.

Arthroscopy

The arthroscope can be used to directly visualize the PCL and other intraarticular structures before a final decision on surgical repair [23]. The high incidence of associated injuries and the difficulty in making an accurate diagnosis make this a useful adjunct in planning a course of treatment [19]. It is difficult to visualize arthroscopically the entire PCL because the ACL lies directly in front of it and a portion of the PCL is extraarticular. A posteromedial portal or a standard anterolateral portal and a 70° fore-oblique arthroscope can be used to improve visualization [19, 23]. An intact ligament of Wrisberg or Humphrey lying within the same synovial sheath as the torn PCL can easily be

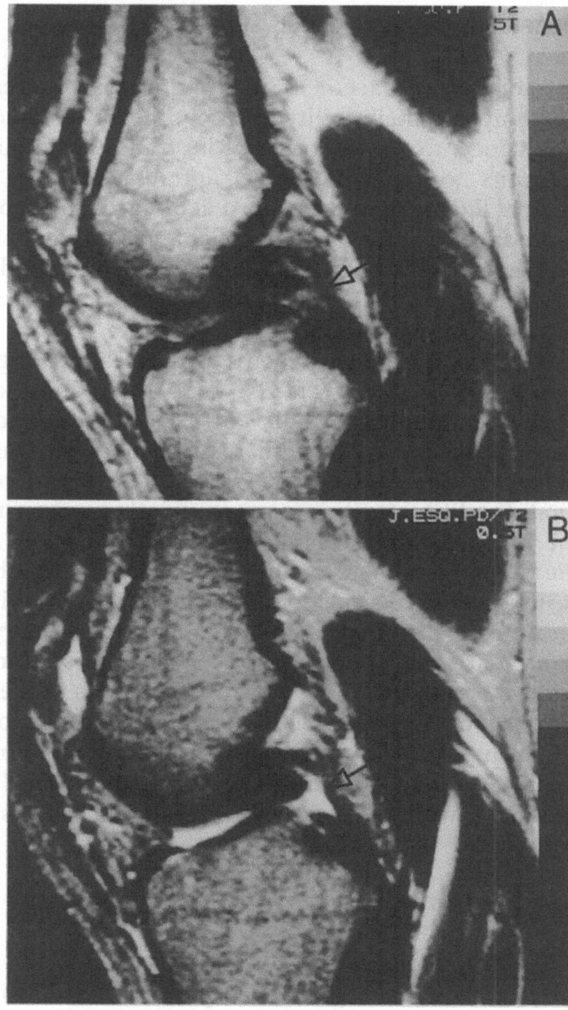

Fig. 2A,B. PCL tear. **A** Proton density weighted (T1) image shows thickening and increased signal intensity proximally within the PCL. **B** T2-weighted image clearly defines complete separation with edema in gap (*open arrowheads*). (Courtesy of RM Liga das Senhoras Catolicas Magnetic Resonance Imaging, Curitiba, Brazil)

interpreted as an intact PCL [12]. It is important to note that the normal ACL appears arthroscopically to be lax in PCL-injured knees. This "pseudolaxity" is caused by tibial subluxation posteriorly when the knee is flexed 90°. This phenomenon, along with the impression of PCL being intact arthroscopically (as the result of healing behind the ACL) may lead the inexperienced surgeon to proceed with unnecessary ACL reconstruction in PCL-injured knees [52].

General Approach to PCL Surgery

Intraoperative placement of PCL grafts is an attempt to duplicate the original anatomy and their ability to accept the physiological forces present in the knee. The concept of isometry is at best an oversimplification. PCL reconstruction attempts to reproduce both anterolateral and posteromedial fibers of the normal ligament, and the normal biomechanical behavior of the PCL cannot be reproduced by any of the available reconstruction techniques to date. To optimize femoral site graft placement, one must understand the normal anatomic relationships of the intercondylar notch, as well as the effect of anterior and posterior femoral tunnel placement [46, 51].

There is a relative paucity of data concerning the *in vivo* or *in vitro* placements of a PCL graft. To date the authors know of one study on postoperative clinical results of PCL reconstructions that has quantified anteroposterior laxity with instrumented measurements [46]. Two very important relationships of the PCL must be considered when choosing a graft site. First, the PCL attaches more superiorly on the roof of the femoral notch than the ACL, as can be easily visualized during any ACL reconstruction. Second, the tibial attachment of the PCL is clearly 1 cm below the joint line, which is easily visualized on an MRI coronal view. Similar to the ACL, the femoral location is the primary determinant of isometry, and the most critical determinant is not the location of anteroposterior placement along the intercondylar roof but rather how close one was to the intercondylar roof on the 12 o'clock position [26]. This position is located towards the roof of the intercondylar notch, on average 11 mm from the junction of the notch with the trochlear roof. The tibial site should be placed at the normal anatomic attachment, which is clearly very posterior and distal (Fig. 3).

Graft Selection

PCL graft biomechanics differ little from ACL reconstruction. Indeed, biomechanical issues, graft incorporation tendencies, durability, and types of tissues considered are the same for each reconstruction. Some PCL reconstructions require that the graft be brought over the back of the tibia and subjected to a very tight turn. Also, the geometry of the tunnel placement requires a longer graft, which extends from the medial femoral condyle to below the posterior ridge of the tibia and runs anteriorly through a hole in the metaphysis of the tibia [67]. In contrast to the many animal and clinical studies on ACL reconstruction, very few concern PCL reconstruction. It would be safe to assume, however, that the principles for each are the same, except for the limitations just noted. The role of the prosthetic replacement of the PCL was defined as limited, if any [41]. Synthetic materials such as polytetrafluoroethylene, coated carbon fiber, and ultrahigh molecular weight polyethylene have been considered as options either in a prosthetic or augmentation role. The occurrence of abrasion, rupture, and intraarticular debris, and inflammation has challenged their merit and narrowed the in-

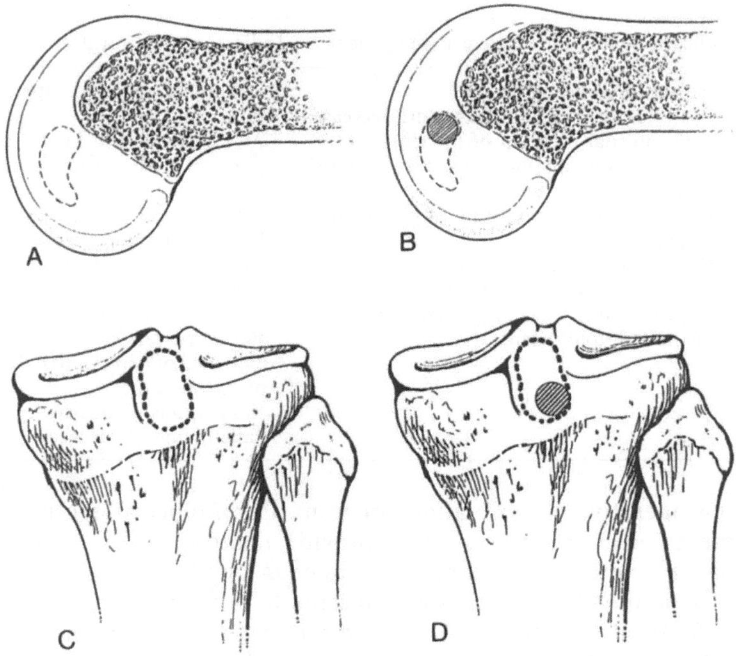

Fig. 3. A,C Outline of tibial and femoral insertion sites of the PCL. **B,D** Ideal tunnel location to reproduce the anterolateral component of the PCL if a single bundle reconstruction technique is used. (Adapted from [30a])

dications for their use [91]. In PCL reconstruction the continuous posteriorly directed stresses placed on the graft provide (in theory) a rationale for augmentation or prosthetic replacement. Reports in the literature have both supported and refuted the benefits of this device. There is at least one synthetic ligament (Ligament Advanced Reinforcement System, LARS; Arc-sur-Tile, France) for which good results have been reported in ACL/PCL reconstruction at 2- and 5-year follow-up [46]. We refer the reader to Chap. II.4 for further details.

Surgical Indications

The rationale for surgery is based on the premise that restoration of PCL function by intraarticular ligament reconstruction reestablishes knee biomechanics, thereby significantly improving function and preventing further joint damage [80]. Careful patient selection is the critical first step toward achieving these goals. Patients with chronic, isolated PCL injuries are treated with an aggressive therapy program and activity limitations (Table 3). PCL reconstruction surgery is a viable option for those patients who do not satisfy these criteria, or who have found nonoperative management to be unsuccessful. Patient factors that must be taken into consideration include age, financial and job constraints, expectations, and the ability to participate in the

Table 3. Decision making in PCL treatment

Indications for nonsurgical management	Indications for surgical management
* Isolated PCL injury Posterior drawer < 10 mm and decreases with internal rotation of the tibia No associated ligamentous, meniscal, or articular surface injury	* PCL and associated capsuloligamentous injury ACL or medial or lateral collateral ligament, posterolateral corner (arcuate/popliteus)
* No or minimal symptoms of giving way or pain	* Symptoms of giving way or pain despite focused nonoperative treatment
* No evidence of progressive articular cartilage deterioration on radiographs	* Static virus alignment, lateral joint space opening on single-leg standing AP radiograph
* Negative bone scan of patellofemoral or medial compartments	* Gait abnormality Varus, posterolateral thrust
	* Evidence of progressive degenerative changes on serial scans
	* Nonresponsive quadriceps reinforcement
	* Patellofemoral disease

postoperative rehabilitation program. Chronological age limits for knee ligament reconstruction surgery provide relative guidelines only. Symptomatic patients who are less than 50 years of age, and who wish to continue an active athletic life-style are reasonable candidates for surgical treatment.

Surgical treatment is indicated for chronic PCL knees with associated capsuloligamentous injuries resulting in markedly abnormal tibial translation and rotation (Table 3). Posterolateral instability is the most common associated problem requiring surgical attention. Posterolateral reconstruction and/or valgus osteotomy should be considered for patients with a varus mechanical axis, a significant varus thrust with walking, and asymmetric lateral joint space opening on single-leg standing AP roentgenograms. Surgical intervention should be entertained if patients develop pain consistent with early arthritis, or if serial bone scans reveal evidence of progressive degenerative changes. Attention should be given to patellofemoral pain which appears to restrain athletic activities much earlier than medial compartment arthritis and retards rehabilitation programs. There is more controversy concerning surgical treatment of isolated PCL lesions, either acute or chronic; nevertheless, our tendency is to treat these lesions surgically before associated secondary laxities develop, or anterior knee pain impairs sports practice. One should not wait 5–15 years for classic roentgenological OA features to make a decision.

Acute PCL Injury

A review of the literature fails to provide any specific guidelines in the treatment of acute isolated PCL tears. The relative infrequency of PCL tears makes it difficult to accumulate long-term follow-up data in a large series of patients. Conflicting opinions exist concerning nonoperative and operative treatment [6, 15, 36].

The treatment plan for a complete rupture of the PCL is determined by the physical examination and the functional level of that knee. Fowler [20] and

Bergfeld [6] defined an isolated PCL rupture as the following: (a) straight abnormal posterior instability of the tibia on the femur of 10 mm or less without any varus-valgus instability in extension or at 20° of knee flexion and a normal ACL, (b) no abnormal external rotation of the tibia or recurvatum, and (c) a decrease in the amount of posterior displacement with internal rotation of the tibia on the femur. Both authors have demonstrated excellent functional results even in competitive professional and amateur athletes using rehabilitation alone without the use of bracing. If a patient undergoes a thorough rehabilitation trial and continues to have functional symptoms of abnormal posterior instability with activities of daily living and/or sports, he is considered a candidate for PCL reconstruction. Ruptures of the PCL associated with grade III injuries to the medial complex [medial collateral ligaments (MCL) and posteromedial corner] or lateral complex [lateral collateral ligament (LCL) and posterolateral corner] in the acute setting require surgical intervention. Not infrequently the PCL should be reconstructed, either with an autograft central third bone-patellar tendon-bone or with an allograft bone-patellar tendon-bone or Achilles tendon. A prosthetic ligament or a synthetic ligament augmentation device is not indicated at the moment, except in a research setup.

Nonoperative treatment is generally recommended for grades I and II posterior laxity (i.e., tears with mild to moderate posterior tibial subluxation). With complete tears causing posterior displacement of 1 cm or more the correct decision is more difficult [15]. A controversy that arose in the literature involves the treatment of the acute tears of the substance of the PCL [4]. Of patients achieving good quadriceps strength 88% had good or excellent results, compared with only 14% good to excellent results when poor strength was reported [3].

Isokinetic testing of patients [59] showed that in patients who fully returned to sports and were completely satisfied with their outcome mean torques were greater then 100% of the uninvolved quadriceps. The most important factor determining outcome in a conservatively treated knee is the presence or absence of other ligamentous laxities [68]. The patient's inability to separate out functional outcome compared with objective testing is reported [20], and acceptable functional stability does not necessarily require absolute static stability. Functional outcome on a group of 43 patients followed up for 6.3 years showed good to excellent results in 12 of 14 with unidirectional instability whereas only 14 of 29 with multidirectional instability showed similar standings [83]. The belief that all isolated PCL injuries do well has been recently disputed [43]. Out of 40 patients 65% stated their activity level was limited, and 90% complained of knee pain, despite a 99% average isokinetic muscle strength compared to the contralateral side. On radiographic evaluation an increasing degenerative change was seen with increased intervals from time of injury.

At present most authors recommend nonoperative treatment for acute isolated PCL injuries and reserve surgical reconstruction for symptomatic chronic PCL injuries, acute bony avulsions, and combined injuries [36]. Acutely, isolated PCL injuries can be managed by splinting the knee in extension until the pain subsides, followed by early motion. It is essential that rehabilitation of PCL injuries emphasize quadriceps strengthening. Successful compensatory mechanisms in a PCL-deficient knee demand extremely strong quadriceps muscles contracting much earlier than normal in the gait cycle [59].

Authors' Preferred Method of Treatment

Whenever the PCL is torn and is associated with a bony avulsion, it is strongly suggested to reduce and fix the fragment with a screw to reduce the morbidity associated with laxity and malunion of such a fracture. The surgical exposure is usually the posterior approach, and the fragment is reduced and fixed with one 4.5-mm cortical interference screw. Smaller fragments that do not hold a screw can be fixed using metal wires through the tibial metaphysis and held in place in the anterior cortex, but we seldom see such small fragments. Results of ORIF using either method are usually good and restore to a great extent the posterior stability.

When the tear of the PCL is interstitial, there are several parameters to be followed in spite of many surgical techniques available to date. It is now accepted that a primary repair of an interstitial tear of the PCL gives results as poor as those for the ACL. We suggest conservative treatment for "isolated" PCL lesions with less than 10 mm posterior drawer. These patients must be oriented in the beginning to physical therapy to control the inflammatory phase and the quadriceps atrophy; the second stage of their treatment includes a progressive reinforcement of the quad and proprioception. They must be seen regularly in the office to keep track of posterior laxity/instability and patellofemoral pain. We regard this approach as a "safe" one, but we believe that no gold standard exists so far, and that the subject is open to discussion.

Combined PCL lesions (posterolateral corner, ACL) usually give posterior laxities greater than 10 mm. We recommend reconstruction of the various structures since these knees are usually unstable and symptomatic and to avoid degeneration of the articular surfaces. Some authors prefer early surgery after trauma, expecting greater knee stability due partly to joint rigidity. We prefer to await the resolution of the inflammatory phase after trauma and reconstruct the ACL-PCL after 6–8 weeks, when the knee's range of motion is back to normal, along with a repair or a reconstruction of the posterolateral corner. Meniscal lesions are also repaired, if possible. We are currently using a synthetic ligament and hardware (LARS) to reconstruct the PCL using an anterior approach, and early results are encouraging. Extracapsular structures lesions (MCL, LCL, posteromedial corner) are rarely repaired in the acute phase because when the central axis of the knee is restored, these structures are stabilized and heal. Graft choice for PCL reconstruction is discussed in Chap. II.2.

The use of the patelar tendon to reconstruct the PCL has some pitfalls, such as anterior knee pain, transitory quadriceps weakness, and eventual postoperative failure. To avoid such problems, a contralateral graft may be taken. No matter which graft is taken to reconstruct isolated or combined PCL lesions in an acute setting, the primary goal is to restore the knee's biomechanics, with great care to reconstruct the PCL-posterolateral corner associated lesions.

Chronic PCL Injury

Symptoms often associated with chronic posterior knee instability are weakness or decreased functional capacity of the knee. The patient sometimes has pain with excess use, especially in the patellofemoral area. Often symptoms are caused by an associated intraarticular pathological condition such as degenerative changes. The knee does not tend to give way as it does with ACL injuries. If the patient has symptoms that can be attributed to posterior tibial subluxation, surgical reconstruction should be considered. It is important to determine preoperatively whether the tibia can be passively reduced to a normal anterior position. Otherwise, it is difficult to reduce the tibia at the time of surgery unless extensive dissection is carried out. It may be advisable to document the absence of other intraarticular pathological conditions causing the symptoms by arthroscopy before reconstruction.

Although some authors have reported satisfactory results with nonoperative treatment of isolated PCL injuries [43], other investigators have reported good subjective but only fair objective results [36]. A recent follow-up study of nonoperatively treated isolated PCL injuries suggests that significant activity-related pain and degenerative changes (especially in the medial compartment) is common with this form of treatment [83]. PCL injury has a poor clinical outcome when associated with rotatory laxity, quadriceps weakness, meniscectomy, patellar chondrosis, and gross laxity [33]. Mueller [55] recommends late repair of PCL injuries by dissecting out the ligament through scar tissue and then treating it as a primary repair. He states that the proximal and distal stumps of the PCL retain their vascular pedicle, and that the ligament does not atrophy and retract as does the ACL. We have no experience with such late repairs of the PCL.

Dynamic augmentation by transfer of the medial head of the gastrocnemius has been recommended [53]. The results of dynamic reconstruction using a bone block transfer of the origin of the medial head of the gastrocnemius have been published [36]. Insall and Hood believe this acts as a "dynamic" reconstruction with the tibia moving forward on the femur with active plantar flexion of the foot. Other authors disagree and do not recommend this procedure [64]. Other dynamic transfers, such as the popliteus tendon [50], "double tendon" (gracilis and semitendinosus) [89], and semimembranosus [77] have not become popular and sacrifice important structures. Hughston reported good subjective but discouraging objective results using the medial head of the gastrocnemius [33].

Surgical reconstruction of chronic posterior knee instability has been described using a number of methods [77]. Kennedy and Grainger [44] recommended transfer of the semitendinosus tendon through femoral and tibial drill holes. Hey-Grooves [30] in 1920 and Trickey [85] in 1980 reported on the same type of reconstruction but used both semitendinosus and gracilis tendons. Clancy et al. [12] described a method using free patellar tendon graft with intact tibial and patellar bone attachments. They reported results of PCL reconstruction in 13 patients with chronic posterior instability with an average follow-up of 31 months [47]. The overall objective and functional results were good or excellent in 11 knees. The advantages reported by Clancy for using this method are bone-to-bone healing and adequate tensile strength of the graft. Moreover, a two-bundle reconstruction using two femoral tunnels reproduces more closely the biomechanics of the PCL (Fig. 4).

Authors' Preferred Method of Treatment

Prevention of degenerative arthritis as a sole indication for surgery has not been clearly demonstrated in the literature to date. The rationale for surgery is based on the premise that intraarticular ligament reconstruction reestablishes knee biomechanics, significantly improves function and prevents further joint damage. Careful patient selection is the critical first step necessary to achieve these goals.

If a patient undergoes a through rehabilitation trial and continues to have functional symptoms of abnormal posterior instability with activities of daily living and/or sports, he is considered a candidate for PCL reconstruction. Moreover, a "hot patella" in serial bone scans is a strong argument for PCL reconstruction. There are several difficulties found in chronic PCL reconstruction. It is difficult to reproduce the broad femoral attachment of the PCL when using one single autograft, such as the patellar tendon or the hamstrings. It is clear for us that one should try to reconstruct the PCL using two bundles, an anterolateral tight in flexion and another posteromedial tight in extension. If the surgeon uses a large allograft (Achilles or patelar tendon), it may then be possible to reproduce this large attachment using a single bundle.

Reconstruction of the PCL can either be performed arthroscopically [48] or by arthrotomy [5, 47, 81]. The use of an autograft bone-patellar tendon-bone usually requires an arthrotomy because it is very difficult to turn the bone block into the bone tunnels within the knee arthroscopically. Techniques to reconstruct the PCL arthroscopically frequently use an Achilles tendon allo-

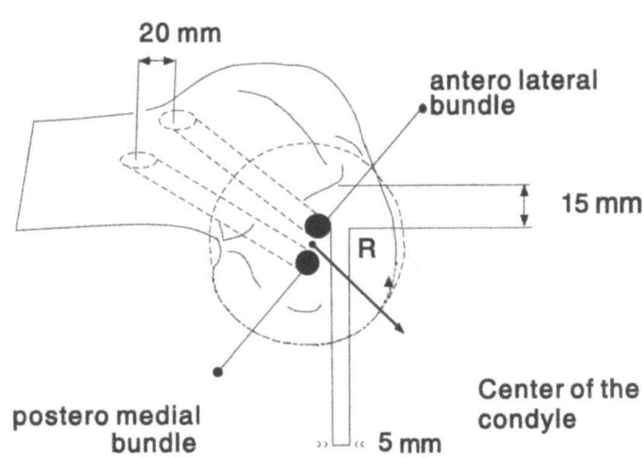

Fig. 4. Ideal femoral tunnel location to reproduce the anterolateral and posteromedial components of the PCL if a double bundle reconstruction technique is used. (Reproduced with permission from [45a])

graft and use the more compliant tendinous portion of the graft to navigate the bone tunnels within the knee. A theoretical decrease in patellofemoral symptoms and extensor mechanism morbidity would be anticipated using an arthroscopically performed allograft reconstruction rather than the traditional arthrotomy and autograft PCL reconstruction. However, no comparative data are available, and one procedure cannot be recommended over the other. The principles of PCL reconstruction require secure fixation of the graft and accurate tunnel placement with documentation on radiographs of the course of the tibial tunnel before final overdrilling of the guide pin. Furthermore, for isometric placement, the femoral tunnel should be high on the roof of the intercondylar notch rather than on the wall of the medial condyle [26]. Postoperatively it is recommended that the knee be placed in extension, which tightens the posterior capsule and minimizes the amount of posterior sag. The knee is held in extension for a few weeks, depending on the degree of additional trauma to the collateral ligaments and the stability of graft fixation at surgery. Olecranization of the patella does not prevent posterior tibial displacement and might even contribute to it [66].

Our choice for PCL reconstruction so far would be a contralateral bone-patellar tendon-bone graft, split in half at its proximal femoral insertion, using an anterior approach. An extraarticular augmentation on the posterolateral corner can be done in patients with a 15+mm posterior laxity, once this procedure reduces the stress on the PCL. It is also important to note that such a laxity usually follows secondary stabilizer's laxities. We are currently testing a new synthetic ligament that has had some good results in France. This ligament (LARS,) is polyester-made. It comes in various number of fibers, but the one proposed for PCL reconstruction (PC 60), is made up of 60 fibers and two distinct portions: a 35-mm intraarticular portion which contains only longitudinal fibers and a cylindrical knitted structure which is intended for the bony tunnels. Laboureau proposes a two-bundle reconstruction, which provides 120 fibers with an approximate tensile resistance of 5000 N. Although there is only scarce clinical support for this implant, some objective results at 2- and 5-year follow-up are encouraging: 73% of patients with 5 mm or less posterior drawer using dynamic radiographic measurement at 200 N (Telos, Austin and Cess, Fallston, Md.) at 5-year follow-up and only 14% occasional knee effusion [46].

Stability over time is a reasonable question. Artificial ligaments have the reputation to fail sooner or later. In Laboureau's series evaluated at 18–24 months, the author found no deterioration of mechanical quality which one would expect with a break [46]. The PCL is extraarticular and located in a highly vascularized area [69]. This favors the formation of a powerful fibrous healing and ensure results over time. It seems that if posterior laxity is corrected immediately, and if this correction is maintained for a sufficient time, the posterior fibrosis would be organized and oriented. This theory is illustrated by histological studies on a knee dislocation after a severe trauma [46]. We are presently proceeding with a multicentric clinical study in Canada to test this ligament, but our results are too preliminary to recommend the widespread use of this prosthesis.

Combined PCL Lesions

ACL and PCL Injuries

For combined ACL and PCL injuries we attempt to reconstruct the PCL first. This is due mainly to the difficulty to correctly place and tension both grafts at a time. Another problem are rehabilitation programs, once one must combine antagonist muscle strenghtening at the same time. We use a double-bundle technique and prefer to base both of our tibial tunnels on the anteromedial tibial cortex, separating the two tunnels by approximately 3 cm. We routinely use semitendinosus-gracilis tendons autografts and/or LARS prosthesis for reconstructing the PCL and patellar tendon-bone autografts for ACL reconstructions as needed. Tunnel and graft preparation and graft passage are done in a routine manner. The PCL graft is secured, being careful to recreate the normal step-off of the medial femoral condyle. The delayed ACL reconstruction is carried out and then fixed using an interference screw placed endoscopically for the femoral tunnel and an interference screw to secure the graft in the tibial tunnel. In cases of combined PCL and LCL injuries, we prefer to reconstruct the LCL with semitendinosus tendon autograft or a LARS ligament. We prepare bone plugs and place it in a 9-mm tunnel in the proximal fibula with a 7-mm interference screw. MCL are usually left unrepaired. However, significant MCL laxity (more than 3+) can be treated with primary repair alone or in combination with augmentation with allograft if an approach to take semitendinosus-gracilis tendons grafts has already been performed, and devascularization has occurred. These combined injuries are often associated with knee stiffness [40].

ACL reconstruction in combination with PCL reconstruction demands careful consideration of tissue source and the implications of its use. In this instance it is beneficial to be familiar with both bone-patelar tendon-bone and hamstring techniques. Options include any combination of autogenous or allogenous bone-patelar tendon-bone or hamstring tendons. Allografts or contralateral autogenous grafts may also be used.

Posterior Cruciate Ligament and Posterolateral Deficiencies

In patients with chronic PCL and posterolateral (PL) deficiencies and varus alignment, a progressive varus thrust often becomes apparent as both the primary and secondary lateral restraints are rendered ineffective. In these situations axial realignment must be considered as the primary procedure, with soft tissue reconstruction as a secondary and perhaps later concern [10].

Chronic posterolateral instability of the knee is frequently associated with other ligamentous injuries and may even be iatrogenic in origin if important stabilizers are used to correct other instabilities. Hughston stated that this condition is "easily missed, misdiagnosed, and mistreated" [34]. The posterolateral structures include the LCL (fibular collateral ligament), the fabellofibular ligament (short collateral ligament), the popliteus muscle, the arcuate ligament, the iliotibial band, and the lateral head of the gastrocnemius muscle [68, 70]. Posterolateral instability continues to challenge the orthopedic surgeon despite improved knowledge of the anatomy and biomechanics of the posterolateral corner and an advanced understanding of the clinical presentation and physical examination of these complex knee instability patterns. Functional disability manifested by giving-way episodes with activities of daily living and alterations of gait are much more common when PCL injury is associated with posterolateral capsuloligamentous injury. The resultant posterior tibial subluxation, increased adduction moment, and medial concentration of the joint reaction force may contribute to the development of arthrosis. The fundamental question continues to be whether early surgical intervention can prevent the deleterious consequences of a chronic PCL deficiency.

While there are a number of different technical methods of valgus osteotomy, the end-point of all these is the transfer of the mechanical axis to the lateral aspect of the knee. Careful preoperative planning that includes allowance for a lateral thrust is present on X-ray films, calculating the size of wedge, as well as meticulous attention to principles of osteotomy surgery, and the need for early range of motion and full protected weight bearing are mandatory. Following osteotomy, if the patient continues to have functional instability, soft tissue reconstruction is considered [87]. Laboureau reported on a series of 20 patients with PCL and PLC instability; 16 were treated with a simultaneous reconstruction. Each had less than 5 mm posterior draw at 200 N at 90° of knee flexion postoperatively. Two failures were due to failures on the posterolateral corner reconstruction [45].

Reconstruction of the PCL alone occasionally imparts increased PL stability, depending on the degree of initial PL injury. Evaluation of PL stability on completion of the PCL reconstruction is necessary. If significant PL instability at 30° and 90° remains, a PL reconstruction may be performed. It is important to realize that PL reconstruction alone in the presence of combined PCL/PL injury is likely to fail. When performing these two procedures, the PCL reconstruction should be completed first, ensuring that the femorotibial joint is reduced. The PL reconstruction follows.

Posterior Cruciate Ligament and Medial Collateral Ligament

Medial collateral ligament lesions heal most often. Therefore, one could postpone PCL reconstruction for 6–8 weeks to allow such healing and avoid knee stiffness and loss of motion after surgery. If still necessary, this combination of injuries can then be addressed surgically. The axial alignment must be examined specifically for valgus, and if this exists, bony realignment should be considered before performing any soft tissue procedure. The PCL reconstruction should be completed first, confirming that there is anteroposterior reduction. Following this, there is often a notable increase in medial stability, in which case the medial procedure may be deferred. If medial instability remains, however, a medial reconstruction may be performed.

Rehabilitation

Rehabilitation of the PCL reconstructive or nonoperative patient is greatly dependent on dynamic quadriceps stability. The biomechanics of the PCL and PL during various exercises are not well understood; however, research is being performed to advance the clinical management following these injuries. The clinician must realize that tremendous tibiofemoral shear forces are created during various knee exercises, in both closed and open chain. In particular, there are tremendous stresses applied to the PCL during OKC-resisted knee flexion. The clinician must also realize the role of the hamstrings during most closed chain exercises; therefore the authors recommend an early program emphasizing isolated open chain quadriceps strengthening progressing to closed chain drills once adequate quadriceps strength has been established. The PCL rehabilitation program can no longer be thought of as an ACL rehabilitation program "turned around" [88]. The principles of rehabilitation include joint motion and stability as well as muscular endurance and strength. These should be considered during the immobilization phase, the surgical and postsurgical phase, and continued through the early healing, late healing, and final healing stages. Conservative and surgical treatment programs for functional rehabilitation are essentially the same, except for the time factors involved. The times of completion of different levels of activity are generally longer in the surgical program [49].

Conservative PCL Treatment

Partial PCL injuries with a firm end-point on posterior drawer and only a few millimeters of pathological posterior displacement that decreases with internal tibial rotation can show good early functional results when treated by rehabilitation. Rehabilitation consists of the immediate control of inflammation and swelling by rest, ice, compression, and elevation, and the judicious use of a nonsteroidal anti-inflammatory medication. Protected weight bearing is allowed, as is active motion of the knee. Because active quadriceps activity is an antagonist to posterior laxity, it is critical to concentrate on progressive resistance exercises for the quadriceps. A brace is optional but is recommended for associated partial collateral ligament injury. The expected return to sports by the athlete is 1–3 months.

Surgical PCL Treatment

In rehabilitation the initial emphasis is on acquiring a full range of motion. Muscular conditioning and proprioceptive retraining follow. In most instances full protected weight bearing in extension is appropriate in the first postoperative weeks, and as gait returns to normal, full unprotected weight bearing may begin. In the immediate postoperative period, options include no immobilization and devices such as cooling splints or range-of-motion braces. As far as safe muscle activity is concerned, the horizontal components of the antagonist muscles are balanced at 22° of flexion; moreover, simultaneous contraction of all muscle groups can unload the cruciate ligaments entirely at flexion angles above 22° [58].

Complications

Some complications after ligamentous reconstruction are not common, but do exist: infection, difficulty in attaining full range of motion, difficulty with static stability, hemarthrosis [74], and problems with graft harvesting, placement, and fixation. It seems wise to recommend the use of a wide tourniquet cuff using pressure no greater than twice the systolic arterial pressure and for the least amount of time necessary [54, 82]. The patient believed most at risk is one with capsular defects caused by capsular ligamentous ruptures. Peek [60] reported the case of a compartment syndrome complicating an arthroscopic procedure.

Patellar tendon graft can be damaged by partial or complete transection, most commonly at the inferior pole of the patella. Fracture of a patella or tibial bone block can occur if one is too vigorous in levering the bone block out of the patella or tibia. This can be prevented by making the bone block at least 5 mm in thickness, by using an oscillating saw to cut completely through the outer cortex, and by avoiding undue levering of the bone block [25]. If it should occur, Graf recommends immediate internal fixation. Neither the routine postoperative rehabilitation nor the clinical result should be compromised if stable, anatomic internal fixation is achieved. Anterior knee pain can also be present in patients after bone-patelar tendon-bone graft harvest.

Fixation of the graft, either within the bone tunnels or to the femoral or tibial cortex, is the most likely site of graft failure in the first few weeks of ACL/PCL reconstructions. Pitfalls of interference screw fixation include graft, screw, and tunnel length mismatch, tunnel and screw divergence or convergence, graft advancement or translocation within the tunnel, graft fracture, and suture laceration [2]. Loss of motion following PCL reconstruction differs from motion loss following ACL reconstruction. PCL-reconstructed knees often require an excessive period of time to regain flexion. Insertion of the graft in the wrong location or tensioning of the graft with the knee in extension can result in excessive tensioning of the graft in knee flexion, resulting in loss of motion [37]. Peroneal nerve or vascular injury can occur [38], especially in knees with combined injuries involving the posterolateral corner. Most neuropraxies resolve in time (up to 18 months).

Late laxity following PCL reconstruction is common. Almost all PCL reconstructions develop some mild laxity. However, similar to others, we have found that this is less of a problem clinically. Problems with laxity can be exacerbated, however, with divergent tunnel placement and combined ligament injuries that are not adequately addressed at the time of surgery.

References

1. Andrews JR, Edwards JC, Satterwhite YE (1994) Isolated posterior cruciate ligament injuries. History, mechanism of injury, physical findings, and ancillary tests. Clin Sports Med 13(3):519–530
2. Bach BR (1989) Potential pitfalls of Kurosaka screw interference fixation for anterior cruciate ligament surgery. Am J Knee Surg 2:76–81
3. Balkfors B (1982) The course of knee ligament injuries. Acta Orthop Scand 198 [Suppl]:1–99
4. Barton TM, Torg JS, Das M (1990) Posterior cruciate insufficiency: a review of the literature. Sports Med 1:419
5. Berg EE (1995) Posterior cruciate ligament tibial inlay reconstruction. Arthroscopy 11(1):69–76
6. Bergfeld JA (1990) Diagnosis amd nonoperative treatment of acute posterior cruciate ligament injury. The annual meeeting of the AAOS, Instructional Course, no 208. New Orleans, 8–13 February 1990
7. Berquist TH, Ehman RL (1990) The knee. In: Berquist TH (ed) MRI of the musculoskeletal system, 2nd edn. Raven, New York
8. Butler DL, Noyes FR, Grood ES (1980) Ligamentous restraints to anterior-posterior drawer in the human knee: a biomechanical study. J Bone Joint Surg Am 62:259–270
9. Castle TH, Noyes FR, Grood ES (1992) Posterior tibial subluxation of the posterior cruciate-deficient knee. CORR 284:193–202
10. Clancy WG (1988) Posterior cruciate ligament injuries. Symposium presented at the annual meeting of the American Academy of Orthopedic Surgeons, Atlanta, 8 February
11. Clancy WG (1989) Repair and reconstruction of the posterior cruciate ligament. In: Instructional Course lectures, AOSSM Annual Meeting, Traverse City, Michigan, pp 141–155
12. Clancy WG Jr, Shelbourne KD, Zoellner GB, Keene JS, Reider B, Rosenberg TD (1983) Treatment of knee joint instability secondary to rupture of the posterior cruciate ligament. Report of a new procedure. J Bone Joint Surg Am 65(3): 310–322
13. Clancy WG Jr, Sutherland TB (1994) Combined posterior cruciate ligament injuries. Clin Sports Med 13(3):629–647
14. Covey CD, Sapega AA (1993) Injuries of the posterior cruciate ligament. J Bone Joint Surg Am 75(9):1376–1386
15. Dandy D, Pusey R (1982) The long-term results of unrepaired tears of the posterior cruciate ligament. J Bone Joint Surg Br 64:92–94
16. Dejour H, Walch G, Peyrot P (1988) The natural history of rupture of the posterior cruciate ligament. Fr J Orthop Surg 2:112–120
17. DeMeo PJ, Bergfeld JA (1992) Posterior cruciate ligament injuries. Mediguide Orthop 11:1–6
18. Durselen L, Claes L, Kiefer H (1995) The influence of muscle forces and external loads on cruciate ligament strain. Am J Sports Med 23(1):129–136
19. Fanelli GC, Giannotti BF, Edson CJ (1994) The posterior cruciate ligament arthroscopic evaluation and treatment. Arthroscopy 10(6):673–688
20. Fowler PJ, Messieh SS (1987) Isolated posterior cruciate ligament injuries in athletes. Am J Sports Med 15(6):553–557
21. Fukubayashi T, Torzilli PA, Sherman MF, Warren RF (1982) An in vitro biomechanical evaluation of anterior-posterior motion of the knee: tibial displacement, rotation and torque. J Bone Joint Surg Am 64:258–264
22. Girgis FG, Marshall JL, Al Monarem ARS (1975) The cruciate ligaments of the knee joint. Anatomical, functional and experimental analysis. Clin Orthop 106:216–231
23. Gold DL, Schaner PJ, Sapega AA (1995) The posteromedial portal in knee arthroscopy: an analysis of diagnosis and surgical utility. Arthroscopy 11(2):139–145
24. Gollehon DL, Torzilli PA, Warren RJ (1987) The role of the posterolateral and cruciate ligaments in the stability of the human knee: a biomechanical study. J Bone Joint Surg Am 69:233–242

25. Graf B, Uhr R (1988) Complications of intra-articular anterior cruciate reconstruction. Clin Sports Med 7(4):835–848
26. Grood ES, Hefzy MS, Lindenfield TN (1989) Factors affecting the region of most most isometric femoral attachment. I. The posterior cruciate ligament. Am J Sports Med 17(2):197–207
27. Grood ES, Noyes FR (1988) Diagnosis of knee ligament injuries. Biomechanical precepts. In: Feagin JA (ed) The crucial ligaments. Churchill Livingstone, New York
28. Gross ML, Grover JS, Bassettt LW, Seeger LL, Finerman GAM (1992) Magnetic ressonance imaging of the PCL: clinical use to improve diagnostic accuracy. Am J Sports Med 20:732–737
29. Harner CD, Livesay GA, Kashiwaguchi S, Fujie H, Choi NY, Woo SL (1995) Comparative study of the size and shape of human anterior and posterior cruciate ligaments. J Orthop Res 13(3):429–34
30. Hey-Groves EW (1920) The crucial ligaments of the knee joint: their function, rupture, and the operative treatment of the same. Br J Surg 7:505
30a. Harner CD, Livesay GA, Choi NY, et al (1992) Evaluation of the sizes and shapes of the human anterior and posterior cruciate ligaments: a comparative study. In: Transactions of the 38th Annual Meeting of the Orthopaedic Research Society, Washington, p 123
31. Hirshman HP, Daniel DM, Miyasaka K (1990) The fate of unoperated knee ligament injuries. In: Daniel DM, Akeson WH, O'Connor JJ (eds) Knee ligaments: structure, function, injury, and repair. Raven, New York, pp 482–503
32. Hodler J, Haghaghi P, Trundell D, Resnick D (1992) The cruciate ligaments of the knee: correlation between MR appearance and gross and histological findings in cadaveric specimens. Am J Roentgenol 159:357–360
33. Hughston JC, Degenhardt TC (1982) Reconstruction of the posterior cruciate ligament. Clin Orthop 164:59–77
34. Hughston JC, Jacobson KE (1985) Chronic posterolateral rotatory instability of the knee. J Bone Joint Surg Am 67:351–359
35. Hunter JC, Chapman JR (1995) Isolated avulsion of the posterior cruciate ligament: an uncommon dashboard injury. Am J Roentgenol 164 (5):1190
36. Insall JN, Hood RW (1982) Bone-block transfer of the medial head of the gastrocnemius for posterior cruciate insufficiency. J Bone Joint Surg Am 64:691–699
37. Irrgang JJ, Harner CD (1995) Loss of motion following knee ligament reconstruction. Sports Med 19(2):150–159
38. Jackson DW, Proctor CS, Simon TM (1993) Arthroscopic assisted PCL reconstruction: a technical note on potential neurovascular injury related to drill bit configuration. Arthroscopy 9(2):224–227
39. Jakob RP, Warner JP (1990) Lateral and posterolateral rotatory instability of the knee. In: Jakob RP, Staubli H-U (eds) The knee and the cruciate ligaments. Springer, Berlin Heidelberg New York, pp 463–494
40. Johner R, Ballner PM, Roggė D, Burch HB, Jakob RP, Tscherne H (1992) Dislocation of the knee. In: Jakob RP, Staubli H-U (eds) The knee and the cruciate ligaments. Springer, Berlin Heidelberg New York, pp 307–317
41. Johnson RJ, Beynnon BD, Nichols CE, Renstrom PA (1992) The treatment of injuries of the anterior cruciate ligament. J Bone Joint Surg Am 74(1):140–151
42. Kannus P, Bergfeld J, Jarvinen M, Johnson RJ, Pope M, Renstrom P, Yasuda K (1991) Injuries to the posterior cruciate ligament of the knee. Sports Med 12:110
43. Keller PM, Shelbourne KD, McCarrol JR, Rettig AC (1993) Nonoperatively treated isolated posterior cruciate ligament injuries. Am J Sports Med 21(1):132–136
44. Kennedy JC, Grainger RW (1976) The posterior cruciate ligament. J Trauma 7:367–377
45. Laboureau JP (1993) Laxités postero-externes. 3rd European Congress of the Eastern France Orthopaedics and Trauma Society, Strasbourg, June 1993
45a. Laboureau JP (1993) Posterior cruciate ligament reconstruction. LARS booklet for synthetic PCL replacement. Arc-sur-Tille, France
46. Laboureau JP, Bercovy M, Derick GH (1993) Plastie à 2 faisceaux du ligament croisé posterieur. Resultats comparés plastie autogène – plastie artificielle. 1st European Congress of Orthopaedics, Paris, April 1993

47. Lipscomb AB Jr, Anderson AF, Norwig ED, Hovis WD, Brown DL (1993) Isolated posterior cruciate ligament reconstruction. Long-term results. Am J Sports Med 21(4):490–496
48. Malek MM, Fanelli GC (1993) Technique of arthroscopically assisted PCL reconstruction. Orthopedics 16(9):961–966
49. Markey KL (1991) Functional rehabilitation of the cruciate-deficient knee. Sports Med 12(6):407–17
50. McCornmick WC, Bagg RJ, Kennedy CW, Leukens CA (1981) Reconstruction of the posterior cruciate ligament: preliminary report of a new procedure. Clin Orthop 118:30–34
51. Miller MD, Harner CD (1995) The anatomic and surgical considerations for posterior cruciate ligament reconstruction. Instr Course Lect 44:431–440
52. Miller MD, Harner CD, Koshiwagushi S (1994) Acute posterior cruciate ligament injuries. In: Fu FH, Harner CD, Vince KG (eds) Knee surgery, 1st edn. Williams and Wilkins, Baltimore, pp 749–768
53. Moore HA, Larson RL (1980) Posterior cruciate ligament injuries: results of early surgical repair. Am J Sports Med 8:68
54. Moore MR, Garfin SR, Hargens AR (1987) Wide tourniquets eliminate blood flow at low inflation pressures. J Hand Surg (Am) 12A:1006–1011
55. Muller W (1983) The knee: form, function and ligament reconstruction. Springer, Berlin Heidelberg New York, pp 246–248
56. Nonweiler DE, Schenck RC Jr, DeLee JC (1993) The incomplete bicruciate knee injury. A report of two cases. Orthop Rev 22(11):1249–1252
57. O'Connor JJ (1993) Can muscle cocontraction protect knee ligaments after injury or repair? J Bone Joint Surg Br 75(1):41–48
58. O'Connor JJ, Biden EN, Bradley J, Fitzpatrick D, Yopung S, Kerkshaw C, Daniel DM, Goodfellow J (1990) The muscle stabilized knee. In: Daniel DM, Akeson WH, O'Connor JJ (eds) Knee ligaments: structure, function, injury, and repair. Raven, New York
59. Parolie JM, Bergfeld JA (1986) Long-term results of nonoperative treatment of isolated posterior cruciate ligament injuries in the athlete. Am J Sports Med 14(1):35–38
60. Peek RD, Haynes DW (1984) Compartment syndrome as a complication of arthroscopy: a case report and a study of interstitial pressures. Am J Sports Med 12:464–468
61. Prietto MP, Bain JR, Stonebrook SN, Settlage RA (1988) Tensile strength of the human posterior cruciate ligament (PCL). In: Orthopaedic Research Society, 34th annual meeting, Las Vegas. Trans Orthop Res Soc 13:195
62. Racanelli JA, Drez D Jr (1994) Posterior cruciate ligament tibial attachment anatomy and radiographic landmarks for tibial tunnel placement in PCL reconstruction. Arthroscopy 10(5):546–549
63. Race A, Amis AA (1992) Mechanical properties of the two bundles of the human posterior cruciate ligament. Trans Orthop Res Soc 17:124
64. Roth JG, Bray RC, Best TM, Cunning LA, Jacobson RP (1988) Posterior cruciate ligament reconstruction by transfer of the medial gastrocnemius tendon. Am J Sports Med 16:21–28
65. Rubinstein RA Jr, Shelbourne KD, McCarrol Jr, VanMeter CD, Retling AC (1994) The accuracy of the clinical examination in the setting of posterior cruciate ligament injuries. Am J Sports Med 22(4):550–557
66. Rungee JL, Fay MJ, Deberardino TM (1995) Olecranization of the patella. Orthopedics 18(1):27–34
67. Sapega AA, Covey DC (1994) The biomechanics of femoral and tibial posterior cruciate ligament graft placement. Clin Sports Med 13(3):553–559
68. Satku K, Chew CN, Seow H (1984) Posterior cruciate ligament injuries. Acta Orthop Scand 55:26–29
69. Scapinelli R (1963) Studies on the vasculature of the human knee joint. Acta Anat 70(3):26–31
70. Seebacher JR, Inglis AE, Marshall JL, Warren RF (1982) The structure of the posterolateral aspect of the knee. J Bone Joint Surg Am 64:536–541
71. Shelbourne KD, Rubinstein RA Jr (1994) Methodist sports medicine center's experience

with acute and chronic isolated posterior cruciate ligament injuries. Clin Sports Med 13(3):531–543
72. Shenck RC, Grood ES, Noyes FR, Fishman EK (1992) Computerized stress tomography of posterolateral instability of the knee. Am J Sports Med 5(4):202–209
73. Skyhar MJ, Schwartz E, Warren RF, Torzilli P, Ortiz G (1993) The effects of sectioning the PCL and posterolateral complex on the articular contact pressures within the knee. J Bone Joint Surg 75:694–699
74. Small NC (1993) Complications in arthroscopic surgery of the knee and shoulder. Orthopedics 16(9):985–988
75. Smidt GJ (1973) Biomechanical analysis of knee flexion and extension. J Biomech 6:79–92
76. Smith KL, Daniels JL, Arnocczky SP, Dodds JA, Cooper TG, Gottschalk A (1994) Effect of joint position and ligament tension on the MR signal intensity of the cruciate ligaments of the knee. J Magn Reson Imaging 4(6):819–822
77. Southmayd WW, Rubin BD (1980) Reconstruction of the posterior cruciate ligament using the semimembranosus tendon. Clin Orthop 150:196–197
78. Speer KP, Spritzer CE, Bassett FH III, Feagin JA Jr, Garrett WE Jr (1992) Osseous injury associated with acute tears of the anterior cruciate ligament. Am J Sports Med 20:382
79. Sudasna S, Harnsiriwattanagit K (1990) The ligamentous structures of the posterolateral aspect of the knee. Bull Hosp Jt Dis 50:35–40
80. Swenson TM, Harner CD (1995) Knee ligament and meniscal injuries. Current concepts. Orthop Clin North Am 26(3):529–46
81. Thomann YR, Gaetcher A (1994) Dorsal approach for reconstruction of the posterior cruciate ligament. Arch Orthop Trauma Surg 113(3):142–148
82. Thorblad J, Ekstand J, Hamberg P, Gillquist J (1985) Muscle rehabilitation after arthroscopic meniscectomy with or without tourniquet control. Am J Sports Med 13:133–135
83. Torg JS, Barton TM, Pavlov H, Stine R (1989) Natural history of PCL-deficient knee. Clin Orthop 246:208–216
84. Torzilli PA, Veltri DM (1995) Biomechanical analysis of knee motion and stability. In: Nicholas JA, Hershmann EB (eds) The lower extremity and spine in sports medicine. 2nd edn. Mosby, St. Louis, pp 641–673
85. Trickey EL (1980) Injuries to the posterior cruciate ligament: diagnosis and treatment of early injuries and reconstruction of late instability. Clin Orthop 147:76
86. Van Dommelen BA, Fowler PJ (1989) Anatomy of the posterior cruciate ligament: a review. Am J Sports Med 17:24–29
87. Veltry DM, Warren RF (1994) Operative treatment of posterolateral instability of the knee. Clin Sports Med 13(3):615–627
88. Wilk KE (1994) Rehabilitation of isolated and combined posterior cruciate ligament injuries. Clin Sports Med 13(3):649–677
89. Wirth CJ, Jager M (1984) Dynamic double tendon replacement of the posterior cruciate ligament. Am J Sports Med 12:39–43
90. Woo SLY, Hollis JM, Adams DJ et al (1991) Tensile properties of the human femur-anterior cruciate ligament-tibia complex: the effect of specimen age and orientation. Am J Sports Med 19:217–222
91. Woods GW, James SL, Homsy CA, Prewitt, Slocum DB (1985) Synthetics in anterior cruciate ligament reconstruction: a review. Orthop Clin North Am 16:227
92. Zarins B, Fish DN (1995) Knee ligament injury. In: Nicholas JA, Hershmann EB (eds) The lower extremity and spine in sports medicine, 2nd edn. Mosby, New York, pp 825–863

II.8 Polyethylene Fibers as Substitute for Anterior Cruciate Knee Ligament in Goats: A Histological Study

M. Soudry, J. Aunallah, H. Solomon, J.H. Boss, I. Shijrawi, and D.G. Mendes

Introduction

The treatment of an unstable knee resulting from anterior cruciate ligament (ACL) tear is still controversial. Many different procedures for reconstruction have been described in the quest for a solution to this problem. All surgical methods aim at reproducing the properties of the natural ligament as accurately and quickly as possible. The search for the optimal ACL replacement has led researchers in various directions, but the two major groups are those preferring allogenic substitution and those favoring the use of synthetic material. It is well known that autogenic grafts undergo necrosis postoperatively, and that several months are required for revascularization [3, 4]. Allogenic materials are nonviable to start with; moreover they stretch out with time, resulting in loss of stability. A synthetic substitute is expected to take over the mechanical load of the damaged ligament, and the period of immobilization following surgery is remarkably shortened or eliminated. There are a few materials with the mechanical strength to withstand the normal loading of the ACL and the millions of cycles of cyclic loading to which the ACL is subjected. Carbon was expected to be an inducting and conducting material that would transform from a synthetic prosthesis into a biological replacement with natural collagen, but it quickly failed due to its brittleness [7]. Attempts to coat it with gelatin or polylactic acid did not improve its quality, the main drawback being its inability to withstand bending and shear stresses that lead to breakage. Among other materials that have been used, Dacron proved biocompatible but failed at the bone ligament interface due to stress concentration. Other materials approved by the FDA include the Gore-Tex and the Ligament Augmentation Device (LAD).

Gore-Tex is so far the strongest synthetic ligament, made of polytetrafluoroethylene with a tensile strength of 5300 N and a stiffness of 322 N/mm. It is reported to withstand 37 million cycles without reduction in ultimate tensile strength. The LAD prosthesis tensile strength is reported to be 1159 N with stiffness of 36 N/mm and 10% loss in strength following 1 million cycles [6]. Polyethylene, as a biomaterial (ultrahigh molecular weight) is well known in its solid form but not well enough in its fiber structure. In industry the fiber structure has been reported to have good mechanical properties. For this reason polyethylene fibers were been chosen to be tested experimentally as a ligament substitute in goat stifles determine (a) the morphology of the knee joint after their implantation, and (b) the histological response of the knee joint and remote organ and areas after their implantation.

Materials and Methods

As ligament substitute we used high tensile polyethylene fibers of 150 Deniers. These are of elliptic cross-sectionof 15–25 mm in diameter, braided to a thickness of 2 mm diameter (Cendis Medical, Paris). They were sterilized with beta rays and purified through a special desenzymage process. The prosthesis had an average ultimate tensile strength of 1240 N with an average elongation to failure of 16%.

Twenty goats of 40 kg each, and 1 year of age, underwent surgery. General anesthesia was used with endotracheal intubation. We operated on the right stifle and used the left as control. Four knees were operated on the contralateral side as sham procedures.

After excision of the ACL the proximal tibia and the lateral femoral condyle were drilled with a 3.5-mm drill. We aimed to drill in isometric areas as closely as possible and the intrarticular entrances rounded to minimize friction between bone and ligament. The polyethylene ligament was inserted through the tunnels and fixed to bone through staples, and also with interposed solid polyethylene cork between ligament and bone. Closure was performed in layers. The limbs were placed in a bulky supportive cotton dressing. Antibiotics were given 1 h preoperatively, intraoperatively, and 1 day postoperatively. Roentgenograms of the knees were performed after surgery and prior to sacrifice. When the surgical wound healed, the goats were sent back to the farm. They were followed up every 2 weeks for walking mobility and knee stability.

Results

One goat died of an unknown cause. Two goats were noticed to have some limp: one because the knee became infected and the other due to an unexplained weight loss 6 months after surgery. All the remaining 17 goats used their hind limb normally.

Clinical Examination

All goats were found to have a stable knee. However, it was our observation that ACL sacrifice did not destabilize the knee. Roentgenograms on sacrifice did not show a particular pattern of bone reaction.

Morphology

The goats were sacrificed at 3 months, 6 months, and 1 year. At 3 months the ACL was found to be covered with a transparent and vascularized thin layer of soft tissue (Fig. 1). The knee had a normal-looking appearance of synovia and articular cartilage in the femur, tibia, and patella. At 6 months the polyethylene remained intact except in a case in which there was an impingement at the femoral insertion. The synovial membrane and articular cartilage were also found to be macroscopically intact. At 1 year the polyethylene fibers were covered with a thick soft tissue. The articular cartilage and other surrounding tissues were macroscopically sound. In two knees among 17 the polyethylene was found to be torn at the ligament bone junction.

Histology

Processing was performed by the undecalcified technique using the Jung-Polycut microtome of 4-μm thickness. In general, the organization of polyethylene structure showed a stereotypical pattern which was completed within the first 3 months. Thereafter no significant changes took place; no further tissue ingrowth nor thickening of the surrounding fibrous capsule were evident.

The intraarticular portion of the implant (Fig. 2) stained with Von Kossa-McNeal (in which mineralized bone is black and fibrous and granulomatous tissue is violet) was surrounded by a relatively thick layer of fibrous tissue from which oblique fibrous tracks radiate into the central fibrous band, and which is

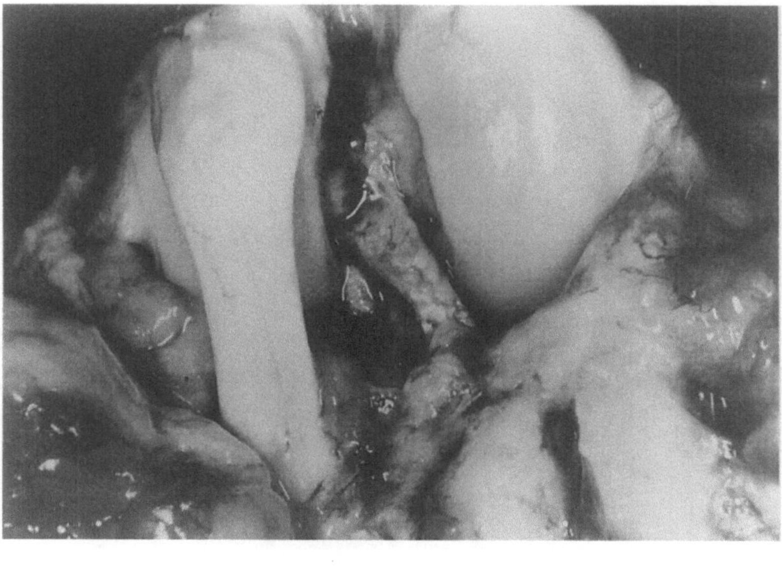

Fig. 1. Goat 201 at sacrifice. The ligament prosthesis is covered with a thin, transparent, vascularized layer of soft tissue at 14 weeks

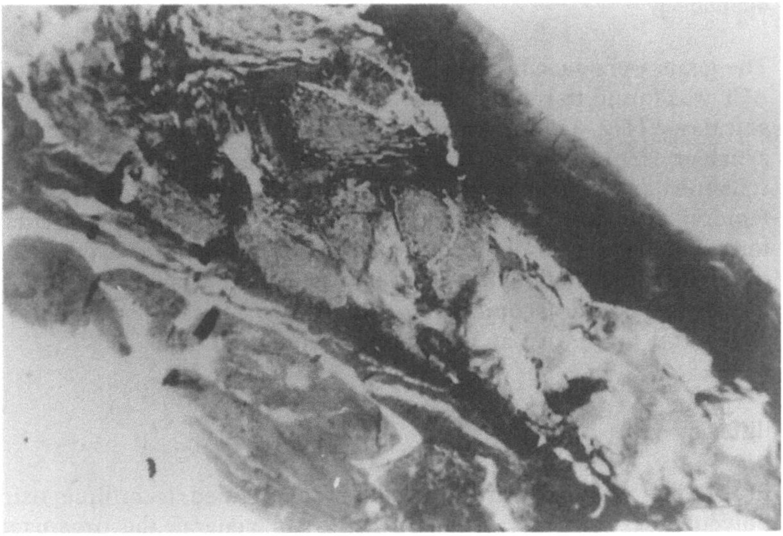

Fig. 2. The intraarticular portion of the ligament prosthesis surrounded by interface membrane intermingled with granulation tissue that is invading in between the bundles

focally lined by hyperplastic synovial cells. The thickness of the capsule is approx. 0.75 mm on each side.

Figure 3 shows the same section, under polarization. The strands of the braid are separated from one another by the oblique fibrous tracks which

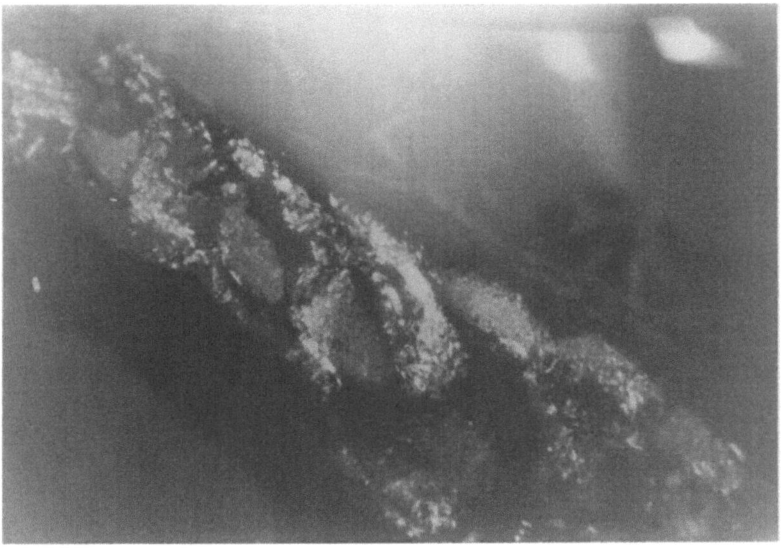

Fig. 3. The same as Fig. 2, under polarized light highlighting the bundles of polyethylene fibers

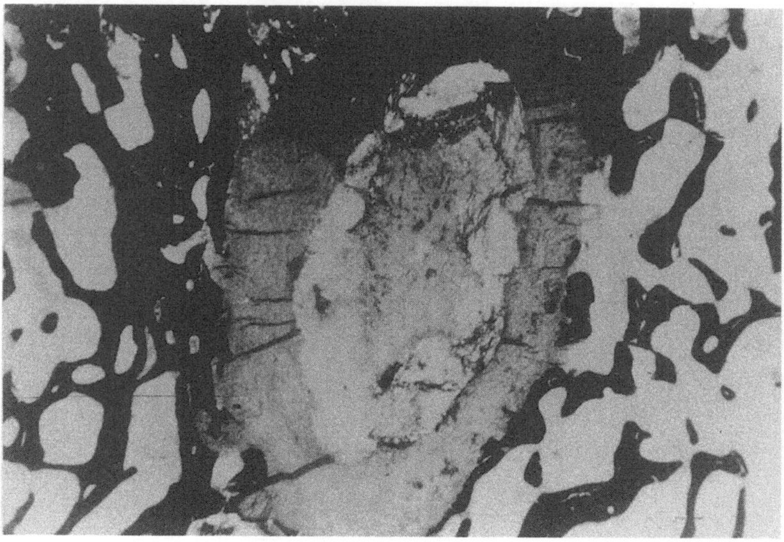

Fig. 4. Transverse section across the body tunnel housing the polyethylene ligament prosthesis

merge to form the central fibrous band. Note that there are no polyethylene breakdown products in the connective tissue. In the transverse intraosseous section of the polyethylene (Fig. 4) we see the same mode of organization as in the intraarticular, with bone surrounding the structure described. Common to both patterns was the formation of a densely textured layer of collagenous tissue in between the polyethylene substitute and the bone. The capsule thickness was almost the same in all goats and equal to the empty space surrounding the prosthesis. The diameter of the bony tunnels increased by about 50% after 3 months of daily activities. This increment did not increase further with time.

There was bone remodeling, as evidenced by the bone formation with increased osteoblastic and osteoclastic activity. The bone was partially sclerotic. Figure 5 presents the same picture under polarized and compensated light showing the fibrous interface membrane, central fibrous band, and oblique fibrous tracks within the ring-shaped polyethylene braid. No polyethylene breakdown products are evident in the interface membrane or in the central fibrous band.

In the area of bone remodeling the outermost aspect of the fibrous membrane merged with each other; the continuity of the small bony trabeculae was interrupted by connective tissue. Surfaces of the trabeculae not abutting the interface membrane was smooth and covered by an osteoid seam and numerous cuboidal osteoblasts. The opposite surface was uneven, and many Howship lacunae as well as osteoclasts are evident. In the area of transition of capsule into prosthesis, there was streaming of fibrous tissue between polyethylene and encircling of individual fibers.

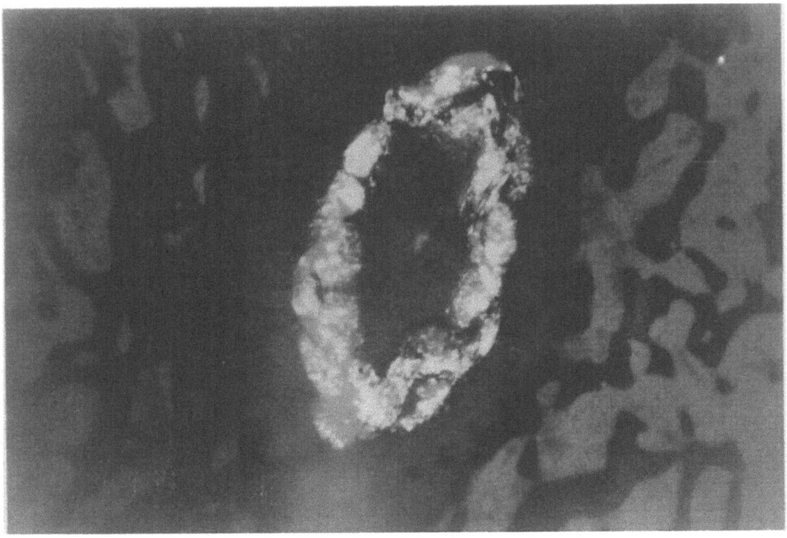

Fig. 5. The same as Fig. 4, under polarized light, highlighting the polyethylene fibers

On the prosthetic aspect of fibrous capsule polyethylene fibrils were surrounded by fibrous tissue with some individual polyethylene fibers engulfed by mono- and polykaryonic macrophages. The synovia showed minimal synovial reactivity, generally not differing significantly from the control knee. Lymphatic glands from ipsilateral inguinal area were examined and found free of polyethylene, as were the organs of the reticuloendothelial system, lungs, liver, and spleen.

Discussion

There are numerous reports in the orthopedic literature on experience with the use of various synthetic ligaments including carbon fiber, Dacron, polytetrafluoroethylene, and polypropylene. The major drawback of carbon fiber has proven in our experimental work to be its fragility on bending and shear stresses[7]. Although osseointegration of the ligament to bone was observed in a number of cases, the continuous inflammatory reaction in the new ligament, the reactive degeneration of the articular cartilage and the soft tissues as a reaction to the wear particles, and the persistent inflammatory reaction in the new ligament itself prevented clinical success of this new material in spite of the stability that was achieved in the knee joint.

Claes et al. [2] showed that carbon fiber and Dacron ligaments cause the formation of granulomas in the fibrous capsule and synovial membrane as a reaction to the wear particles of the material. Lopez et al. [5] described the histological findings following replacement of the ACL with the synthetic Dacron ligament. They noted an inflammatory response in the fibrous capsule encapsulating the prosthetic ligament and granulomatous response to the wear particles of the material and only minor amount of the new collagen fibers. With the passage of time the fibrous tissue looked more mature. McPherson et al. [6] studied the Kennedy LAD device and found poorly organized fibrous tissue around the prosthesis. After 24 months postsurgery they noted ingrowth that anchored the ligament within the tibial and femoral canal.

In our project we studied a synthetic, high-tensile polyethylene which was expected well to withstand loading and friction and to minimize the formation of wear particles that cause a granulomatous inflammatory response. The morphological findings were indeed that the polyethylene fiber ligament wear particles are rare, causing a few localized granulomas, in contrast to the severe granulomatous response to wear particles of bulk polyethylene joint prosthesis. The synovial lining of the knees that received the implant showed no granulomatous response. Rather, there was only a very mild nonspecific reaction with normal articular cartilage and menisci in most of the goats. The fact that the ligament prosthesis was enveloped by fibrous tissue and not by granulation tissue emphasizes the inertness of the material. This confirms the biocompatibility and biofunctionality of prosthetic ligaments made of polyethylene fibers.

All the synthetic ligaments that we implanted in goat stifles had a fibrous anchorage in the tiny tunnel through a fibrous capsule which enveloped the ligament anchored by Sharpey-like fibers to the new cortical bone of the walls of the tunnel. The fibrous capsule created a composite with the prosthetic ligament by the invasion of collagen in between the bundles of the polyethylene fibers forming a specific kind of configuration.

Conclusion

In conclusion, our experimental model in goats shows that:

- When impingement was not present between bone and ligament, gross morphology did not alter.
- Implantation of the polyethylene in the form of fibrils did not exert significant untoward effects on the soft and hard tissues of the knee joint. The extent and intensity of the body's granulomatous response to polyethylene was localized, scant, and nondestructive.
- There was negligible detritic material, which means that there was no evidence of fragmentation of polyethylene fibers.
- In this model the polyethylene substitute was not osseointegrated within the intraosseous channel. Fixation was limited to inner periphery of capsule and connective tissue bands from it in between the bundles of polyethylene. It seems that the initial discrepancy between ligament diameter and oversized canal led to increased resorption.

References

1. Bolton CW, Bruchman WC (1985) The Goretex expanded tetrafluoroethylene prosthetic ligament. Clin Orthop 196:186–195
2. Claes L, Burri C, Neugebauer R, Piehler J, Mohr W (1983) Animal experiments for comparison of various alloplastic materials in ligament replacement. In: Burri C, Claes L (eds) Alloplastic ligament replacement. Aktuelle Probleme der Chirurgie und Orthopedie, vol 26. Huber, Berne, p 101
3. Frank C, Amiel D, Woo SL-Y, Akeson W (1985) Normal ligament properties and ligament healing. Clin Orthop 196:15–25
4. Friedman MJ, Sherman OH, Fox JM, Del Pizzo W, Snyder SJ, Ferkel RJ (1985) Autogenic anterior cruciate ligament (ACL). Anterior reconstruction of the knee. Clin Orthop 196:9–14
5. Lopez-Vazquez E, Juan JA, Vila E, Debon J (1991) Reconstruction of the anterior cruciate ligament with a Dacron prosthesis. J Bone Joint Surg Am 73:1294–1300
6. McPherson GK, Mendenhall HV, Gibbons DF, Plenk H, Rottman W, Sanford JB, Kennedey JC, Roth JH (1985) Experimental mechanical and histological evaluation of the Kennedy ligament augmentation device. Clin Orthop 196:186–195
7. Mendes DG, Iusim M, Angel D, Rotem A, Grishkan A, Mordochvitch D, Boss Y (1985) Histological pattern of biomechanic properties of carbon fiber augmented ligament tendon. A laboratory and clinical study. Clin Orthop 196:51–60

II.9 Motion-Induced Inhibition of Effective Osseous Fixation of a Polyethylene Filamentous Prosthesis: Histological, Histomorphometrical, and Biomechanical Studies of a High-Tensile Polyethylene Filamentous Braid for Replacement of the Anterior Cruciate Ligament in Goats

J.H. Boss, I. Shajrawi, M. Soudry, D.G. Mendes, R. Schäferh, U. Soltesz, and D. Ulrich

Introduction

Much has been written about the treatment of the anterior cruciate ligament (ACL)-deficient unstable knee. The necessity of reconstructing or replacing the injured or lost ACL is hotly debated. Most surgeons opting for the operative approach choose to implant a patellar tendon (or other collagenous structures) autograft to replace the native ligament. However, this weakens the remaining patellar ligament; furthermore, the simple construct of a grafted tendon or fascia does not remodel into the anatomically complicated and functionally demanding configuration characteristic of the natural ACL [1]. Still, tendon autografting will remain the treatment of choice in the foreseeable future because the clinical results of ACL replacements with synthetic filamentous prostheses are extremely disappointing. Despite discouraging clinical and experimental outcomes, investigators have not given up the search for an optimal ligament replacement by artificial devices.

Synthetic filamentous prostheses serve either as scaffolds for ingrowth of collagenous tissue (formation of a so-called "neoligament") or as permanent replacements. Prerequisites for the success of procedures aimed at permanently replacing the ACL with synthetic filaments include secure anchorage to the osseous bed of the femoral and tibial tunnels and prevention of mechanical attrition of the implants. The histological and histomorphometrical patterns of the anchorage and tissular organization of a high-tensile polyethylene filamentous braid (HTPEFB) in the stifle of the goat have been reported previously [8, 17]. The HTPEFB prosthesis becomes surrounded within the intraosseous tunnels by thick fibrous interfacial membranes which adversely affect the biomechanics of the ACL replacement. In addition, while the HTPEFBs themselves become poorly organized by ingrowing collagenous tissue, they are excessively separated from their osseous bed at the femoral and tibial exits by fibrous tissue. Experiments carried out at our laboratories show interfacial movements upon loading of the implanted HTPEFBs to be primarily responsible for the inhibition of adequate tissular organization.

Materials and Methods

The HTPEFB (marketed by Cendis Medical, France) consists of polyethylene filaments braided to a thickness of 2 mm. The individual filaments are of 150 Denier. They are elliptical in cross-section and 15–25 µm in diameter. The average ultimate tensile strength of the braids is 1240 N. The average elongation to failure is 16%, and the average loss of strength is 57% following 10^7 cycles of loading at 4 N. Following subcutaneous implantation for 3 months, the braid's ultimate tensile strength and elongation to failure are essentially unchanged; they withstand 9×10^6 cycles of loading up to 8 N prior to failure.

We excised the right ACL of 1-year-old goats, weighing about 40 kg. The HTPEFB was inserted through 3.5-mm-wide tunnels drilled in the femur and tibia at isometric points of entry. The braid was fixed to the femoral and tibial cortices with polyethylene corks and two staples. On the third postoperative day the goats were transferred to a farm where they were daily put out to graze in a pasture. They were examined every 2 weeks until killed after 3, 6, or 12 months. At the time of sacrifice the braids along with the surrounding hard and soft tissues were excised. The samples were fixed in formalin for histological evaluation and kept frozen until used for biomechanical testing. Two control groups were studied to compare the findings of ACL replacements with those obtained after implantation of the braids in nonloaded intraosseous and extraosseous sites. In one control group, HTPEFBs were implanted into 3.5-mm-wide tunnels drilled horizontally through the cortex of the distal femoral metaphysis. In the other control group, HTPEFBs were inserted into a subcutaneous pouch. In both control groups the braids were retrieved after 3 months.

The formalin-fixed, undecalcified specimens were cut alternately in the transversal and longitudinal directions of the braids and embedded in methylmethacrylate. Following polymerization at room temperature, the polymethylmethacrylate blocks were sliced on a Buehler low-speed saw. Sections were cut at 4 µm on a Reichert-Jung Polycut microtome and stained by the von Kossa (with McNeal's counterstain), Goldner, and Movat techniques [16].

In addition to the routine microscopical observations, the relationships of the braids to the surrounding hard or soft tissues as well as the connective tissue ingrowth and the ensuing granulomatous response were quantified. Using the Cue-2 image analysis system (Olympus, Lake Success, N.Y., USA) and the C-2 Morphometry Software version 1.7.87, the "volumes" of the fibrous, and granulomatous tissues and that of the polyethylene filaments were determined by calculating the mean of ten measurements obtained in a manner similar to that applied in the quantification of bone volume [3]. Special attention was paid to the stages of image enhancement, segmentation and thresholding, and "removal" of small objects so as to obtain a processed picture as close as possible to the original microscopical appearance.

Results

The organizational pattern and anchorage mode of the HTPEFBs as well as the histomorphometrical findings were essentially identical in the goats killed 3, 6, or 12 months following the ACL replacements. The results are therefore described without further reference to the postoperative intervals. The femoral and tibial tunnels were enlarged by 29%–100%, measuring 4.5–7.0 mm in diameter. The braids were separated from the inner aspects of the bony tunnels by thick, bland, fibrous interfacial membranes, averaging 1.2 mm in width. The longitudinally oriented collagen fibers of the interfacial membranes were attached via Sharpey-like collagen fibers to a ring-shaped "neocortex." This ring-shaped bony shell averaged 0.6 mm in width (range 0.3–1.0 mm), and its bone volume was 80%–90%. The abundant osteoclasts at the inner aspect of the neocortex and the numerous osteoblasts abutting on broad osteoid seams at the outer aspect of the neocortex evidenced ongoing remodeling. The osteoclastic activity accounted for the expansion of the tunnels. The fibrous tissue of the interfacial membranes merged with a circumferential transitional zone composed of giant celled granulomatous tissue with some admixed fibroblasts and collagen fibers (Fig. 1). The transitional zone averaged 0.5 mm in width. The polyethylene filaments occupied 10%–20% of the volume of the transitional zone. Tissular ingrowth into the bulk of the implants was poor, the volume

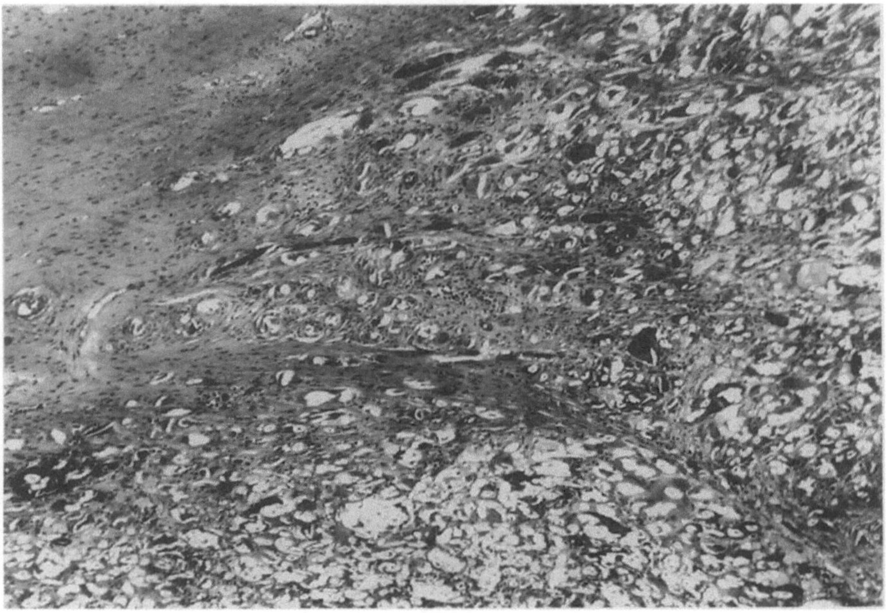

Fig 1. The fibrous tissue of the interfacial membrane merges with the circumferential transitional zone consisting of giant celled granulomatous tissue with some admixed fibroblasts and collagen fibers. The thin, oblique, fibrous tract emerging from the interfacial membrane separates two polyethylene filamentous bundles from one another. Von Kossa stain, ×30

occupied by fibroblasts, collagen fibers, and mono- and polykaryonic macrophages ranging around the value of 10%. Relatively large segments of the braids were unorganized, being devoid of tissue for as long as 12 months postoperatively, and the individual polyethylene filaments were separated from each other by acellular, proteinaceous precipitates. The interfacial membrane of the braids was linked via several thin, oblique, fibrous tracts (Fig. 1) with a rather thick, longitudinally oriented, central, fibrous band, averaging 0.7 mm in diameter (Fig. 2).

The central bands consisted of several intertwining bundles of densely packed collagen fibers, which were oriented mainly in the longitudinal axis of the implant. On average, the polyethylene filaments occupied 30% and 37% of the volume of the oblique tracts and central bands, respectively. A granulomatous transitional zone was also interposed in between the central band and the bulk of the prosthesis (Fig. 3). Discrete granulomas induced by polyethylene wear particlets [7] were extremely rare within the interfacial membranes and the fibrous tissue at the exits of the braids from the intraosseous tunnels.

The organizational pattern of the HTPEFBs within the joint cavity was similar to that in the intraosseous tunnels. The braids were invested by a thick, longitudinally oriented, fibrous coat, which was covered by a synovial-like

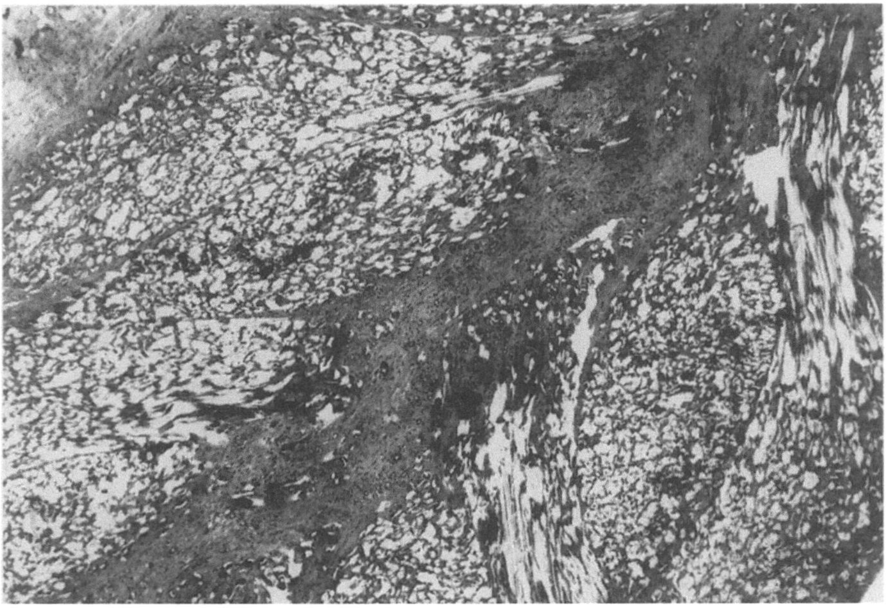

Fig 2. Longitudinal section of an HTPEFB within the femoral tunnel. There is scanty tissular ingrowth into the braid's bundles, the polyethylene filaments of which are separated from each other mainly by a proteinaceous precipitate. Thin, oblique, fibrous tracts compartmentalize the braid and link with the rather thick, central, fibrous band, which is oriented in the direction of the braid. Von Kossa stain, ×30

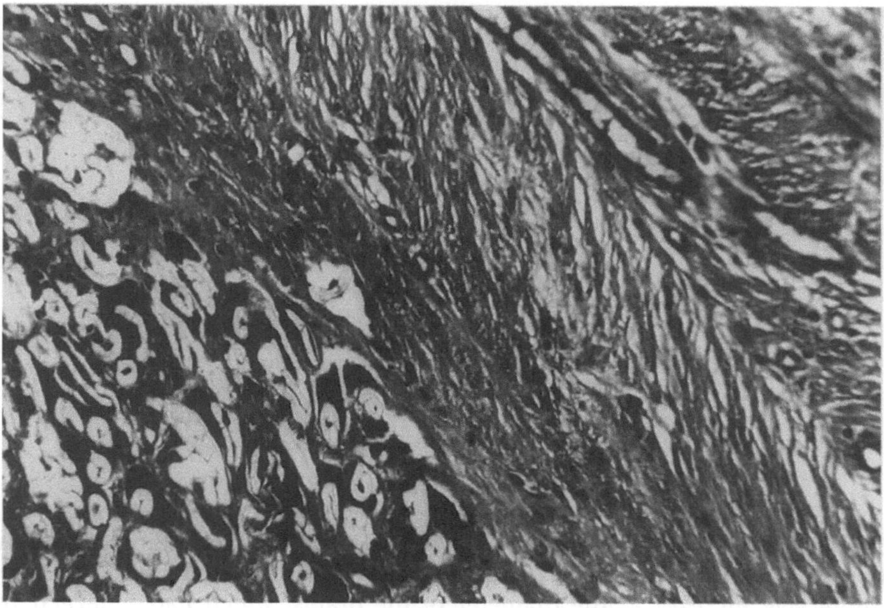

Fig 3. The transitional zone interposed between the densely textured collagenous tissue of the central fibrous band and the bulk of the prosthesis. The fibrous tissue of the band merges with the transitional zone. The mono- and polykaryonic macrophages of the latter are admixed with some fibroblasts and collagen fibers which separate individual or small groups of polyethylene filaments from one another. Von Kossa stain, ×200

layer. These coats, which averaged 1.5 mm in width, corresponded spatially to and were in continuity with the interfacial membranes of the intraosseous portions of the braids. Otherwise, the qualitative and quantitative findings in the intraosseous and intraarticular portions of the prostheses were identical.

The organizational pattern of the HTPEFBs replacing the goats' ACL qualitatively resembled that of braids implanted into either the femoral metaphyseal cortex or inserted into a subcutaneous pouch. However, there were substantial quantitative differences in the host response to the functionally loaded ACL replacement, on the one hand, and the nonloaded intracortical and subcutaneous implants, on the other. The histograms in Figs. 4–6 summarize the results of the histomorphometrical evaluations.

The HTPEFBs which had been horizontally implanted into the distal metaphysis of the femur were enclosed within thin, fibrous interfacial membranes averaging 0.4 mm in width. The osseous tunnels were expanded by 37%–100%, measuring 4.8–7.0 mm in diameter. Neither the inner and outer granulomatous transitional zones nor the oblique fibrous tracts differed from those observed in the ACL replacements. There was ample ingrowth of predominantly fibrous tissue into the bulk of the braids, the volume occupied by the ingrowing tissue averaging 34%. The central fibrous bands were well developed, their diameter averaging 1.2 mm. The polyethylene filaments occupied on average

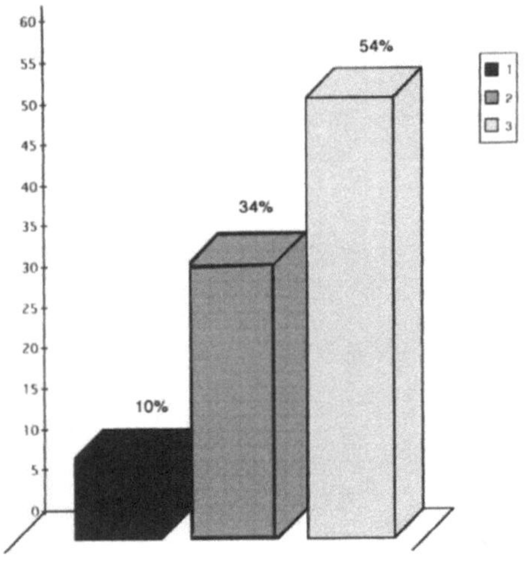

Fig 4. Comparison of mean percentage of tissular ingrowth into the bulk of loaded HTPEFBs replacing the ACL (*1*), nonloaded HTPEFBs implanted into the metaphysis (*2*) and nonloaded HTPEFBs inserted into a subcutaneous pouch (*3*)

7% of the volume of the central fibrous bands. The abundant ingrowth of fibrous tissue into the braids, and the thick central fibrous bands caused an impressive increase in the diameter of the braids, which had enlarged from their original 2 mm to 4.5–6.5 mm. Contrasting with the purely fibrous nature of the central bands of the implants replacing the goats' ACLs, some irregularly shaped and variously sized, anastomosing osseous trabeculae occasionally crisscrossed within the fibrous tissue of the central bands of the transcortically implanted braids.

Thin fibrous capsules, averaging less than 0.2 mm (range, 0.10–0.23 mm), enclosed the subcutaneously inserted HTPEFBs. The capsules were composed of few concentrically arranged fibrocytes and collagen fibers. As with the

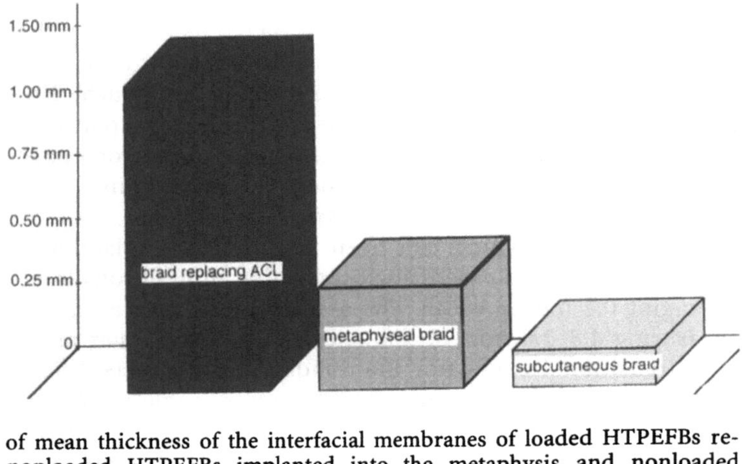

Fig 5. Comparison of mean thickness of the interfacial membranes of loaded HTPEFBs replacing the ACL, nonloaded HTPEFBs implanted into the metaphysis and nonloaded HTPEFBs inserted into a subcutaneous pouch

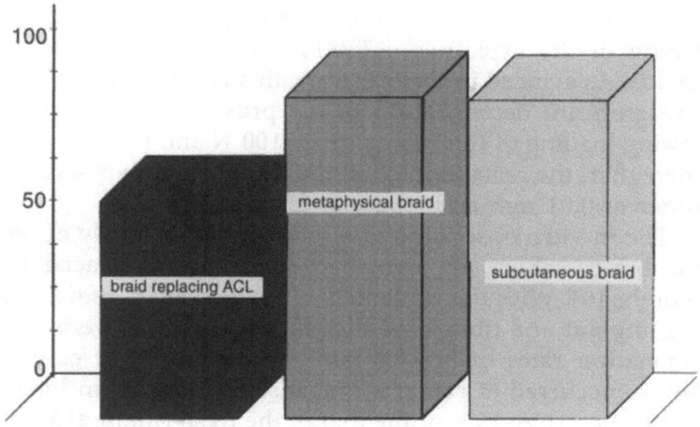

Fig 6. Comparison of the averaged percentage of fibrous tissue within the central fibrous bands of HTPEFBs replacing the ACL, nonloaded HTPEFBs implanted into the metaphysis and nonloaded HTPEFBs inserted into a subcutaneous pouch

nonloaded braids implanted into the femoral metaphyses, the subcutaneous implants were also greatly expanded, averaging 5.5 mm in diameter (range, 4.0–6.3 mm). No transitional zone had developed at the braid-capsule junctions. Some macrophages and many fibroblasts, producing abundant collagen fibers, had invaded the bulk of the HTPEFBs and separated individual or small groups of filaments from one another, such that on average the tissue occupied 54% (range, 38–82%) of the implants' cross section. Well-developed oblique fibrous tracts, compartmentalizing the braids, were attached to the central fibrous bands. In at least some samples, the diameter of the central fibrous bands could not be measured precisely because the extensive tissular ingrowth would have required an arbitrary delineation of the boundary between the fibrous tissue within the braids and the central fibrous bands. In four specimens in which the boundary was more or less clearly demarcated, the diameter of central fibrous bands averaged 1.1 mm. The polyethylene filaments occupied an average of 8% of the volume of the central fibrous bands.

Biomechanical evaluations were carried out on the stifles of goats of non-operated (control) goats and the stifles of animals with a braid-replaced ACL. The ramp setup at loads in the range of F_{max}=500 N and F_{max}=5 kN was used for the relaxation and tear tests (at 45° of flexion), respectively. The goats' natural ACLs were stiffer than the polymeric prostheses and displayed no creep at maximum loads. In addition, the hysteresis of the natural ACL was considerably lower than that of the braids. In the tear tests (under controlled displacement of 0.6 mm/s) the load increased without load drop up to at least 540 N (range, 540–1120 N), where damage to the braids was first noticed. The corresponding displacements ranged from 3.0–5.2 mm. Maximum force was reached after a further load increase (F_{max}=1100–1360 N). All the measured maximum loads exceeded the value of 1.1 N at displacements ranging from 5.1 to 8.0 mm.

The load-elongation curves (up to 50 N) disclosed quite significant elongation of the explanted HTPEFBs when compared with the fresh (unused) braids. As evinced in the tear tests, this decrease in the braids' stiffness was due to significant deterioration of the prostheses during the service period. Following loading of the braids to 50–100 N and keeping the load constant at 5 N thereafter, the relaxation of all the explanted braids was exceedingly low (in the order of 0.01 mm after the first few seconds).

The in vitro observations disclosed two distinctly different failure modes of the braids which, in vivo, had successfully replaced the goats' ACL for 6 months [9]. With the explanted stifles held at 45° of flexion and the load axis of the implant and tibia kept aligned, the tear tests were maintained to failure at elongation rates of 0.3–0.6 mm. In a minority of cases, so-called prosthetic failure occurred at rather low loads. The maximum load achieved (F_{max}) was 150 N, i.e., 10%–15% of the that of the fresh (unused) HTPEFBs. In most cases, so-called fixation failure expressed loss of the implants' anchorage in the femoral or tibial tunnels at loads which were considerably smaller than those withstood by the explanted stifles. During the tests, the implants were pulled out of their tunnels. The maximum loads achieved were 140–520 N and the corresponding displacements were 11–30 mm. There was no significant difference in the fixation strength of the braids in the femoral and tibial tunnels (140–520 N and 180–500 N, respectively). Following the loss of anchorage, the implants were removed from the stifles, fixed with two grips, and retested. The failure loads of the withdrawn braids ranged from 530 to 1250 N.

At the completion of the in vitro tests, the fibrous tissue investing the braids was peeled away. Most but not all samples exhibited many torn polyethylene filaments close to the braids' exit from the femoral tunnel, corresponding anatomically to the site at which the braids rubbed against the intercondylar notch during locomotion. Despite the wear and tear of the polyethylene filaments during service, neither detritic synovitis nor a prosthetic debris-induced granulomatous response in the fibrous tissue at the exit from the femoral tunnel were detected.

Discussion

The clinical results of the replacement of the goats' ACL by a HTPEFB prosthesis were favorable. There was no overt change in the animals' behavior. Clinically the stifles appeared to be stable for the 3- to 12-month observation period [9]. However, both the morphological and biomechanical evaluations of the retrieved specimens revealed unsatisfactory results. First, tissue had hardly grown into the bulk of the retrieved braids. Second, the implants were separated from their bony beds by thick, fibrous interfacial membranes. Third, the majority of the implants revealed in vitro insufficient strength of fixation to the bone, demonstrating their weak anchorage within the intraosseous tunnels. In addition, after peeling away the investing fibrous tissue, wear and tear of the polyethylene filaments were detected at the part corresponding to the braids' contact with the bone of the intercondylar notch.

Polyethylene in bulk form is a biocompatible material [5]. Mechanical causes rather than compositional inadequacy, therefore, plausibly account for the inefficacy of the ACL replacement by HTPEFB. There is no evidence of wear debris-induced granulomatous reactions [9] within the bone-braid interfacial membrane or synovium because the polyethylene filaments virtually do not break down into small particles. However, in the wake of enhanced osteoclastic activity at the innermost aspect of the newly formed neocortex at the periphery of the interfacial membrane, the femoral and tibial tunnels expand to a disastrous degree. Fibrous tissue apparently occupies the evolving gap incrementally at the implant-bone junction. In the absence of a "small-particles disease" [10], which often frustrates orthopedic surgeons concerned with artificial joints, the recruitment and stimulation of the osteoclasts cannot be attributed to bone-resorbing mediatory substances synthesized by macrophages ingesting polymeric wear debris.

Fibrous tissue luxuriantly grew into the bulk of the subcutaneously inserted HTPEFBs and a very thin, fibrous capsule surrounded the implants. Fibrous tissue grew fairly well into the bulk of the braids implanted into the distal femoral metaphyseal cortex. At this site, the implants were enclosed within relatively thin fibrous interfacial membranes. It is reasonable to assume that the braids were not burdened functionally by tension at the metaphyseal locale, although they were perhaps subject to some compressive forces. On the contrary, the functionally loaded HTPEFBs replacing the goats' ACL remained practically unorganized by tissular ingrowth for as long as 1 year postoperatively. In addition, they were separated from their bony beds by a remarkably thick fibrous interfacial membrane, the breadth of which comprised as much as one-third of the diameter of the expanded intraosseous tunnels. Common to the braids replacing the ACL implanted into the metaphyseal cortex or inserted into the subcutis was the formation of the oblique fibrous tracts and central fibrous bands, the dimensions of which did not vary significantly in the three locations. In the nonloaded situation, fibroblasts grew amply in between the individual filaments, between the several bundles constituting the braid, and into the formerly empty center of the braid. Hence, the braided construct appeared to be suitably conducive to creeping fibroblastic substitution of the interstitial fluid postoperatively accumulating in the interstices of the filamentous implant. This being the case, braided designs may be the optimal architecture of devices planned to replace tendons or ligaments. The ample, longitudinally oriented collagen fibers of the central fibrous bands are likely to carry a critical part of the tension forces.

The loaded HTPEFBs replacing the goats' ACL were attached to and separated from the bony shells of the femoral and tibial tunnels by thick interfacial membranes. Since the braids were probably not osseointegrated during the initial healing phase and were not attached to the bone by an enthesislike junction, the effective length of the implants is the distance between the corks affixing the braids to the femoral cortex and the tibial cortex. The relatively long implant acts during locomotion as both a string and a spiral such that interfacial movements are all but an inevitable consequence. The goats were active, and interfacial motion was thus operative as of the immediate postimplantation period; as a result there was no motionless period during which the bone could have adapted to the implant. The adverse effects of interfacial movements on the efficacious integration of intraosseous devices are well known [11]. The same laws evidently govern bone-implant relationships irrespective of differences in the chemical composition of diverse devices. When replacing patients' ACL with the irritative and granuloma-inducing carbon filamentous prosthesis, osseointegration of the occasional implant does take place under the best of circumstances, provided a stable interfacial situation prevails [6]. The interfacial movements inhibit new bone formation and also stimulate macrophages. The mediatory substances of macrophagic sources that activate bone resorption and stimulate fibroblastic proliferation and collagen synthesis include leukotrienes, prostaglandins, interleukins, osteoclast-activating factor, tumor necrosis factor, collagenase, elastase and growth factors [2, 4, 12, 14, 18]. Macrophages abound in the transitional zone of the ACL replacements. Motion-induced macrophagic activities conceivably participate in the bone resorption and fibrous tissue formation that is evident histologically within the femoral and tibial tunnels.

The morphologically ascertained failure of the tissular organization and anchorage of the braids within the intraosseous tunnels readily explains the results of the biomechanical tests. The low in situ fixation strength is due to inadequate attachment of the implants to the bone within the tunnels. This low fixation strength, corresponding to fixation failure at forces of 140–180 N, has been found to be the limiting factor of the majority of ACL replacements. It compares unfavorably with the fixation strength of the goats' native ACLs, which tolerate loads in excess of 1 kN.

The enthesis constitutes the anatomically and functionally effective load-transferring mechanism at the normal ligament-bone junction. As with other synthetic filamentous prostheses, and in contrast to collagen fiber devices [15], an enthesis is not reestablished following replacement of the goats' ACL with the HTPEFB. In contrast to an enthesis, a fibrous interfacial membrane separates the bone from the polymeric implant. It is not surprising, therefore, that the artificial ligaments of the explanted stifles fail in the in vitro tests more often at their bony insertion site than within their midsubstance.

Until bioengineered artificial ligaments are available, the significance of a motion-free interfacial situation prevailing during the initial postoperative period cannot be overemphasized. Whether reconstituted collagen fiber–synthetic filamentous composites or growth factors-releasing devices allow for the formation of enthesis-like junctions attaching implants to the bone is a moot question. The HTPEFB has obvious advantages over carbon filamentous implants, which have all but been banished from everyday orthopedic practice. The prohibitive brittleness of the carbon filaments in bending is responsible for the unacceptably severe and painful detritic synovitis which patients commonly report to their surgeons. Polyethylene debris-induced granulomas do not develop in the synovial membrane following replacement of the goats' ACL with the HTPEFB prosthesis because there is no intraarticular dissemination of small breakdown particles. Carbon filaments elicit an ongoing inflammatory and granulomatous response which inhibits the formation of mature collagenous tissue such that a decade or so postoperatively the fibrous tissue within carbon braids or tows is of the nature of young scar tissue [13]. As far as we know, the HTPEFB implant is the only device conducive to the formation of oblique fibrous tracts and a central fibrous band which are likely at least partially to carry the loads exerted on the artificial ligament. On the other hand, the absence of secure bony fixation because of the interposition of the thick fibrous interfacial membrane makes the HTPEFB unsuitable for ACL replacement in clinical practice. The avoidance of interfacial movements during the early healing phases and engineered formation of an enthesis-type bone-implant junction are the outstanding and challenging goals of research in implantology of synthetic filamentous ligaments.

References

1. Amis AA, Dawkins GPC (1991) Functional anatomy of the anterior cruciate ligament. Fibre bundle actions related to ligament replacements and injuries. J Bone Joint Surg Br 73:260–267
2. Amstutz HC, Campbell, P, Kossovsky N, Clarke IA (1992) Mechanisms and clinical significance of wear debris-induced osteolysis. Clin Orthop 276:7–18
3. Bar Shira-Myamon B, Coleman R, Cohen A, Steinhagen-Thiessen A, Silbermann M (1989) Age-related bone loss in the lumbar vertebrae of CW-1 female mice. Calcif Tissue Int 44:36–45
4. Bonfield TL, Colton E, Marchant RE, Anderson JM (1992) Cytokine and growth factor production by monocytes/macrophages on protein preabsorbed polymers. J Biomed Mater Res 26:837–850
5. Boss JH, Shajrawi I, Aunullah J, Mendes DG (1995) The relativity of biocompatibility. A critique of the concept of biocompatibility. Isr J Med Sci 31:203–209
6. Boss JH, Shajrawi I, Mendes DG (1991) Intraosseous anchorage of the carbon fiber composite ligament by bony ingrowth. J Appl Biomater 2:241–242
7. Boss JH, Shajrawi I, Mendes DG (1993) Histological patterns of the tissue reaction to polymers in orthopaedic surgery. In: Cohn D, Kost J (eds) Biomedical polymers. Elsevier, London, pp 11–17
8. Boss JH, Shajrawi I, Mendes DG (1994) The nature of bone-implant interface: the lessons learned from implant retrieval and analysis in man and experimental animal. Med Prog Technol 20:119–142
9. Boss JH, Shajrawi I, Soudry M, Anullah J, Solomon H, Mendes DG (1994) Studies on a novel anterior cruciate ligament polyethylene fiber prosthesis: the histomorphological pattern of organization and bony anchorage of a polyethylene fiber prosthesis in the stifle of the goat. Clin Mater 15:61–67
10. Dorr LD, Bloebaum R, Emmanuel J, Meldrum J (1990) Histological, biochemical and ion analysis of tissue and fluids retrieved during total hip arthroplasty. Clin Orthop 261:82–95
11. Goodman S, Aspenberg P (1992) Effect of amplitude of micromotion on bone ingrowth into titanium chambers implanted in the rabbit tibia. Biomaterials 13:944–948
12. Maquire JK Jr, Coscia MF, Lynch MH (1987) Foreign body reaction to polymeric debris following total hip arthroplasty. Clin Orthop 216:213–233
13. Mendes DG, Soudry M, Levine E, Silbermann M, Shoshan S, Gross U, Shajrawi I, Boss JH (1992) The collagen composition of the carbon fiber composite ligament: an electronmicroscopical, biochemical and immunohistological study. J Long-Term Effects Med Impl 1:305–319
14. Olson EJ, Kang JD, Fu FH, Georgescu HI, Mason GC, Evans CH (1988) The biochemical and histological effects of artificial ligament wear particles: in vitro and in vivo studies. Am J Sports Med 16:558–570

15. Panni AS, Denti M, Franzese S, Monteleone M (1993) The bone-ligament junction: a comparison between biological and artificial ACL reconstruction. Knee Surg Sports Traumatol Arthroscopy 1:9–12
16. Schenck RK, Olah AJ, Hermann W (1984) Preparation of calcified tissues for light microscopy. In: Dickson GR (ed) Methods of calcified tissue preparation. Elsevier, London, pp 184–214
17. Shajrawi I, Aunullah J, Soudry M, Solomon H, Mendes DG, Boss JH (1993) Quantification of the tissue response to a polyethylene prosthesis of the ACL in the goat: a histomorphometric study. Orthop Int Edn 1:455–460
18. Spilizewski KL, Marchant RE, Anderson JM, Hiltner A (1987) In vivo leukocyte interactions with the NHLBI-DTB primary reference materials: polyethylene and silica-free polydimethylsiloxane. Biomaterials 8:12–17

Part III
Future Trends: Tissue Engineering, Regeneration

III.1 Artificial Materials and Structures for Extracellular Matrix Scaffolding of Orthopedic Tissues: A Review

C. Chaput, C.H. Rivard, A. Selmani, and L'H. Yahia

Introduction

Tissue-engineering materials and approaches have recently been proposed as the future direction for tissue and organ regeneration. Ideal surgical reconstructions would replace biological tissues like autogenous tissues. In many cases autogenous reconstructions are limited by the nonoptimal availability of usable autogenous tissues or by donor site morbidity. Allografts are susceptible to increasing risks of infection and to transmission by infectious diseases or agents. Alloplastic materials (synthetic implants) offer no potential for direct infectious agents transmission but increase the susceptibility to indirect infections and inadequate long-term immunological interactions with the host and in some cases are plagued with clear failures, rejections, and uncertain long-term functioning. As a consequence, alternative materials and therapies have been investigated with the hope of benefiting from the advantages of living biological tissues or from those of artificial materials and structures.

A New Step in the Biomaterials Science

This corresponds to the more recent developments in the gradual evolution of biomedical materials for surgical implants and devices. Biomaterials science can be divided into four main phases:

- Industrial materials in ad hoc devices provided few benefits to medicine and surgery.
- "Inert" industrial materials in bioengineered devices were proposed to replace injured tissues (e.g., Dacron vascular grafts, Charnley hip prosthesis).
- Bioengineered materials were obtained, using bioengineered designs, and "bioactive" devices were proposed to replace "inert" ones (e.g., hydroxyapatite coatings, collagen-based implants).
- The current developments of "tissue-engineered" implants propose the synergy between artificial and biological materials to create "biohybrid" devices, stimulate tissue regenerations rather than tissue replacements, and promote the controlled and natural delivery of bioactive molecules by using immunoisolated cells. Multidisciplinary research groups, biotechnical industries, and governmental agencies now target the future clinical applica-

tions and impacts of tissue-engineering technologies: living skin transplantation, bone healing, organ function substitutions, nerve, cartilage, ligament, tendon, blood vessel, and dental soft-tissue regenerations will be achieved in the next decade.

Toward a Controlled and Assisted Tissue Regeneration?

Tissue regeneration is a new strategy for replacing or repairing biological tissues and for organ-function substitutions or aids [8, 76, 121]. Therapeutic needs are real, for example, for the dermal tissue reconstruction and the regeneration of meniscus, cartilaginous organs, bone and articular cartilage, vessels, trachea, abdominothoracic tissues, and ligaments and tendons. The engineering of biological tissues is not limited to biological tissue repair but also extends in particular to the cellular transplantation and to the sick or deficient organs: diseases of the liver [36, 37, 49, 85, 119, 169] and pancreas [38], diabetes, Parkinson's disease [30, 65], and deficiencies in neurotransmitters [1, 2, 72, 77, 80, 89, 109, 111, 123, 139, 144, 174–176, 178, 179, 181, 185].

Human cells can be retransplanted immediately or used in contributing to recreate tissue equivalents that can be applied in the repair or replacement of diseased or injured biological tissues. Nevertheless, if cells can proliferate in vitro as in vivo and create new tissues, they can construct by themselves and then organize and structure the neotissues required for the repair or replacement in a two or three-dimensional, precise, and desired shape. Cellular and tissue transplants are therefore urgently needed as potential temporary structures to serve as skeletons for the three-dimensional formation of neotissues in vitro or in vivo. Advances in molecular biology, cellular biology, immunology, and biomaterials engineering have led to great progress in tissue engineering and culture. The most promising area of application is no doubt the combination of cellular transplants and biocompatible materials; here cells must be placed in synthetic materials similar in form to the organs or tissues [121]. Acellular matrix constructions do not necessarily mean the induction, direction, or facilitation of the cascade of events during the remodeling in vivo of tissues and organs. A more promising approach is the manufactured scaffold enrichment in vitro by cellular culture in these scaffolds. These cells in culture allow the addition of extracellular components by cellular biosynthesis and thus can create the conditions required for adequate recognition and remodeling in vivo [8, 10, 11].

Techniques for Cellular and Tissue Transplants

Concepts and Definitions

The term "tissue engineering" is relatively new and continues to evolve. A recently proposed definition is [76]: the application of principles, engineering, and scientific methods of life to the comprehending of structure-functional relationships of pathological and normal mammalian tissues, and for developing biological substitutes to reestablish, maintain, or improve tissue functions. The basic hypotheses rest on using living cells alone or living cells and inert or natural extracellular components for biological substitute [76]. This approach is therefore the opposite of that in simple inert, artificial implants. Subcomponents and areas of interest have been proposed to promote a technological development (Symposium on Tissue Engineering, UCLA Symposium, Keystone, Calif., USA, April 1990). Research subjects in biological tissue engineering include the growth of cells in a three-dimensional system and the development of carriers and vehicles for the reintroduction of living cells, the development of therapeutic systems to deliver proteins, growth factors or hormones, new methods of culture for cellular types considered "recalcitrant", and the expression of transgenic proteins for cellular transplants. Additional areas are the limitation of immunogeneicity for transplants and the development of biosensors to understand the cellular behavior during the regeneration processes both in vivo and ex vivo, and of biomarkers to monitor the tissue regeneration.

Liver and pancreas were certainly the preferred organ targets during the materials and methods development for the cellular transplantation [36, 38,

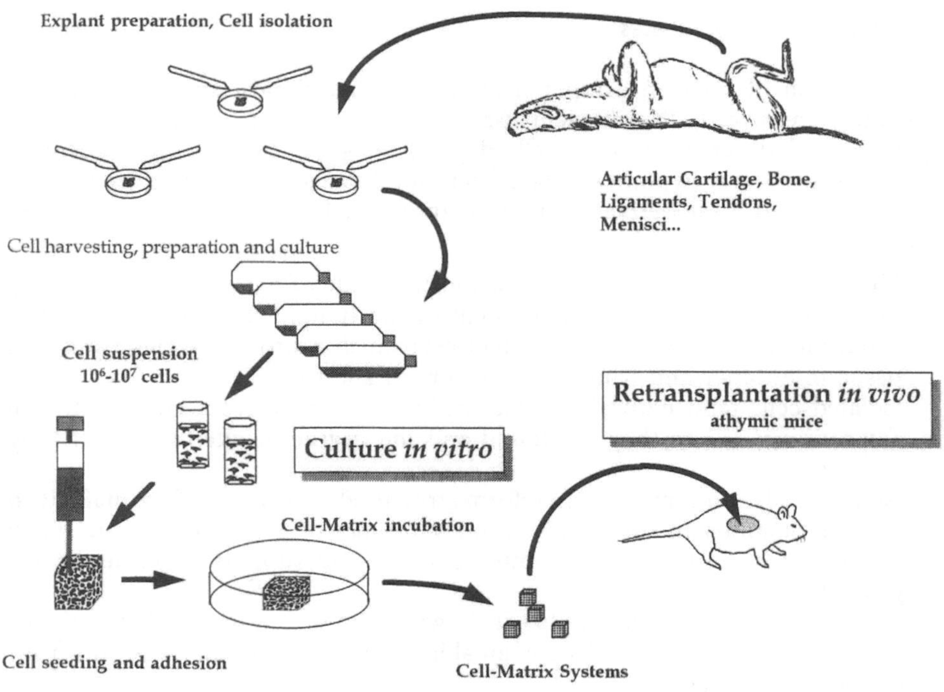

Fig. 1. Tissue-engineering approaches often promote the combination of in vitro cultures of biohybrid tissue equivalents and in vivo retransplantation, remodeling, and regeneration. A scenario illustrating tissue engineering of spinal structures

49]. Liver tissue regenerates very well and cellular transplants can be made quite rapidly from weak quantities of donor tissues. The initial size is regained only in several weeks, and hepatropic growth factors are known and available and can be used. There is a real therapeutic need to treat liver diseases or deficiencies; in 1989 nearly 30 000 deaths in the United States were directly associated with diseases of the liver, but only 2160 liver transplants were available and used. In 1991 liver metabolic disorders, cirrhosis, or cancer were responsible for 205 000 surgical operations in the United States. Cellular transplants represent a real benefit and therefore offer an alternative as the most attractive for the transplant of organs. Surgical intervention to transplant an organ may be premature and delayed for a long time. Urgent factors can thereby be managed or eliminated. The first available organ does not have to be taken; this can be postponed until the transplant is effective and complete. Finally, transplant immunogenicity can be controlled by cellular selection, and the need for medication (immunodepressing drugs) can be reduced. Cellular transplantion by direct cell injection at ectopic sites has already been achieved for several organs [87, 168]. Nevertheless, targeted transplant structuring will rapidly require the use of supports and guides in association with isolated cells, especially for more hierarchically structured tissues, and used in more important biomechanical contexts.

Observations and Hypotheses

The survival of cells on ectopic sites, as with the clinical success, has considerably improved. The efficient use of the donor tissues and the survival, growth, and differentiation of cells depend on good culture and transplant conditions. Several hypotheses have been put forward concerning the cellular transplant, and these derive from observations made on both cells and tissues [87, 164]:

- Biological living structures are in a constant state of dynamic balance. They undergo remodeling and replacement of the functional tissue to the contribution and renewal of component cells. Such a process varies from one tissue to another and from one structure to another.
- Isolated cells tend naturally to reform in vitro structures according to the environment where they are found and the degree of alteration that they undergo.
- Isolated cells can remodel and reform structured neotissues. They make it in a limited manner in suspension in the substance of the mature tissue. They begin without intrinsic organization and have no support to guide the regeneration.
- The tissue cannot be implanted in large volumes due to problems of nutrition (1.0–3.0 ml) and cells are not able to survive more than 100 μm from the nearest capillary.
- The cellular form is determined by the cytoskeletal components, and the attachment to scaffolds plays an important role in the cellular division and in the differentiation of functions. Dissociated cells placed in mature tissues

as suspension and without cellular attachment cannot find sites of attachment and determine their polarity. This may limit the total number of implanted cells that can remain viable to organize, proliferate, and function.

Moreover, as Bell [8, 10, 11] points out, cells are organized in vitro in three dimensions, as they are in vivo, and express morphological and biochemical profiles very different from those observed during cultures in vitro in monolayers but similar to those in vivo. Three-dimensional cultures are therefore important in analyzing cellular behavior at the level of biological tissues and are increasingly important for tissue reconstruction [62].

Today tissue regeneration is based on the attachment of cells isolated on artificial supports whose structural properties and surface allow the guidance, organization, and growth of cells, and whose configuration allows a cellular surviving in vivo by simple distribution. After implantation it is hoped that the growth of the cellular mass is accompanied by a vascularization of the cell-biomaterial complex, and that these two mechanisms can be influenced by external factors to the site: growth factors, hormones, proteins, and adequate stimuli. The modeling of tissue substitutes created by three-dimensional culture in collagen gels infiltrated by fibroblasts was proposed by Jain et al. [84]. This study allows an understanding of the concentration of constraints in the biological composite and contributes to knowledge of tissue growth, development, remodeling and of the healing of the soft tissue substance.

Cellular retransplantion and tissue regeneration are operations in vitro and in vivo that differ in methodology and technique depending on the cellular type, the processed organ, and the tissue or artificial extracellular matrix. Various methods of tissue regeneration have been presented since the simple and pure reconstruction in vivo, simple cellular seeding in vitro, and transplantion to the various levels of reconstruction in vitro with subsequent retransplantion. Figure 1 summarizes the research scenario to evaluate the in vitro and in vivo culture and retransplantation procedures.

Materials and Structures of the Artificial Extracellular Scaffolds

Scaffold Biomaterials

Permanent natural and synthetic biomaterials have been developed and used for many years in medicine and surgery. However, nonbiodegradable polymeric structures are of little or no informational content for cells, cannot be remodeled, and do not adequately transfer the biomechanical stresses to the native tissues. In addition, they are subjected with time to the physical alterations, wear, and debris formations associated with potential immunological reactions, and ultimately provide poor artificial systems for long-term tissue reconstruction and regeneration.

Bioabsorbable or remodelable materials offer alternatives for designing a new generation of improved and biocompatible surgical implants. Many years of investigations have provided experimental data to use the available bioab-

Table 1. Artificial materials for scaffoldings of orthopedic and neutral tissues

	Artificial scaffoldings for tissues
Natural and synthetic biomaterials used or experimented for scaffolds	▶ Type I collagen ▶ Collagenlike polymer ▶ Elastinlike polymer ▶ Fibronectinlike polymer ▶ Lamininlike polymer
	• Poly(lactic-co-glycolic) acid (PLGA, Vicryl) • Poly(vinyl alcohol) (PVA, Ivalon) • Poly(L-lactic) acid (PLLA) • Poly(lactic acid-co-lysine) with arginine–aspartic acid–lysine sequences • Poly(glycolic) acid (PGA) • Poly (DTH-carbonate) • Poly (ethylene oxide) (PEO) • Polyurethanes: LDI (lysine diisocyanate)–PCL (polycaprolactone) • Poly(2-hydroxyethyl methacrylate) (pHEMA) and relative copolymers • Poly(diethyl glycol methacrylate) (pDEGMA) • Poly(2-acrylamino-2-methyl sulfate propane) (pAAMPS) • Poly(2 hydroxyl-L-glutamate) (pHEG)
	✓ Animal and human biopolymers ✓ Secreted extracellular matrix; animal and human tissues

Table 2. The basic requirements for biomaterials in scaffolding materials for orthopedic tissues

Basic requirements
• Biodegradable and biocompatible (bioabsorbable) Nontoxic materials
• Biocompatible degradation by-products Nontoxic degradation products
• Low immunogeneicity
• Physical properties similar to those of the tissue being duplicated (e.g., biomechanics, density)
• Easily processed in complex shapes Transformable into foams, sponges, filaments or threads Suterable
• Capable of mobilizing cells, of promoting the cell attachment
• Remodelable and recognized by the tissue cells
• High information content materials for cells (ECM molecules)
• Specific properties in accordance with the tissue to be duplicated

sorbable materials with more efficiency: the initial mechanical performances, biocompatibility of the materials and associated degradation by-products, surface chemistry, biodegradation kinetics, and effects are among key parameters for selecting adequate absorbable materials. Scaffolding devices and constituent materials must meet some basic requirements for biomaterials or devices devoted to tissue engineering:

- As noted above, cells need scaffolds or templates to regenerate tissues in vitro and in vivo.
- It is not evident that cells stimulated by adequate scaffolds can remodel then and reform a tissue or an association of tissues.
- Scaffolds may be repopulated by cells before implantation, or if they are appropriate, they may mobilize host cells, or cells may migrate into them and repopulate them after implantation.
- Materials that are not remodelable by cells are poor scaffolding materials.
- Frozen and cryopreserved allografts can be considered as natural scaffolds with low immunogenicity but are not effective as commercial solutions for tissue substitutes. Biomimetic design of such materials may be attractive.
- Instructive scaffolds with favorable molecular diversity and supramolecular structure may stimulate tissue regeneration [9].

Generally of low informational contents for cells are bioabsorbable synthetic polymers [polylactic acid (PLA), polyglycolic acid (PGA), polyglactine (PLGA), poly(vinyl alcohol; PVA), polyurethanes (PU), polycarbonates], cross-linked peptides (unnatural amino acids), and recombinant proteins (collagenlike, elastinelike polymers) [60, 134, 173]. Naturally occurring materials such as animal or plant biopolymers (e.g., collagens, elastin, laminins, fibronectins, GAGs, proteoglycans, alginic acids), secreted biomatrix (EHS tumor cell matrix), and animal and human tissues (e.g., dermis, dura, peritonium, porcine heart valve, ligament and tendon grafts) are of high informational content for cells [25–29, 41, 48, 59, 67, 69, 75, 116, 130, 142, 151–153, 165]. Both these low and high informational content materials have been used to bioengineer living cell/material implants; however, man-made polymers, cross-linked peptides, recombinant proteins, and biopolymers have been preferred for the recent tissue-engineering developments [27, 28, 39, 55]. Collagen, collagen-proteoglycans, collagen-glycosaminoglycans (GAGs), hydroxyapatite, calcium carbonate, poly-L-lactic acid (PLLA), PGA, PLGA, PVA , poly(ethylene-oxide), and PUs have been preferentially selected and transformed to culture three-dimensional tissue equivalents for skin, dermal tissue [9–11, 41, 43, 50, 51, 74, 140, 180–182, 158], ligament [24, 26, 29, 68, 145, 170], cartilage, meniscus [12, 13, 34, 44, 46, 53, 66, 91, 93, 94, 104, 133, 135, 136, 154, 155, 147, 165], and bone regeneration [78, 82, 104, 108, 124, 127, 159, 161, 162, 166].

Scaffold Structures

Material transformation is another key parameter for producing artificial scaffolds for tissue culture: customizable three-dimensional highly porous structures must be processible in various shapes and sizes to duplicate tissue geometric appearance. The regeneration of cartilaginous organs, for example, of ears or noses, will require special shapes and dimensions. As a consequence scaffolds must be easily processible, keep their shape intact during the culture and implantation, and withstand surgical handling. Various techniques have been developed to produce three-dimensionally, highly porous polymeric structures: a limited number of structure types provide bioabsorbable shape-customizable porous systems. Tables 1 and 2 summarize the polymeric biomaterial and basic requirement for artificial scaffolding materials of orthopedic tissues.

Foams and sponges are rather experimental while textiles are of commercial origin in most cases. These two types of porous structures differ in terms of pores or similipores. Sponges and foams consist of pores of various forms and sizes, interconnected or not [47, 61, 92, 132]. Textiles are porous but do not have real pores, rather interstices created by the fibers or filaments arrangement are behind this similiporosity. Porosity and permeability concepts of the material textiles have been analyzed by Guidoin et al. [73] and Pourdeyhimi [131]. The two structural types are used in creating artificial extracellular matrices, but other polymeric structures are also available.

Textile Constructions

Textile constructions were among the first products used to support tissue culture and reconstruction processes: textile and nonwoven fibrous structures have interstitial porosity characteristics, high surface area/volume ratios, and oriented constituent basic materials (fibers, filaments) [37, 79, 131]. Another advantage is that simple absorbable textile products are readily available from biomedical companies [22, 23, 40, 74, 105, 118, 119, 122]. However, extreme customization of such textile structures in research laboratories is difficult since the textile technology requires special equipment and expertise not common in nontextile research laboratories [67, 88, 99].

Textiles are well known in the biomedical area, especially for prostheses and implants of soft biological tissues (sutures, arteries, valves, ligaments and tendons, cardioabdominal walls and diaphragms). There are several types of textile structures: one-dimensional for fibers, filaments, threads, and yarns, two- and three-dimensional for the knitted, woven, nonwoven, felts, braided, and plaited structures. These textile constructions have been used in implantable devices, most commonly permanent, except for sutures and some reinforcing implants. There are thus a number of textile products commercially available from the textile or biomedical industry, although textiles made of biodegradable polymer are more limited commercially in terms of product diversity.

Using Textiles for Cell-Seeded Implants

In 1988 Vacanti et al. [164] experimented with cellular transplantations by using a biodegradable polymeric textile. They used the experimental spinning of filaments in polyanhydride and polyorthoester, and used also sutures of Vicryl, a PLGA 910 (90GA/10LA), copolymer of PGA and PLA. Synthetic sutures were dissociated to form filamentous strands, while individual filaments were attached partially to form a fibrous sheets on which Vacanti et al. [164] cultured various cellular types isolated from the pancreas, liver and intestine of animals. The fibrous sheets served as supporting EC matrices to cells during the culture, then the whole was reimplanted to the level of the organ targets. In the same manner, Vacanti et al. [163] used a suture of (90GA/10LA) coated of PLGA 370 (30GA/70LA) and stereate of calcium for the culture and the transplantation of chondrocytes in the rat (articular cartilage). The utilization of artificial fibrous structures continued with experimentations of Cima et al. [36, 163] that evaluated the cellular transplantation from matrices in Vicryl and Dexon. For Vicryl these are braided sutures that are unbraided and dissociated to form a constituted fibrous strands with filaments with a diameter of 14 μm. The second matrix in Dexon was commercial and constituted by fibers 15 μm in diameter. Hyalin chondrocytes taken from the animal were cultivated there, and the neotransplant was implanted subcutaneously in nude mice. The viability and efficiency of the cellular transplantion were initially promising [36]. The reconstruction of the articular cartilage by cellular transplants is of great interest, and von Schroeder et al. [167] applied a matrix of PLA to its reconstruction in the rabbit. In 1992 cartilaginous organs were studied in culture, reconstruction, and transplantation. Vacanti et al. [160] used nonwoven PGA fibers with a diameter of 14 μm and carved them into rectangular forms of same sizes. They then used a mixture of PLA and PLGA in solution for coating/impregnation of rectangular forms of nonwoven PGA. The final structure had the shape of a human ear and was composed of a nonwoven PGA core coated and bonded with PLA-PGLA polymers. This meadow-formed matrix served as support for the culture of hyalin chondrocytes of animals. The artificial matrix/cell was then implanted subcutaneously to the dorsal level in rats. The tissue regeneration in vivo seems effective in the short term, but long-term studies must be continued. In 1993 Freed et al. [64] introduced the reconstruction of cartilage by using a PGA nonwoven of a surface density approximately 8.5 mg/cm^2 and with fibers of 14 μm in diameter. Culture and reconstruction studies were achieved in vitro by the mean of bovine cells (chondrocytes). Kim et al. [91] repaired artificial bony defects created in the rat model and applied a nonwoven PGA (interstices between fibers: 75–100 μm). Vacanti et al. [162, 163] worked on the culture of osteoblasts, and to the regeneration of the trabecular bone on PGA substrata (diameter of fibers: 15 μm) and proposed the construction of two distinct orthopedic tissues (bone/cartilage) by double-seeding and simultaneous culture on a nonwoven support (interstices between fibers: 75–100 μm). Grande et al. [71] attempted to reconstruct the articular cartilage, and evaluated several artificial EC matrices with a nonwoven PGA (size of pores: 200 μm) and an open woven in Vicryl (size of pores: 210 μm). They determined that the culture in vitro of articular cartilage is allocated by the type of substratum, and the best result was obtained with a matrix in collagen.

Recent Development of Biotech Textiles

Cima et al. developed the fibrous structure (textiles), making it easily available for the cellular transplantation, and improved the textile substrata as artificial EC matrices [36, 37]. Mikos et al. [118–120] developed a method to build of genuinely three-dimensional textile structures; this method had been inspired by that presented previously by Vacanti et al. [160]. It consisted of a thermochemical bonding of a fibrous nonwoven PGA structure by PLA solutions and relied upon the solubility of the PLA and nonsolubility of the PGA in methylene chloride. They tested their new structures for the culture in vitro of hepatocytes and for the reconstruction of the liver. The study of Mikos et al. [118–120] was probably one of the ultimate stages in the work on porous and biodegradable synthetic matrices as a basis for textiles. Felts and nonwoven, knitted, and woven fabrics or meshes applied to the cellular culture and to the tissue reconstruction have been the most often used available structures to the commercial level. Biodegradable surgical fabrics (e.g. of Davis & Geck, Somerville, N.J., USA) are the preferred material of investigators. The development of specific biodegradable textiles to the culture in vitro and to the reconstruction in vivo was, again, necessary, especially for the three-dimensional scaffolding of neotissues.

Recent advances have been made in applying textile technology to the specialized processing of textile fabrics for tissue scaffolding. Textile manufacturers and cell biologists are now working closely together to produce fabrics and felts with the specific and adjustable properties required by tissue culture and regeneration. Production lines for biodegradable textile materials have been proposed, particularly for nonwoven (felts) constructions: fiber orienting, cutting, crimping, needle-processing, and heat-press treatment for the design of special matrices for "biotech textiles." With such textile production, similar to that in the cotton industry, scientists and manufacturers refining textile materials for tissue engineering are approaching the question of "how to make cartilage in a textile factory" [188, 189].

Porous Alloplastic Matrices

Alternative Artificial Scaffolds

Foams and sponges have been proposed as alternatives to textiles [52–54, 101, 105, 122, 156]. Nearly every polymeric material can be processed into porous structures such as foams or sponges. This has been carried out experimentally in research laboratories and industrially by biotechnological companies. Foams and sponges can be customized in terms of three-dimensional shapes, sizes, porosity networks, polymer organizations, etc. [47, 61]. The simplest foam designs are easily fabricated in research laboratories with few chemical skills or equipment and provide effective scaffoldings for cell and tissue cultures. More refined, uniform, or regular structures require more specialized facilities. Phase-separation processing of polymeric foams have recently allowed the incorporation of biochemicals in porous polymeric materials. Polymeric foams and sponges have been studied extensively in recent tissue-engineering studies on tissue culture or organ equivalents. Artificial extracellular matrices can be manufactured in the laboratory, are genuine porous structures, and are similar to foams and sponges. A number of experimental techniques have recently been developed for their production. Synthetic extracellular matrices, both two- and three-dimensional and porous, are in fact similar to foams and sponges of collagen only in terms of micropore structures and macro/microscopic aspects.

Processing Porous Polymeric Foams

In 1990 Groot et al. [46] proposed porous and biodegradable polymeric matrices for the reconstruction of meniscal tissues. Matrices were produced from PU on the basis of lysine diisocyanate, PU on the basis of polycaprolactone, and PU on the basis of methylene diphenyldiisocyanate estane. For highly porous structures two techniques were proposed: freeze-drying/salt-leaching and polymerization in situ, and "salt-leaching." In the former the polymer is dissolved in a solvent, and a nonsolvent is then added to facilitate phase separation. A soluble and crystalline product in the water, generally of sodium chloride, citrate, or saccharose, is added to the solution; it is then poured into a mussel and preserved at −15°C. Passage within a freeze-dryer connected to a pump allows elimination of the initial solvent under weak pressure (500 mbar). Depending on the point of fusion of the solution and the type of evaporation of the cosolvent, the specimen remains coagulated and a supplementary cooling is necessary. The solution is then submerged into distilled water to eliminate salts or saccharose crystals. The latter technique applies especially to the PU. Prepolymers are diluted with a solvent to homogenize it with the salt or saccharose crystals, and the solvent is then eliminated by evaporation. Crystals of the polymer/crystals mixture are eliminated by successive washings in baths of distilled water. The obtained foams are isotropic, and the size of pores varies with the techniques and materials. It may be less than 50 μm or greater than 100–300 μm, with tunnels in systems of a diameter of 50 μm. Morphologies and microporous internal structures of foams are variable enough to obtain from porous equipment sufficient diversity. Groot et al. [46] used the following technique of porous polymeric structure manufacture by freeze-drying: the polymer is dissolved in a solvent system, the solution is cooled until the solvent freezes, and the solvent is eliminated by being placed in vacuum. The technique was applied by Elema et al. [58] to produce four meniscal implants of different porosities, and foams can be strengthened by means of PLA fibers. The study focused on the implantation of different porous structures: PU/PLA (foam), PU (foam), and PU/PLA strengthened by PLA fibers. The authors observed among other things that reinforcement by PLA fibers delays biodegradation of the porous matrix (100–300 μm) and stops invasion and growth of the fibrocartilaginous tissue. In 1991 Klompmaker et al. [93, 94] experimented with the same porous synthetic matrices in PU and PLLA. Implants were made from PU foam or PLA/PU foam to strengthen by fibers of PLLA. The relative proportions of pore sizes 0–50/100–300 μm were 67%/33% for PU/PLLA and 59%/41% for PU alone. Porous matrices were formed by growing such it formed the natural meniscus, and the implantation in vivo demonstrated real capacities of fibrocartilaginous reconstruction.

A similar technique (with sodium chloride) was used by Bruin et al. [19] in manufacturing bilayer membranes for skin regeneration. The porous matrix was constituted of PU based on an ester of glycolide-co-e-caprolactone and a urethane of lysine diisocyanate. The inferior layer represented the zone of dermal regeneration, was obtained by soluble particles in water (NaCl), and proved to be macroporous. Beumer et al. [14, 15] also used a porous scaffold of multiple layers to produce dermal substitutes after cellular seeding. The inferior layer in PLLA was macroporous and achieved by crystals of anhydrous sodium citrate (salt-leaching). The superior layer was obtained from Polyactive, an elastomeric material of ether block/ester segmented [polyorthoester/poly(butylene terephthalate)].

In 1993 Freed et al. [64] developed porous matrices in PLA, particularly in PLLA. The method of manufacture consisted of "particulate-leaching/solvent-casting" in which the polymer and crystals of sodium chloride were placed into solution. The polymer solution/crystals were carefully mixed and homogenized, and the solvent was then eliminated by simple evaporation. After passage under void (100 mmHg), the solvent was totally eliminated, and the resulting matrix was submerged in an important volume of water distilled to eliminate crystals of salt. This resulted in a porous support, and the maximal pore size was measured to be 515 μm. Porous PLA matrices were then applied in vitro to the culture of chondrocytes and were implanted subcutaneously and dorsally in the rat. Freed et al. [64] found potential advantages in using polymeric porous and biodegradable supports rather than natural material. Alloplastic materials allowed the manufacture of three-dimensional matrices for specific clinical applications and the selection of polymeric materials that degraded similarly to the secretion of fibrocartilaginous matrices. There was a reduction of the inflammatory response of the host by the implantation of cells cultured on the support and no inflammatory response once the support degraded. These authors also coated PLLA foams with collagen and noted the formation in vitro and in vivo of neocartilage. They expressed the hope of being able to use this technique for orthopedic reconstruction.

Wald et al. [169] used a similar technique ("salt-leaching/solvent-casting") to manufacture porous laminated PLLA discs 1.35 cm in diameter and 0.5 cm thick. The size of pores was 126 and 166 μm. Porous discs were then associated to form a system of feeding and cellular transplant using a catheter. Cellular cultures used hepatocytes, and the distribution within the polymeric matrix was studied using blue-colored polystyrene microbeads (1–10 μm) and staining by acidic orange 8. Injections of polystyrene allowed determination of the uniformity of the porosity within the structure and some parameters of manufacture. The manufacture and the porous foam association by the process of lamination were new stages in developing artificial three-dimensional matrices and were presented by Mikos et al. [119]. Lamination for biodegradable and porous supports of PLA and PLGA was by "salt-leaching/solvent-casting" and consisted essentially of solvent fusion. This technique allowed the pasting and connecting of polymeric porous matrices that evolved a three-dimensional structure. Mikos et al. [119] presented several multicomponent foams connected by lamination and applied to the reconstruction of the nose and in processing of the metacarpal phalanx. Techniques of contour cartography and geometric reconstruction of the foam scaffold have also been employed for the construction and assemblage of polymeric three-dimensional constructions of particular forms.

Polymeric Hydrogels

Micro-/macroporous hydrogels and gels have been investigated in various materials to provide templates for regeneration. Acrylic and amino acid hydrogels have been evaluated particularly for neural, ligament, and tendon regeneration.

Hydrogels are absorbent polymers able to absorb very great volumes of water or aqueous solutions in a predictable and controlled manner [138] while keeping intact their mechanical integrity. When they are dived in aqueous media, hydrogels inflate and some of their properties change, but in no event do they undergo dissolution. Hardness, permeability, and transparency are modified by contact with the aqueous medium. Hydrogels are the most often cross-linked polymeric materials, and thermoplastic hydrogels are generally cross-linked polymers but with linear molecular structures, which renders them thermoplastic and soluble in some organic solvents [17]. Biomedical applications can be diversified in implanted pharmacological systems, arterial substitutes, bandages, coating for biosensors, and soft-tissue substitutes, but the main commercial application is in contact lens [17, 110]. The absorption of water allows limiting the reactions of friction with tissue components, and in a more general manner they do not generate adverse reactions (inflammation, formation of thrombus). They are therefore nontoxic if the monomers that may be present in the hydrogel are extracted, are chemically stable, and possess a weak surface tension with aqueous environments. This latter point renders them potentially attractive for applications with blood contact, since the biocompatibility with the blood is directly connected to the thromboresistance and to the weak interfacial tensions [110]. Other properties of hydrogels include swelling to the aqueous system contact, the delivery of substances soluble in the water and incorporated in the hydrogel, and transportation and retention of fluids and their hydrophilicity.

Acrylic Hydrogels

Principal polymeric hydrogels were derived of 2-hydroxyethyl methacrylate (acrylic; HEMA) and 1-vinyl-2-pyrrolidone. The possiblity of using hydrogels of poly(2-hydroxyethyle methacrylate) (pHEMA) as simple substrata for cellular cultures has been studied by Lyndon [110]. The effect of the chemical composition and specific grouping presence in hydrogels on the cellular cultures has been analyzed by Smetana et al. [148, 149], especially for pHEMA, the copolymers of HEMA and diethyl glycol methacrylate-, poly(diethyl glycol methacrylate), the copolymer of HEMA and *N*-/1-ethyl-2-pyrrolidone/acrylamide, the copolymer of HEMA and diethyl aminoethyl methacrylate, and poly(2-acrylamido-2-methyl sulfate propane). For porous hydrogels pHEMA is preferred and permits porous constructions with variable pore sizes. HEMA and pHEMA were used for the first time in 1934 by Woodhouse [179] and evaluated as biomaterials in 1951 by Wichterle [172]. There have been few clinical evaluations for such materials, however, except for ocular systems, especially for soft and absorbent contact lenses. In 1993 Chirila et al. [31] presented hydrogels of HEMA as porous substrata for cellular and tissue transplants. The pHEMA hydrogels polymerized in mass are considered as nonporous, while those polymerized in solution are either microporous or macroporous. For polymerization in solution the increase in the quantity of dilution in the mixture of monomers produced an increase in pore size. The water was a solvent of the HEMA and a nonsolvent of the pHEMA, and its concentration in the initial mixture of monomers was critical for the porosity. Normally the hydrogel is homogeneous for a weak concentration in water and is a porous foam (micropores 10–100 nm, macropores 100 nm–1.0 μm). If this proportion is greater than 45%, the pHEMA hydrogel becomes heterogeneous and resembles to real sponges. Interconnected canals are formed by the separation of the monomers during polymerization, and the porosity was created by spaces between the particles of polymers in solution. The size of particles is 2–5 μm and that of canals 40–80 μm. There are therefore two types of porous hydrogels for biomaterials: the foam pHEMA and the sponge pHEMA [31]. The tissue evaluation of these hydrogels (pHEMA) has been carried out beginning in 1967 in Czechoslovakia, and cellular invasion was observed to a depth of 500 μm. As a rule, the greater the porosity, the stronger the cellular invasion is, and the weaker the fibrous encapsulation (in vivo). Effects of porosity on the behavior of the hydrogels were studied by Sprincl et al. [150], who demonstrated that the homogeneous hydrogels are encapsulated while the heterogeneous or hydrogels foams are penetrated by cells and tissues. The porosity was directly related to the initial water presence, and the cellular invasion predominance and neovascularization were observed for hydrogels with more than 70% of water. Sponges in hydrogels are being experimented with in various areas, especially for mammary prostheses, for the artificial pancreas, and in neurology. In addition, Chirila et al. [31] evaluated in vitro and in vivo a sponge of pHEMA as a component of a keratoprosthesis. In the same manner, a hydrogel of pHEMA was introduced in a composite artificial tendon by Davis et al. [45]. Woerly et al. [175–179] used the hydrogels of pHEMA and poly(-glycol methacrylate) (pGMA) for intracerebral implantations: the hydrogels were porous with a very regular microscopic and structured aspect. Foams were considered hyperporous with 50% or more porosity (void) and hypoporous with 50% or less: pHEMA hydrogels showed 50%–85% porosity (void) and were rather hypoporous, and their average pore size was reduced, not exceeding 7 μm, with a maximal size of approximately 100 μm. pGMA hydrogels were variable, with approximately 60%–65% of void. The pGMA hyperporous hydrogels possessed 50% pores of more than 10 μm, but other hydrogels were generally hypoporous, and pores did not exceed 6–7 and 13–15 μm in size. The maximal pore size here was also approximately 100 μm. Woerly et al. [111, 174–178] have analyzed the morphology and microstructure, the interaction of the porous hydrogels with collagen, the composition in polar chemical groups, and the ionic composition have been for the hydrogels of pHEMA and especially of pGMA. The hydrogel p(HEMA) has very often been associated with other polymers and synthetic or natural structures for the production of implants. pHEMA-collagen hydrogels were evaluated in vivo at the rat to the level of their biocompatibility [86]. pHEMA-collagen hydrogel was analyzed for applications in pharmacological liberation. p(HEMA) was used for the confection of composites with silicone. Composite implants p(HEMA)-silicone were studied by Cifkova et al. [35] and Lopour et al. [106]. In ligamentotendinous surgery Migliaresi et al. [117] and Davis et al. [45] proposed implants made partly of p(HEMA): polyester + p(HEMA) [117] and PLA + polycaprolactone/p(HEMA) [45].

Other Hydrogels

Various other hydrogels have also been presented by investigators for applications in three-dimensional cultures. For example, Larsen et al. [100] evaluated in vivo (mouse) a commercial elastoviscous hydrogel on the basis of hyaluronans. The hylan was implanted intradermally and subdermally in mice. As for polymeric porous nonhydrogels structures, the hydrogels are being developed as culture supports and living transplants for tissue regeneration. Their future development and possible applications will continue to be of great importance. Their application is principally in porous synthetic equipment for tissue reconstructions, despite a great complexity of manufacture.

Other Artificial Constructions

Porous hollow fibers, hierarchically oriented hollow structures, smooth, nonporous or porous microbeads, and microtubules are special polymeric products developed for tissue culture and organ function substitution by cellular engineering. Extracorporeal function substitutions using cells and substratum currently used similar structures, but made of nonabsorbable polymers such as polysulfone.

Table 3. Experimental use of artificial material and structures for tissue cultures and transplantations in Orthopedics (cartilage and bone)

Tissue/animal model	Materials structures	Reconstruction procedures (in vitro/in vivo)	Ref.
Articular cartilage			
Calf (cartilage, shoulder)	28 fibers of PLGA 910 coated with PLGA 370 and calcium stereate (Vicryl). * Diam. fibers: 14 μm * Dist. interfibers: 0–200 μm	Culture in vitro with chondrocytes (3–24 days) Implantations (nude mice), subcutaneous (8–81 days)	[163]
Rabbit (cartilage, femoral condyle)	Porous PLA matrices, commercial origin	Direct implantations of the PLA matrices (12 weeks) for articular defect repair	[167]
Rabbit (cartilage, femoral condyle)	Porous PLA matrices, commercial origin PDLA; P(D,L)LA or PLLA	Implantations of matrix/perichondrocytes for articular defect repair (sacrifice of animals at 6 weeks)	[34]
Calf (cartilage, shoulder)	Nonwoven/felt of Vicryl (PLGA), commercial origin	Cultures in vitro with chondrocytes (0–6 weeks) Implantations (nude mice), subcutaneous, after 10 days culture, (0–9 months)	[3, 7, 36]
Rat (cartilage)	Nonwoven/Felt of PLLA and PGA, commercial origin	Cultures in vitro of chondrocytes. Implantations (rats) in periostal cavities, after seeding of chondrocytes	[82]
Rabbit (cartilage, knee joint)	Nonwoven of PGA (thick 2.0 μm, pores 200 μm) Nonwoven/felt of Vicryl (pores 210 μm) Porous matrices of type I collagen (pores 100–300 μm), commercial origin	Adhesion of chondrocytes during 2 h, closed-loop recirculation cultures for 7, 14, 21, and 35 days, then histological analyses	[71]
Bovin/Rat (chondrocytes/fibroblasts)	3D matrices of type I collagen (pores 1.0–100 μm)	Cultures in vitro of chondrocytes (articular cartilage) and fibroblasts (tendons) (1–14 days)	[66]
Rabit (articular cartilage)	3D matrices of type I collagen, commercial origin	Cultures in vitro of chondrocytes (I weeks) Implantations (rabbit) of cell-polymer systems (4–24 weeks)	[13]

Calf (articular cartilage, shoulder)	PGA fibrous mesh bonded with PLLA PGA mesh thickness: 500 μm Interstitial dist.: 75–100 μm	Seeding and culture of chondrocytes (changing cell densities) for 1 weeks Implantation of a cell-matrix; subcutaneous in dorsal regions (nude mice); sacrifice at 3, 6, and 12 weeks	[133]
Bovine FemoroPatellar Groove (FPG) Anterior ribs. Human costal cartilage	PGA fibrous mesh (fiber diam. 14 μm) PLLA foam (NaCl diam. 100–150 μm)	In vitro: seeding of animal FPG and rib chondrocytes In vivo: implantation of human/animal FPG and rib cell-matrix	[64]
Bovine (articular cartilage, knee joint)	Spun PGA fibers (diam. 13–15 μm), PGA felts	Dynamic culture in vitro for 8 weeks; histological and biochemical analyses.	[63]
Rabbits (femoral trochlea)	Solid Polyethylmethacrylate (PEMA) + liquid Tetrahydrofurfuryl methacrylate (THFMA)	Implantation of PEMA-THFMA; in situ polymerization, rabbits were sacrificed at 3 and 6 weeks	[135]
Rabbits (cartilage, ribs)	Anhydrous PLA cylinders	1. Mixed with a concentrated cell solution (10×10^6 cells/ml) 2. Implantation in the medial femoral condyle of rabbits (6 weeks)	[42]
Rabbits (periostum, tibia)	Nonwoven PGA + RGD peptide treatment	1. Seeding and culture within the PGA scaffold (2 weeks) 2. Implantation (intertrochlear groove defect) in rabbits (sacrifices at 8 and 10 weeks)	[136]
Bovine (cartilage)	Hydrophilic and hydrophobic porous high-density polyethylene (HDPE) substrates (pore size 115–135 μm)	Seeding and culture of chondrocytes (7 and 21 days); microscopic and biochemical tests	[104]
Bone regeneration			
Mice (Induction os/cartilage)	Polyanhydres (PCPP-PCPM), compressed discs	Bone induction in vivo by delivery of osteogenic proteins, intramuscular (9–16 days)	[108]

Table 3 (*Contd.*)

Tissue/animal model	Materials structures	Reconstruction procedures (in vitro/in vivo)	Ref.
Calf (Bone, Shoulder)	Nonwoven of PGA, commercial origin (thick 100 μm, interstice 75–100 μm)	1. Culture of osteoblasts (0–2 weeks) 2. Implantations (nude mice), subcutaneous (9 weeks)	[162]
Calf (Bone, shoulder)	Nonwoven of PGA, commercial origin (thick 100 μm, diam. fibers 15 μm)	1. Culture of osteoblasts (0–2 weeks) 2. Implantations (nude mice), subcutaneous, (3–18 weeks)	[161]
Rat (RBMC, bone)	Type I collagen gel, commercial origin	1. Cultures in vitro of osteoblasts 2. Implantations as femoral implant	[159]
Rat (RBMS, bone)	• 3D collagen matrix • Cellulose acetate filter (pores 0.45 μm)	Cultures in vitro of osteoblasts (14 days)	[107]
Mice (cartilage/os)	Type I collagen sponges, + hormones, IGF-ß, insulin, TGFß-1, rhPTH	Cultures in vitro of chondrocytes and osteoblasts (48h–3 days)	146
Osteoblastlike cells (ROS17/2.18)	PLLA foam Polyphosphoester (BPA/PP)	Culture in vitro (72 h)	105
Rats (tibia, bone)	PLA/PGA blend (50:50)	Direct implantations in tibia (7, 14, 21, 28, and 42 days)	78
Rats (bone marrow)	Porous hydroxyapatite substrate (pore diam. 230 μm)	1. Soaked in a bone marrow cells (5×10^8 cells/ml) solution 2. Implantations in rats; subcutaneous (1–4 weeks)	127

Table 4. Experimental use of artificial materials and structures for tissue cultures and transplantations in orthopedics (menisci)

Tissue/ animal model	Materials/structures	Reconstruction procedure (in vitro/in vivo)	Ref.
Menisci			
Dogs (menisci)	Experimental porous matrices LDI-base polyurethane PCL-base polyurethane PLLA + PCL-base polyurethane	Implantations (dogs), midsubstance of meniscus T-shaped lesions	[46]
Dogs (menisci)	1. PU/PLLA (95/5% w/v) + PLLA fibers (25% w) (sugar 100-300 μm + 1,4-dioxane) 2. PU/PLLA (95/5% w/v). (NaCl 100-300 μm + dioxane/trioxane) 3. PU/PLLA (95/5% w/v). (NaCl 100-300 μm + dioxane/trioxane) 4. PU (100% w) (sugar 100-300 μm + dioxane/hexane/water)	Implantations (dogs), midsubstance of medial meniscus T-shaped lesions	[58]
Dogs (menisci)	1. PU/PLLA (95/5% w/v) + PLLA fibers (10% w):67% (0-50 μm) and 33% (100-300 μm) 2. PU (100% w) (pores 83%):59% (0-50 μm) and 41% (100-300 μm)	Implantations (dogs), midsubstance of medial meniscus T-shaped lesions (9-56 weeks)	[93,94]
Dogs (menisci)	Collagen porous implants	Implantations (dogs), medial meniscus (3–12 months)	[101]

Tissue Engineering in Orthopedics

Bone and Cartilage Cultures

Bone Formation

Bone repair is performed approximately one million times annually in the United States (approx. 1 300 000 surgeries for joint replacement, bone grafts, fixations, and facial surgery, 1991). Autogenous and allogenous bone grafts are currently used to accelerate bone ingrowth; however, autografts are limited by donor site morbidity and contour irregularities, and allografts may be associated with cell-mediated immunoresponses in relation to the presence of transplantation alloantigens and pathogens. Metals and ceramics can also be used to support bone ingrowth. Demineralized bone powder stimulates bone ingrowth in animals or humans but does not induce bone tissues at most nonbony sites. The use of bioabsorbable polymeric scaffolds (PGA) has recently been developed to induce bone formation in vitro and in vivo. Osteoblasts and osteocytes were isolated from parent tissues, seeded, and adhered to the polymeric structure and were observed to retain a round configuration in vitro. Osteocalcin was systematically added in vitro in cell culture medium. Neoformed osteoid tissues were observed in polymeric implants seeded with osteoblasts and osteocytes and transplanted in animals. In a first phase (6 weeks) the tissues neoformed in vivo were described grossly and histologically as being composed principally of cartilage tissues, but isolated osteoids were observed at the periphery of the implants near the blood vessels invasion. With time, bone formation and ingrowth increased throughout the tissue-polymer implant. Finally (20 weeks), gross bone formations were present in most polymeric scaffolds seeded with either osteoblasts or osteocytes; trabecular bone formations were observed as encapsulating islands of cartilagelike tissues. Step-by-step evolution of the different cells was also observed at the trabecular bone-cartilage interfaces: hypertrophied cells in the basophilic areas of cartilage tissues, smaller cells in eosinophilic environments near the hypertrophied cells, and then osteocytes in the organized bone and near the small cells. Hypocellular bone marrow was present in the bone formations at 20 weeks, and lamellar bone was also reported [99].

Cartilage Formation

Cartilage repairs and replacements also represent approximately one million surgical operations annually in the United States (approx. 997 000 surgeries for resurfacing, chondromalacias patellae, arthritis, fingers and small joints, osteochondritis, and meniscal replacements, 1991). Replacements are generally performed using cartilage transplantation or implantations of polymer and metal devices. Tranplantations consist in removing healthy cartilage samples, shaping them as required, then reimplanting them where needed. Donor cartilage tissues are quantitatively limited and qualitatively difficult to shape in appropriate three-dimensional implants. Permanent prostheses or devices are commonly associated with resulting infections, host-prosthesis interface failures, and nonadaptive behaviors in relation to the changing stresses in situ. The development of viable living cartilage equivalents has been stressed from isolated chondrocytes and absorbable polymeric scaffolds, and recently evaluated for knee joint cartilage repairs. PGA, poly(lactic-coglycolic) acid polyesters have been transformed under either textile or foam structures and associated with chondrocyte cell lines [42, 63, 71, 136, 161, 162]. Chondrocytes isolated from animal cartilage tissues, then seeded and cultured in absorbable PGA supports, provide histologically and biochemically the signs of neocartilage formation. Mature chondrocytes, sulfated glycoaminoglycans, and type I and II collagen were observed in such cultured cartilage equivalents. Studies have recently been reported for cartilage equivalent formation: human, bovine, and rabbit chondrocytes were seeded and adhered in various polymer scaffolds (PLGA, PLLA, and PGA), then were reimplanted subcutaneously in nude mouses or rabbits for 6–9 months. Biochemical and histological examinations have shown that cartilage tissues of adequate quality (mature chondrocytes, sulfated GAGs, type I and II collagen) have grown and formed within the polymeric substrata. Moreover, the polymeric supports have been totally degraded and dissolved in vitro; however, the neoformed cartilages have kept the initial three-dimensional geometry of the scaffolds. Cartilage developments by intensifying the production of polymeric scaffolds and culture in bioreactors are currently under investigation. Knee joint cartilage replacements with "tissue engineered" cartilage implants have recently been studied in rabbits. Clinical trials for cartilage replacements by tissue engineering will start soon in reconstructive surgery [99].

Isolated cells from parent bone and cartilage tissues and absorbable polymeric scaffolds have been the main raw materials in tissue engineering bone and cartilage neotissues in vitro and in vivo. Improvements of the tissue engineering and culture are currently being developed by designing appropriate supporting materials and culture bioreactors. Tissue engineering techniques and materials are needed in this new area for the regeneration of biological tissues. Another field to be developed in the near future targets the use of bioactive molecules for a better control and management of both in vitro and in vivo culture and regeneration processes [98, 103, 113, 125, 141, 146]. These bioactive molecules include bone morphogenic proteins (BMP), hormones, and growth factors, and they will be certainly involved in the culture of tissue equivalents. BMP and transforming growth factor-β (TGF-β) are currently under study for bone healing and for bone and cartilage formation. BMP helps the formation of cartilage, bone, and bone marrow while TGF-β augments bone growth. Other osteoinducting biomolecules offer promising strategies to assist the tissue regeneration. Table 3 summarizes some studies representative of articular cartilage and bone reconstruction or tissue culture. Table 4 reports the main regeneration studies for menisci.

Ligaments and Tendons Engineering

Compared to cartilage, meniscus, or bone, few studies have been proposed to culture or reconstruct ligament and tendon tissues in vitro or in vivo through bioabsorbable scaffolds, but it must be noted that ligament and tendon surgery represented only 123 000 operations in the United States in 1991. In the 1980s PGA, PLA, and polydioxanone filametous implants were implanted directly in vivo to reconstruct the ligaments of the knee joint, shoulder, or ankle of animals and patients [97]. However, this did not result in extensive use of bioabsorbable reconstructive devices for ligaments and tendons. Novel bioabsorbable materials have recently been selected and developed and are under study for temporary artificial ligaments and tendons [7]. In another way the development and culture of tissue-engineered ligament tissues have still attracted only little interest and limited practical results in comparison to cartilage, fibrocartilage, and bone.

On the other hand, the first attempt to use isolated cells seeded and cultured onto ligament artificial substitutes was proposed by Brody et al. [16]. They used nonabsorbable Dacron Stryker ligament prostheses as scaffolds for cells, seeded them with canine skin fibroblasts, and cultured in vitro the cells until the contact growth inhibition was reached, then reimplanted the skin fibroblasts seeded-Dacron Stryker ligament as anterior cruciate ligament (ACL) substitute in dogs. Gross examinations and histological analyses at 4 and 8 weeks postoperative demonstrated that the seeded Dacron devices were characterized by more uniform and abundant surrounding connective tissue and lower giant cell responses. The study showed that the seeding and culture in vitro of isolated fibroblasts is possible in three-dimensional scaffolds such as artificial ligaments. This technique also permits a better encapsulation of the synthetic ligament by connective tissues and reduces foreign body reaction toward the alloplastic implant; however, it does not result in long-term restoration in vivo of a ligament neotissue with appropriate compositional or structural quality.

Direct Implantation of Collagen Scaffolds

Type I collagen fibers is a major solid component of ligaments and tendons, and collagen is therefore the material preferred for building artificial ligament scaffolds for in vivo or in vitro ligament cultures. In the late 50s natural fibers derived from collagenous materials were still available (Collafil or Mareva) for general uses, but they disappeared thereafter. However, type I collagen was reintroduced in medicine and surgery for tissue engineering of skin equivalents, and type I collagen fibers and textiles are now under production as resorbable biomaterials and under experimentation for various applications in tissue engineering. Collagen can be processed into a great variety of products (e.g., sponges, films, membranes, fibers, flakes) and "biological textiles" (collagen textiles) are now produced specifically for regenerating or reconstructing soft-tissues such as ligaments and tendons [26, 29, 32, 89, 170].

The canine ACL was used as an in vivo model to reconstruct ligamentous tissues through direct implantations of type I collagen fibrous scaffolds. Collagen fibers or threads were processed from bovine tendons, transformed on a customized in-line thread making apparatus, cross-linked using glutaraldehyde vapors, para-formaldehyde/glutaraldehyde solutions and 1-ethyl-3-(3-dimethyl amino propyl) carbodiimide hydrochloride (water soluble carbodiimide). Cross-linked collagen threads were successively braided (4 threads×4×4×3) to build a 576 plies fibrous structures. A 600-ply fibrous structure was also produced by winding on a fixture. Both 576- and 600-ply collagen fibers materials were used as ligament scaffolds for the canine ACL, sterilized by ethylene oxide gas, implanted in dogs by lateral arthrotomy, and mounting the scaffold by the over-the-top technique. The scaffolds were fixed using two titanium staples, proximally and distally [22].

At 12 weeks postoperatively synovial fluid and bony samples, proximal, midsubstance and distal samples specimens of the implants were excised for histological analyses. Formations of collagenous ligament neotissues were observed, with bone attachments at the collagen scaffold ends. Remodeling of the collagen scaffolds clearly depended on the location. Near the staple attachment sites the collagen threads were still intact with a nonorganized cel-

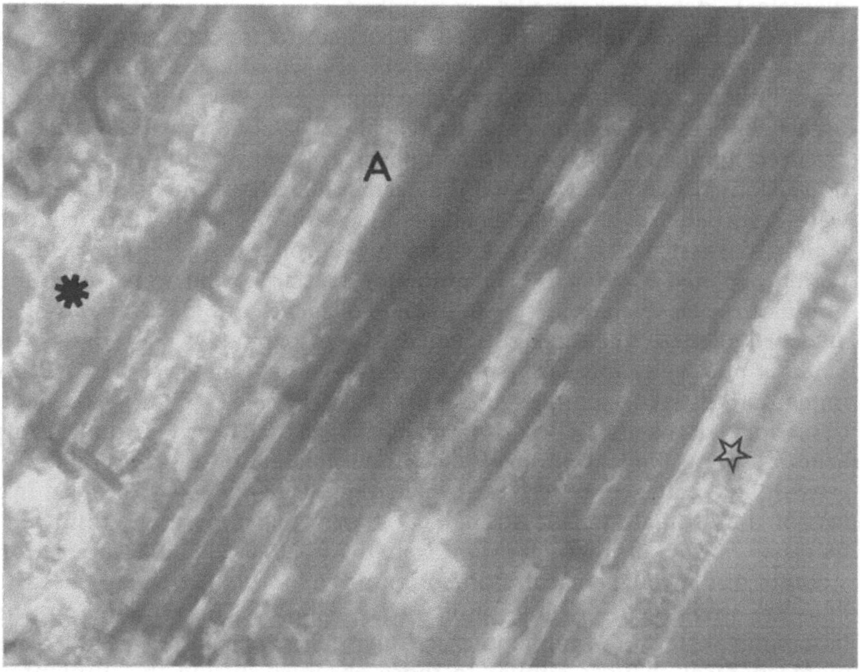

Fig. 2. Carbon tow structures give a good example of the oriented fiber scaffolds used to receive tissue ingrowth and formation. Unfortunately, carbon fibers tend to break down and fragment in nonresorbable particles and debris. Due to clinical complications, carbon fiber ligament devices have been withdrawn. Sirius red 3A staining, polarized light. Carbon fiber bundle (carbon fibers *in black*); ✻, unorganized collagenous tissues; ☆, more organized collagenous bundles (signs of crimps)

lular ingrowth. In the midsubstance large collagenous bundles were observed and oriented along the implant long axis. At the tibial junction in the joint capsule the collagen scaffolds interacted with cartilaginous prominences, suggesting a potential natural refixation. At the microscopic level collagen fibrils were present, showing normal sizes and orientation along the ligament axis. The size distribution of the fibril diameter was slightly broader for the neoligaments than for the contralaterals, but appeared quite similar. However, the organization level in the neoligament midsubstance was not as high as in the controlateral ligament tissues [22].

Fibroblast-Seeded Collagen Scaffolds

Collagen artificial scaffolds were used to reconstruct Achilles tendons and ACL ligaments in animal models [26, 29, 57, 68]. However, collagen ligament scaffolds were not used for direct implantations, but previously seeded with fibroblasts [79, 114, 115], and then cells-collagen materials were retransplanted in rabbits. Insoluble bovine collagen was processed into collagen fibers (60 μm diameter), dehydrothermally cross-linked, and treated with cyanamide or glutaraldehyde products. Collagen fiber scaffolds consisted of composite structures made by assembling about 200 aligned collagen fibers within a non-cross-linked collagen matrix (type I collagen gel). Bovine type I collagen was used to process either the cross-linked fibers or non-cross-linked gel.

Direct implantations of such composite collagen scaffolds for Achilles tendon reconstruction showed that the cyanamide cross-linked collagen implants were resorbed by 10 weeks postoperation and replaced with oriented neotendon tissues. Glutaraldehyde cross-linked collagen materials resorbed less rapidly, elicited a persistent inflammatory reaction, and were surrounded by fibrous tissues. Cyanamide cross-linked collagen scaffolds after remodeling in vivo provided tensile strengths equal to approximately 66% that of normal tendons and quite similar to those of autogenous materials. In a same way implantations of cyanamide cross-linked collagen scaffolds gave neotendon tissue formations after 4 weeks implantations as ACL reconstruction implants. From a biomechanical point of view, the implantation of acellular cyanamide cross-linked collagen materials (initial strength: 12–25 MPa) resulted in a clear loss of the ultimate tensile strength at 4 weeks postoperative, followed by an increase in the ultimate tensile strength at 20 weeks post-operation. The strength of the collagen scaffold/neotissue at 20 weeks (17 MPa) was still lower than the strength of normal rabbit ACL (45 MPa). The remodeling and neotissue ingrowth on the implanted acellular collagen scaffolds seem to result in biomechanical changes similar to those observed during the implantation of ligament autografts. As in the Achilles tendon reconstruction, glutaraldehyde cross-linked collagen scaffolds for ACL reconstruction demonstrated less satisfactory results than did cyanamide cross-linked collagen materials.

As in vascular surgery, fibroblast seeding and culture processes with artificial implants were first initiated on permanent polymeric implants (Dacron materials) but did not result in concrete benefits due to the nonremodeling of the artificial support. Collagen fiber scaffolds, high/low informational content for cells, and remodelable materials, were used to seed and culture ligament and tendon fibroblasts. Collagen substrata generally support good proliferation of fibroblasts in comparison to synthetic substrata, but the fibroblast behavior (proliferation and matrix component biosynthesis) on collagen or other materials scaffolds is also influenced in vitro by the type and source of fibroblasts. On same type I collagen scaffolds fibroblasts from ACL, patella tendon, or synovium do not proliferate or synthesize similarly: ACL fibroblasts do not show high proliferation rates in comparison to fibroblasts from Patellar tendon or synovium but have higher collagen biosynthesis rates than fibroblasts from Patellar tendon and synovium. The fact that ligament or tendon fibroblasts seeded, adhered, and proliferated in vitro on collagen scaffolds remain viable after being reimplanted in situ opens the way to producing fibroblast-seeded ligament analogues for research or clinical applications.

Nerve regeneration has been studied using three-dimensional oriented matrices made of collagen, polymeric hydrogels, and foams. Porous collagen sponges have also been used as three-dimensional systems to regenerate tendons: collagen-GAG matrices are already available and have been studied for

Table 5. Experimental use of artificial materials and structures for tissue cultures and transplantations in orthopedics (ligaments and tendons)

Tissue/ animal model	Materials/structures	Reconstruction procedures (in vitro/in vivo)	Ref.
Ligaments/tendons			
Goat (ACL ligament)	Reconstitued cross-linked collagen fibers Braided implant	Direct implantations in goats (3–11 months)	[29]
Rabbit (Achille's tendon)	Matrix of pHEMA/PCL hydrogel + PLLA fibers	Direct implantations in rabbits (45 days)	[45]
Rabbit	PLLA fiber braided implant	Direct implantations in rabbits, subcutaneous (6–48 weeks)	[97]
Rabbit (ACL ligament)	Reconstitued cross-linked Collagen fibers Glutaraldehyde or Cyanamide cross-linking 200–250 fibers fiber diam. 50-60 μm	Direct implantations in rabbits (4–20 weeks)	[89]
Rabbit (Achille's tendon)	Reconstitued cross-linked collagen fibers. Coating of non-cross-linked collagen	Direct implantations in rabbits (5,10 weeks)	[57]
Rabbit (ligament fibroblasts)	Gels of type I collagen + porous polyethylene anchorages	Cultures in vitro of fibroblasts (12 weeks)	[79]
Newborn Calf (shoulders & knee tendons)	PGA mesh PGA filaments from unbraided PGA sutures	1. Cultures in vitro of fibroblasts (1 week) 2. Implantations in athymic mice, subcutaneous dorsal (3–12 weeks)	[20]
Newborn Calf (ACL ligament)	Gels of type I collagen + human partially demineralized bone anchorages	1. Cultures in vitro of fibroblasts. 2. Culture with tensile loads (1–10 g)	[21]
Dog (ACL ligament)	Reconstitued cross-linked Type I collagen fibers; 576-ply braid and 600-ply bundle.	Implantation over-the-top in dogs (12 weeks)	[22]

skin regeneration, nerve regeneration, or meniscus regeneration, and have recently been developed to surgically treat gap injuries of tendons. The tubular porous collagen-GAG sponges were processed from bovine type I collagen and chondroitin-6-sulfate (Shark cartilage). The internal porosity was created by a cooling/freeze-drying procedure, and the sponges were treated dehydrothermally. This technique is well known to produce collagen sponges with various resulting porosities and porous network characteristics: homogenous isotropic pore, heterogenous isotropic pore organizations, with elongated pores radially or not, and the associated combinations. Mean pore diameters in collagen sponges generally vary from 20 to 130 μm.

This collagenous sponge material can sustain ligament and tendon fibroblast seeding and culture, leading to three-dimensional proliferation and diffusion within the entire volume of the scaffold: a 1.0-g collagen sponge can receive at least several hundred million cells after 28–35 days incubation. Figures 3 and 4 show the proliferation of canine ACL fibroblasts injection-seeded in small collagen sponges and polymeric foams as performed in our laboratory [137]. Fibroblasts rapidly proliferate within the geometric form of the collagen sponge scaffold from the geometric center where they were seeded by simple injection to the entire volume of the sponge. Contraction, remodeling, and resorption of the cell-seeded collagen sponge scaffolds lead to volume or geometry integrity changes.

Bioengineered Ligament and Tendon Analogues In Vitro/In Vivo

While collagen has been the preferred resorbable materials for ligament and tendon reconstruction or culture [4], bioabsorbable synthetic polyesters have been used for fibroblasts seeding, culture, and cell-polymer matrix re-transplantation.

Ligamento- or Tendinogenesis

As for skin and vessels, ligamentogenesis has been studied in culture ligament tissue analogues. Collagen gels between human demineralized bony ends were used to support the culture of ACL fibroblasts and the formation of ligament neotissues. Type I collagen solutions (2.0 mg/ml) were seeded with a 20×10^6 fibroblasts previously isolated from ligaments. Cultures of the ligament equivalents were performed with tensile charges of 1–10 g. Such tissue cultures showed that collagen fibers and fibroblasts in the neoligaments are more aligned for cultures performed with tensile loads. Interestingly, ligamentous neotissues showed good penetration within the anchorages, elastin fibers were observed along the ligament neotissues, and type III collagen was present mostly at the anchorage regions [21].

Ligament and Tendon Culture and Regeneration with Absorbable Synthetic Scaffolds

Bovine type I collagen composite matrices have also been used to reconstruct ACL ligaments, and PLLA bonded type I collagen fiber matrices have been build by assembling 500 reconstituted collagen fibers within a PLLA matrix. In vivo direct implantations showed that PLLA-collagen fibers ligament scaffolds resorb slowly and promote fibrous ingrowth. PLLA matrix allowed an increase of the initial scaffold tensile strength in comparison to pure collagen ligament scaffolds [57].

Recent uses of bioabsorbable polyesters have focused on the building of cell-seeded polymer matrices for ligament and tendon reconstructions. PGA sheets (textile nonwoven) were sampled into 40 mm length × 4 mm width specimens (thickness 100 μm) and used as randomly oriented PGA fiber scaffolds (PGA mesh). Dexon PGA sutures were unbraided and used as parallel PGA fiber scaffolds (PGA filaments). All PGA substrata were then seeded with tenocytes isolated from newborn calf tendons (15×10^6 cells/scaffold). Cultures in the PGA scaffolds were performed for 1 week (5% CO_2, 37°C). Tenocytes seeded PGA mesh (n = 45), tenocytes seeded PGA filament (n = 15), and unseeded PGA mesh and filament matrices (n = 2×5) were subcutaneously reimplanted in the dorsum of mature athymic mice for 3–12 weeks [20]. The authors screened the tendon regeneration in vitro/in vivo by gross and histological examinations and mechanical testing.

At 6 weeks the resulting neoformed tissues showed parallel, longitudinally oriented collagen fibers at the periphery regions of the initial PGA scaffold volume and less oriented or organized collagen fibers in the central part of the scaffold:

- At 6 weeks the collagen fibers longitudinal organization of the neotissues within the reimplanted tenocyte seeded PGA scaffolds was more developed when the PGA scaffolds initially showed parallel orientations of the component PGA fibers (PGA fibrous bundles or woven) rather than nonoriented or randomly oriented organizations of the PGA fibers (PGA nonwoven).
- At 8 weeks biomechanical testing demonstrated that the neoformed tissue-PGA scaffold materials have tensile strengths about one-third of that of normal calf tendons. Pure PGA scaffolds showed tensile strength about 0.18 MPa, neotendons excised at 12 weeks about 18.20±9.72 MPa while bovine tendons exhibited tensile strength about 32.80±5.21 MPa.
- However, at 10 weeks the parallel longitudinal organization of the collagen fibers was observed in the entire volume of the PGA scaffold (periphery + center), and appeared very similar to the ones present in normal bovine tendons.

Table 5 summarizes the reconstruction/tissue culture studies for ligaments and tendons.

The Future of Tendon/Ligament Regeneration

The formation of neoligaments or neotendons was long anticipated in vivo when appropriate materials and scaffolds were implanted in situ to receive the cell proliferation and biosynthesis, as well as the ingrowth of tissue extra-cellular matrix materials. However, good-quality neoligamentous or neo-tendinous tissues (quantity, organization) require adequate nutrition from the surrounding environment and biochemical/biomechanical stimulation. It has been shown that in many cases implantation of artificial materials for ligament reconstruction results in chronic foreign-body reactions, in fibrous ingrowth or encapsulation, but also in unorganized collagen fibers and immature liga-

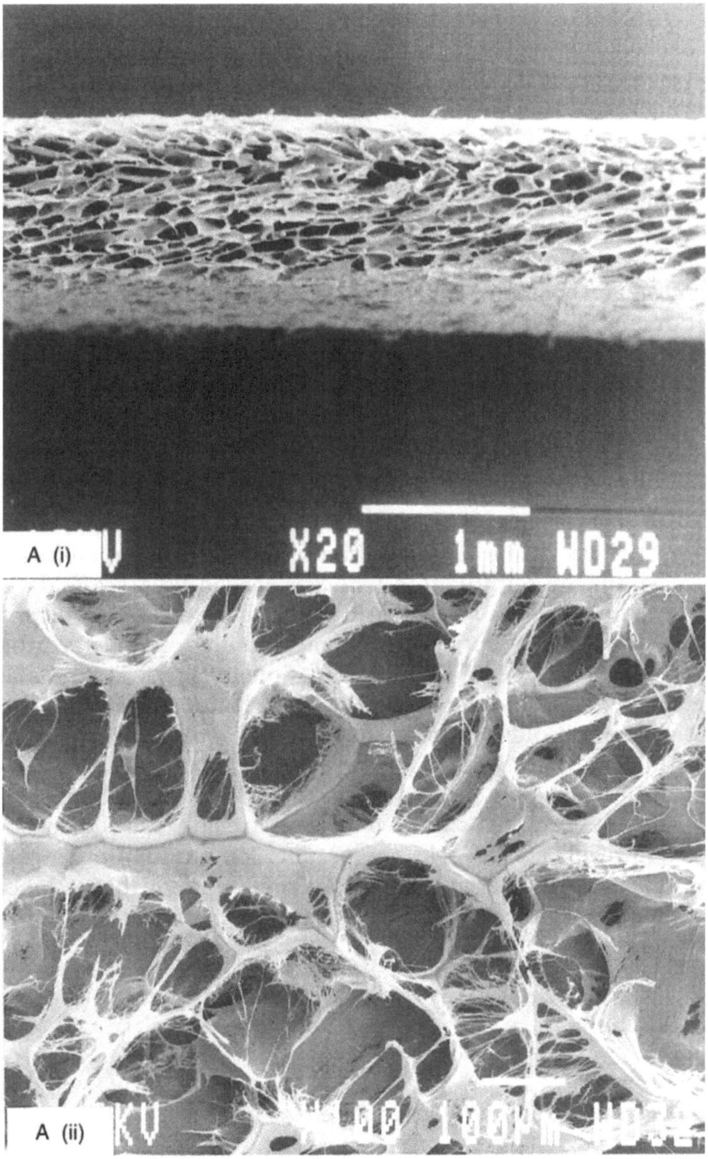

Fig. 3A

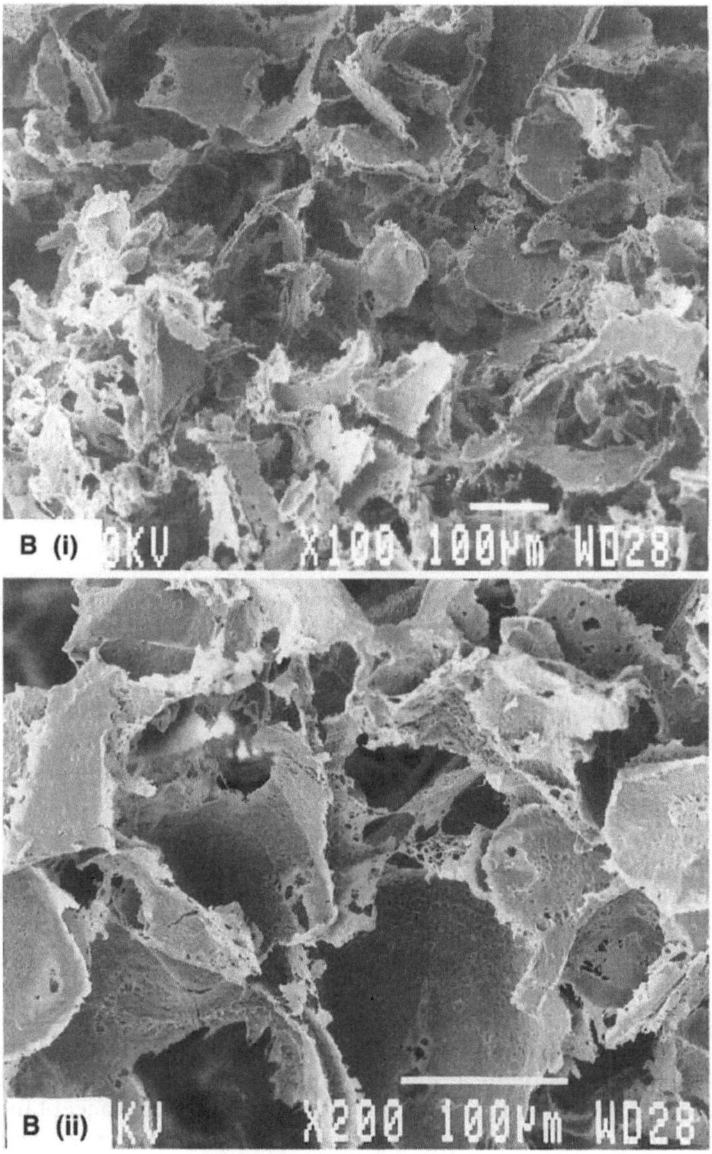

Fig. 3. Examples of bioabsorbable foams and sponges currently used to culture tissue equivalents. **A** Type I collagen (Cellagen) sponges exhibit oriented lamellar porous structures. *Above*, general cross-section, ×20; *below*, surface, ×100. **B** Polymeric foams have recently been preferred to culture and regenerate tissues. *Above*, general view, ×100; *below*, detailed view of the internal polymeric surfaces, ×200

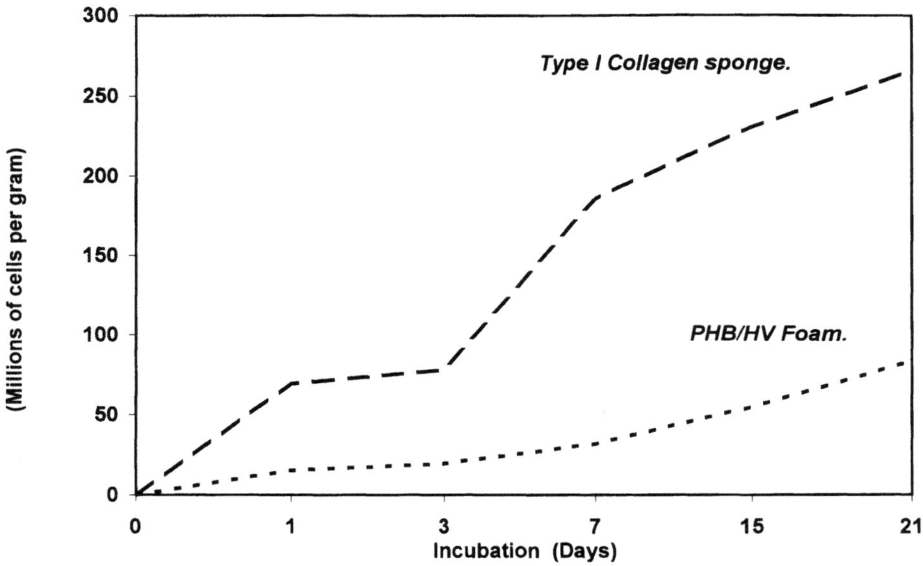

Fig. 4. Canine ACL fibroblast proliferation in three-dimensional lamellar type I collagen (Cellagen) and polyhydroxybutyrate-based structures

ment or tendon neotissues. Scaffolds made of fine, aligned carbon fibers have demonstrated some ability to induce in vivo neotendinous or neoligamentous tissues with attractive tissue quality. Figure 2 shows the collagenous fiber tissue which colonized the carbon fiber tow few months postimplantation as a human ACL reconstruction device. However, fine carbon fibers are extremely fragile and prone to break down in reduced length carbon fibers or carbon particles. In addition, carbon materials were found not to resorb rapidly but almost to migrate up to the lymph nodes. On an other hand, Dacron polyester material did not demonstrate clearly its ability to induce in situ neoligaments or neotissues of appropriate quality. Tissue formations in vivo in Dacron ligament devices more resembled scar or granulomatous tissues.

Biological grafts provide the preferred materials to replace or reconstruct ligaments. Ligament autografts and allografts (tendon or ligament structures) provide good results in terms of in situ biological acceptance, biomechanical performances and long-term biodurability. However, some problems still remain with the use of autografts and allografts: the in vivo remodeling process affects the mechanical properties of the ligament or tendon reconstructions, allografts may be infection or pathogen transmitting agents, sterilization and storage of allografts remain potential problems, the use of autografts suggests secondary morbidity and the removal of functional structures from its site. It is clear, however, that the architecture of tendon or ligament grafts corresponds to optimal structures to restore largely normal tendon or ligament tissues.

Recent research on the ligament or tendon reconstruction and regeneration relied on various targeted different methodological approaches:

- Pure organogenesis of ligament/tendon equivalents, with appropriate stimulations during the culture in vitro.
- Direct implantation of novel bioabsorbable ligament or tendon implants.
- Biomimicking the acellular structure of ligament or tendon biological grafts to reconstruct in vivo the ligament or tendon.
- Culturing in vitro isolated ligament or tendon fibroblasts within oriented artificial bioabsorbable scaffolds, then retransplanting the cell-biomaterial system to complete the ligament or tendon regeneration.

In nearly all cases the reconstruction and regeneration procedures require the development of new bioabsorbable materials (e.g., polymers, proteins), newly oriented or hierarchically structured acellular scaffolds and new methods and materials to culture biohybrid ligament or tendon equivalents.

From our own experience, the fibroblast behavior has been found to depend upon the site (ACL, posterior cruciate ligament or patellar tendon) and the source (animal models, human): stimulation results on human ligament fibroblasts (combined effect of insulin-like growth factor 1 + TGFβ on cell proliferation at 24 h: approx. 30% increase) seem to differ widely from the similar results observed in canine ligament fibroblasts (same conditions: approx. 390% increase). This was observed during growth factors or hormone stimulations of fibroblastic monolayered cultures in culture grade polystyrene 24-well plates. Recent studies have also shown that ligament fibroblasts proliferate well and diffuse through the entire volumes of collagen or polymeric porous materials (Figs. 3, 4) [5, 24, 56, 81, 95, 96, 137]. However, this does not result in the practical regeneration of real living neoligament or neotendon tissues, and efforts must now focus on the culture and formation of three-dimensional neotissues for ligaments and tendons. Improving ligament and tendon cultures requires that basic research studies determine the exact cell behavior, adequate stimulations for the cell proliferation, matrix biosynthesis, and tissue formation for animals and humans.

References

1. Aebischer P, Winn SR, Galletti PM (1988) Transplantation of neural tissue in polymer capsules. Brain Res 448:364–368
2. Aebischer P, Winn SR, Tresco PA, Jaeger CB, Greene LA (1991) Transplantation of polymer encapsulated neurotransmitter secreting cells: effect of the encapsulation technique transaction of the ASME. J Biomech Eng 113:178–183
3. Aguayo AJ, Benfey M, David S (1983) A potential for axomal regeneration in neurons of the adult mammalian nervous system. Birth Defects 19:327–340
4. Allen TD, Schor SL, Schor AM (1984) An ultrastructural review of collagen gels, a model system for cell-matrix, cell-basement membrane and cell-cell interactions. Scan Electron Microsc I:375–390
5. Alper R, Lundgren DG, Marchessault RH, Cote WA (1963) Properties of poly(β-hydroxybtyrate) I general considerations concerning the naturally occurring polymer. Biopolymers 1:545–556
6. Bakker D, van Blitterswijk VA, Hesseling SC, Koerten HK, Kuijpers W, Grote JJ (1990) Biocompatibility of a polyether urethane, polypropylene oxide, and a polyether polyester

copolymer. A qualitative and quantitative study of three alloplastic tympanic membrane materials in the rat middle ear. J Biomed Mater Res 24:489–515
7. Barrow TH (1995) Trends in the synthesis of bioabsorbable polymers medical textiles and biomedical polymers and materials. Proceedings of the Medical Textile Conference, 6–7 September 1995, Clemson, SC
8. Bell E (1991) Viewpoint tissue engineering: a perspective. J Cell Biochem 45:239–241
9. Bell E (1995) Strategy for the selection of scaffolds for tissue engineering. Tissue Eng 1(2):163–179
10. Bell E, Rosenberg M, Kemp P, Gay R, Green GD, Muthukumaran N, Nolte C (1991) Recipes for reconstituting skin. J Biomech Eng 113:113–119
11. Bell E, Scott S (1992) Tissue fabrication: reconstitution and remodeling in vitro. Mat Res Symp Proc, vol 252. Materials Research Society, pp 141–150
12. Ben-Yishay A, Grande DA, Schwartz RE et al. (1995) Repair of articular cartilage defects with collagen-chondrocyte sllografts. Tissue Eng 1(2):119–133
13. Ben-Yishay A, Grande DA, Song EK, Menche D, Pitman M (1991) The use of a cartilage tissue analog for the repair of articular cartilage. 17th Annual Meeting of the Society for Biomaterials, Scottsdale, May 1–5
14. Beumer GJ, van Blitterswijk CA, Bakker D, Ponec M (1993) A new biodegradable matrix as part of a cell seeded skin substitute for the treatment of deep skin defects: a physico-chemical characterization. Clin Mater 14:21–27
15. Beumer GJ, van Blitterswijk CA, Bakker D, Ponec M (1993) Cell-seeding and in vitro biocompatibility evaluation of polymeric matrices of PEO/PBT copolymers and PLLA. Biomaterials 14:598–604
16. Brody GA, Eisinger M, Arnoczky SP, Warren RF (1988) In vitro fibroblast seeding of prosthetic anterior cruciate ligaments. A preliminary study. Am J Sport Med 16(3):203–208
17. Brook MG (1990) Thermoplastic hydrogels. Br J Polymer 23:257–259
18. Brown AF (1984) Neutrophil and monocyte behaviour in three-dimensional collagen matrices. Scan Electron Microsc II:747–754
19. Bruin P, Smedinga J, Pennings AJ, Jonkman MF (1990) Biodegradable lysine diisocyanate-based poly(glycolide-co-caprolactone)-urethane network in artificial skin. Biomaterials 11:291–295
20. Cao Y, Vacanti JP, Ma PX, Ibarra C, Paige KT, Upton J, Langer R, Vacanti CA (1995) Tissue engineering of tendon. Mat Res Soc Symp Proc, vol 394. Materials Research Society, pp 83–87
21. Caron C, Germain L, Goulet F, Auger FA (1994) Tissue-engineering of the anterior cruciate ligament (ACL): comparative contractile properties of fibroblasts. Tissue Engineering Symposium, Taos, New-Mexico
22. Cavallaro JF, Kemp PD, Kraus KH (1994) Collagen fabrics as biomaterials. Biotechnol Bioeng 43:781–791
23. Chaput C, Duval N, Beaupré A, Guidoin R (1990) Les substituts synthétiques du ligament croisé antérieur (LCA): les matériaux et les structures. Annual Meeting, Quebec City Orthopaedic Program, Chteau Frontenac, Quebec City
24. Chaput CJ, DesRosiers EA, Assad M, Brochu M, Yahia LH, Selmani A, Rivard CH (1995) Processing biodegradable natural polyesters for porous soft-materials. II In vitro evaluation of a newly developed biodegradable polyester foam structure as potential tissue-inducing materials. In: Kossovski N, Kossowski R (eds) Advances in materials science and implant orthopaedic surgery II. NATO-Adv Sci Inst (ASI) series E, vol 294. Kluwer, Dordrecht, pp 229–246
25. Chapvil M (1977) Collagen sponge: theory and practice of medical applications. J Biomed Mater Res 11:721–741
26. Chapvil M, Gibeault D, Wang TF (1987) Use of chemically purified and cross-linked bovine pericardium as a ligament substitute. J Biomed Mater Res 21(12):1383–1393
27. Chapvil M, Holusa R (1968) Experimental experiences with the collagen sponge as hemostaticum and tampon. J Biomed Mater Res 2:245–264

28. Chapvil M, Kronenthal RL, van Winckle W (1973) Medical and surgical applications of collagen. In: Hall DA, Jackson DS (eds) International review of connective tissue research, vol 6. Academic, London, pp 1–62
29. Chapvil M, Speer DP, Holubec H, Chapvil TA, King DH (1993) Collagen fibers as a temporary scaffold for replacement of ACL in goats. J Biomed Mater Res 27:313–325
30. Chen LS, Ray J, Fisher LJ, Kawaja MD, Schinstine M, Kang UJ, Gage FH (1991) Cellular replacement therapy for neurological disorders: potential of genetically engineered cells. J Cell Biochem 45:252–257
31. Chirila TV, Constable IJ, Crawford GJ, Vijayasekaran S, Thompson DE, Chen Y-C, Fletcher WA, Griffin BJ (1993) Poly(2-hydroxyethyl methacrylate) sponges as implant materials: in vivo and in vitro evaluation of cellular invasion. Biomaterials 14:26–38
32. Christiansen D, Pins G, Wang MC, Dunn MG, Silver FH (1992) Collageous biocomposites for the repair of soft tissue injury. Mat Res Soc Symp Proc, vol 252. Materials Research Society, pp 151–158
33. Chu CC, Poudeyhimi B, Welsh L (1985) Characterization of morphological and mechanical properties of surgical mesh fabrics. J Biomed Mater Res 19:903–916
34. Chu RC, Couts RD, Yoshioka M, Harwood FL, Monosov AZ, Amiel D (1995) Articular cartilage repair using allogeneic perichondrocyte-seeded biodegradable porous polylactic acid (PLA): a tissue-engineering study. J Biomed Mater Res 29:1147–1154
35. Cifkova I, Lopour P, Vondracek P, Jelinek F (1990) Silicone rubber-hydrogel composites as polymeric biomaterials. I. Biological properties of the silicone rubber-p(HEMA) composite. Biomaterials 11:393–396
36. Cima LG, Ingber DE, Vacanti JP, Langer R (1991) Hepatocyte culture on biodegradable polymeric substrates. Biotechn Bioeng 38:145–158
37. Cima LG, Vacanti JP, Vacanti C, Ingber D, Mooney D, Langer R (1991) Tissue engineering by cell transplantation using degradable polymer substrates. J Biomech Eng 113:143–151
38. Colton CK, Avgoustiniatos ES (1991) Bioengineering in development of the hybrid artificial pancreas transactions of the ASME. J Biomech Eng 113:152–170
39. Coombes AGA, Heckman JD (1992) Gel casting of resorbable olymers. 2. In vitro degradation of bone graft substitutes. Biomaterials 13:297–307
40. Cooper ML, Hansbrough JF, Spievogel RL, Cohen R, Bartel RL, Naughton G (1991) In vivo optimization of a living dermal substitute employing cultured human fibroblasts on a biodegradable polyglycolic acid or polyglactine mesh. Biomaterials 12:243–252
41. Cooperman L, Michaeli D (1984) The immunogenicity of injectable collagen. I. A 1-year prospective study. J Am Acad Dermatol 10:638–646
42. Coutts RD, Yoshioka M, Amiel D, Hacker SA Harwood FL, Monosov (1994) A Cartilage repair using a porous polylactic acid matrix with allogeneic perichondrial cells. Orthopaedic Research Center, New Orleans
43. Dagalakis N, Flink J, Stasikelis P, Burke JF, Yannas IV (1980) Design of an artificial skin. Part III. Control of pore structure. J Biomed Mater Res 14:511–528
44. Davidson JM, Klagsbrun M, Hill KE, Buckley A, Sullivan R, Brewer PS, Woodward SC (1985) Accelerated wound repair, cell proliferation and collagen accumulation are produced by a cartilage-derived growth factor. J Cell Biol 100:1219–1227
45. Davis PA, Huang SJ, Ambrosio L, Ronca D, Nicolais L (1992) A biodegradable composite artificial tendon. J Mater Sci Mater Med 3:359–364
46. de Groot JH, Nijenhuis AJ, Bruin P, Veth RPH, Klompmaker J, Jansen HWB (1990) Use of porous biodegradable polymer implants in meniscus reconstruction. I Preparation of porous biodegradable polyurethanes for the reconstruction of meniscus lesions. Colloid Polymer Sci 268:1073–1081
47. Dehl RE (1984) Characterization of porosity in porous polymeric implant materials. Second World Congress on Biomaterials. 10th Annual Meeting of the Society for Biomaterials, Washington, April 27–May 1
48. DeLustro F, Condell RA, Nguyen MA, McPherson JM (1986) A comparative study of the biological and immunological response to medical devices derived from dermal collagen. J Biomed Mater Res 20:109–120

49. Demetriou AA, Levenson SM, Novikoff PM, Novikoff AB, Chowdhury NR, Whiting J Reisner A, Chowdhury JR (1986) Survival, organization and function of microcarrier-attached hepatocytes transplanted in rats. Proc Natl Acad Sci USA 83:7475–7479
50. Doillon CJ, Silver FH (1986) Collagen-based wound dressing: effects of hyaluronic acid and fibronectin on wound healing. Biomaterials 7:3–8
51. Doillon CJ, Silver FH (1986) Collagen-based wound dressing: control of the pore structure and morphology. J Biomed Mater Res 20:1219–1228
52. Doillon CJ, Silver FH, Berg RA (1987) Fibroblast growth on a porous collagen sponge containing hyaluronic acid and fibronectin. Biomaterials 8:195–200
53. Doillon CJ, Wasserman AJ, Berg RA, Silver FH (1988) Behaviour of fibroblasts and epidermal cells cultivated on analogues of extracellular matrix. Biomaterials 9:91–96
54. Doillon CJ, Whyne CF, Berg RA, Olson RM, Silver FH (1984) Fibroblast-collagen sponge interactions and the spatial deposition of newly synthetized collagen fibers in vitro and in vivo. Scan Electron Microsc III:1313–1320
55. Domb AJ (1990) Biodegradable polymers derived from amino-acids. Biomaterials 11:686–689
56. Doyle C, Tanner EI, Bonfield W (1991) In vitro and in vivo evaluation of polyhydroxybutyrate and of polyhydroxybutyrate reinforced with hydroxyapatite. Biomaterials 12:841–847
57. Dunn MG, Liesch JB, Maxian SH et al. (1994) Fibroblast-seeded collagen scaffolds for ACL reconstruction. Orthopaedic Research Society, New Orleans, pp 36–37
58. Elema H, de Groot JH, Nijienhuis AJ, Pennings AJ, Veth RPH, Klompmaker J, Jansen HWB (1990) Use of biodegradable polymer implants in meniscus reconstruction. II Biological evaluation of porous biodegradable polymer implants in menisci. Colloid Polymer Sci 268:1082–1088
59. Ellingsworth LR, DeLustro F, Brennan JE, Sawamura S, McPherson J (1986) The human response to reconstitued bovine collagen. J Immunol 136:877–882
60. Engelberg I, Kohn J (1991) Physico-chemical properties of degradable polymers used in medical applications: a comparative study. Biomaterials 12:292–304
61. Feldman D, Estridge T (1984) Factors affecting soft tissue ingrowth into porous implants. Second World Congress on Biomaterials. Annual Meeting of the Society for Biomaterials, Washington, April 27–May 1
62. Folkman J, Hochberg M (1973) Self-regulation of growth in three dimension. J Exp Med 138:745–753
63. Freed LE, Vunjak-Novakovic G, Biron RJ, Eagles DB, Lesnoy DC, Barlow SK, Langer R (1994) Biodegradable polymer scaffolds for tissue engineering. Biotechnology 12:689–693
64. Freed LE, Marquis JC, Nohria A, Emmanual J, Mikos AG, Langer R (1993) Neocartilage formation in vitro and in vivo using cells cultured on synthetic biodegradable polymers. J Biomed Mater Res 27:11–23
65. Freed WJ, Poltorak M, Takashima H, LaMarca ME, Ginns EI (1991) Brain grafts and Parkinson's disease. J Cell Biochem 45:261–267
66. Frenkel SR, Pachence JM, Alexander H (1993) Optimization of a cell-seeded collagen implant for cartilage repair. 19th Annual Meeting of the Society for Biomaterials, Birmingham, USA, April 28–May 2
67. Garg AK, Berg RA, Silver FH, Garg HG (1989) Effect of proteoglycans on type I collagen fibre formation. Biomaterials 10:413–419
68. Goldstein JD, Tria AJ, Zawadsky JP, Kato YP, Christiansen D, Silver JH (1989) Development of a reconstitued collagen tendon prosthesis: a preliminary implantation study. J Bone Joint Surg Am 71:1183–1192
69. Gorham SD, Hylan TP, Franch DA, Willins MJ (1990) Cellular invasion and breakdown of three different collagen films in the lumbar muscle of the rat. Biomaterials 11:113–118
70. Graham D, Alexander JJ, Niguel R (1990) Co-culture of aortic endothelial and smooth muscle cells using porous polycarbonate filters. 16th Annual Meeting of the Society for Biomaterials, Charleston, USA, May 20–23

71. Grande DA, Halberstadt C, Schwartz R, Linton J, Naughton G (1993) Evaluation of extracellular matrix scaffolds for engineering of articular cartilage grafts. 39th Annual Meeting Orthopaedic Research Society, San Francisco, February 15–18
72. Guénard V, Valentini RF, Aebisher P (1990) Surface microgeometry of synthetic guidance channels controls the morphology of nerve regeneration. 16th Annual Meeting of the Society for Biomaterials, Charleston, USA, May 20–23
73. Guidoin R, King MW, Marceau A, Cardou A, de la Faye D, Legendre JM, Blais P (1987) Textile arterial prosthesis: is water permeability equivalent to porosity? J Biomed Mater Res 21:65–87
74. Halberstadt C, Anderson P, Bartel R, Cohen R, Naughton G (1992) Physiological cultured skin substitutes for wound healing. Mat Res Soc Symp Proc, vol 252. Materials Research Society, pp 323–330
75. Harris WA, Holt CE, Smith TA, Gallenson N (1985) Growth cones of developing retinal cells in vivo, on culture surfaces and in collagen matrices. J Neurosci Res 13:101–122
76. Heineken FG, Skalak R (1991) Tissue engineering: a brief overview. J Biomech Eng 113:111–112
77. Hoppen HJ, Leenslag JW, Pennings AJ, van der Lei B, Robinson PH (1990) Two-ply biodegradable nerve guide: basic aspects of design, construction and biological performance. Biomaterials 11:286–290
78. Hollinger JO (1983) Preliminary report on the osteogenic potential of a biodegradable copolymer of polylactide (PLA) and polyglycolide (PGA). J Biomed Mater Res 17:71–82
79. Huang D, Chang TR, Aggarwal A, Lee RC, Ehrlich HP (1993) Mechanism and dynamics of mechanical strengthening in ligament-equivalent fibroblast-populated collagen matrices. Ann Biomed Eng 21:289–305
80. Ide C, Tohyama K, Yokota R, Nitatori T, Onodera S (1983) Schwann cell basal lamina and nerve regeneration. Brain Res 288:61–75
81. Inque Y, Kamiya N, Yamamoto Y, Chujo R (1989) Microstructure of commercially available poly(3-hydroxybutyrate-co-3-hydroxyvalerate). Macromol 22:3800–3802
82. Ito K, Fujisato T, Ikada Y (1992) Implantation of cell-seeded biodegradable polymers for tissue reconstruction. Mat Res Soc Symp Proc, vol 252. Materials Research Society, pp 359–365
83. Iversen PL, Partlow LM, Stensaas LJ, Maatamed F (1981) Characterization of a variety of standart collagen substrates: ultrastructure, uniformity and capacity to bind and promote growth of neurons. In Vitro 17:540–551
84. Jain MK, Berg RA, Tandon GP (1990) Mechanical stress and cellular metabolism in living soft tissue composites. Biomaterials 11:465–472
85. Jauregui HO, Gann KL (1991) Mammalian hepatocytes as a foundation for treatment in human liver failure. J Cell Biochem 45:359–365
86. Jeyanthi R, Panduranga Rao K (1990) In vivo biocompatibility of collagen-poly(-hydroxyethyl methacrylate) hydrogels. Biomaterials 11:238–243
87. Jirtle RL, Biles C, Michalopoulos G (1980) Morphological and histochemical analysis of hepatocytes transplanted into syngeneic hosts. Am J Pathol 101:115–123
88. Kato YP, Christiansen DL, Hahn RA, Shieh S-J, Goldstein JD, Silver FH (1989) Mechanical properties of collagen fibres: a comparison of reconstitued and rat tail tendon fibres. Biomaterials 10:38–42
89. Kato YP, Dunn MG, Tria AJ, Zawadsky JP, Silver FH (1991) Preliminary assessment of a collagen fiber ACL prosthesis. 16th Annual Meeting of the Society for Biomaterials, Charleston, May 20–23
90. Kato YP, Silver FH (1990) Formation of continuous collagen fibres: evaluation of biocompatibility and mechanical properties. Biomaterials 11:169–175
91. Kim W, Vacanti JP, Upton J, Vacanti MP, Vacanti CA (1993) The repair of cranial defects in rats with cartilage grown from chondrocytes on synthetic polymer implants. 39th Annual Meeting Orthopaedic Research Society. San Francisco, February 15–18
92. Klein DM, Tyomkin I, Miller B, Rebenfeld L (1990) Pore volume distribution of artificial skin by liquid extrusion analysis. J Appl Biomater 1:137–141

93. Klompmaker J, Jansen HWB, Veth RPH, de Groot JH, Nijenhuis AJ, Pennings AJ (1991) Porous polymer implant for repair of meniscal lesions: a preliminary study in dogs. Biomaterials 12:810–816
94. Klompmaker J, Jansen HWB, Veth RPH, Nielsen HKL, de Groot JH, Pennings AJ (1993) Porous implants for knee joint meniscus reconstruction: a preliminary study on the role of pore sizes in ingrowth and differentiation of fibrocartilage. Clin Mater 14:1–11
95. Knowles JC, Hastings GW (1990) Physical characteristics of "PHB," a degradable polymer. In: Heimke G, Soltész U, Lee AJC (eds) Clinical implant materials, advances in biomaterials, vol 9. Elsevier, Amsterdam, pp 191–198
96. Knowles JC, Hastings GW, Ohta H, Niwa S, Boeree N (1992) Development of a degradable composite for orthopaedic use: in vivo biomechanical and histological evaluation of two bioactive degradable composites based on the polyhydroxybutyrate. Biomaterials 13:491–496
97. Laitinen O, Törmälä P, Taurio R, Skutnabb K, Saarelainen K, Iivonen T, Vainionpää S (1992) Mechanical properties of biodegradable ligament augmentation device of poly(L-actide) in vitro and in vivo. Biomaterials 13:1012–1016
98. Langer R, Moses M (1991) Biocompatible controlled release polymers for delivery of polypeptides and growth factors. J Cell Biochem 45:340–345
99. Langer R, Vacanti JP, Vacanti CA et al. (1995) Tissue engineering: biomedical applications. Tissue Eng 1(2):151–161
100. Larsen NE, Kling MB, Balazs EA, Leshchiner EA (1990) Hylan gel for soft tissue augmentation. 16th Annual Meeting of the Society for Biomaterials, Charleston, USA, May 20–23
101. Li ST, Yuen D, Li PC, Rodkey WG, Stone KR (1994) Collagen as a biomaterial: an application in knee meniscal fibrocartilage regeneration. Mat Res Soc Symp Proc vol 331. Materials Reasearch Society
102. Lim H-S, Han B-K, Kim J-H, Peshwa MV, Hu W-S (1992) Spatial distribution of mammalian cells grown on macroporous microcarriers with improved attachment kinetics. Biotechnol Prog 8:486–493
103. Liu SQ, Ito Y Imanishi Y (1992) Cell growth on immobilized cell growth factor. I. Acceleration of the growth of fibroblast cells on insulin-immobilized polymer matrix in culture medium without serum. Biomaterials 13:50–58
104. LiVecchi AB, Tombes RM, LaBerge M (1994) In vitro chondrocytes collagen deposition within porous HDPE: substrate microstructure and wettability effects. J Biomed Mater Res 28:839–850
105. Lo H, Kadiyala S, Guggino SE, Leong KW (1993) Biodegradable foams for cell transplantation. Mat Res Soc Symp Proc, vol 331. Materials Research Society
106. Lopour P, Vondracek P, Janatova V, Sulc J, Vacik J (1990) Silicone rubber-hydrogel composites as polymeric biomaterials. II. Hydrophilicity and permeability to water-soluble low-molecular-weight compounds. Biomaterials 11:397–402
107. Lowenberg B, Todescan R, Shiga A, Davies JE (1990) Bone formation in three-dimensional collagen matrices in vitro. 16th Annual Meeting of the Society for Biomaterials, Charleston, USA, May 20–23
108. Lucas PA, Laurencin C, Syftestad GT, Domb A, Goldberg VM, Caplan AI, Langer R (1990) Ectopic induction of cartilage and bone by water-soluble proteins from bovine bone using a polyanhydride delivery vehicle. J Biomed Mater Res 24:901–911
109. Lustgarten JH, Proctor M, Haroun RI, Avellino AM, Pindzola AA, Kliot M (1991) Semipermeable polymer tubes provide a microenvironment for in vivo analysis of dorsal root regeneration transaction of the ASME. J Biomech Eng 113:184–188
110. Lyndon MJ (1986) Synthetic hydrogels for cell adhesion studies. Br Polymer J 18:22–27
111. Marchand R, Woerly S (1990) Transected spinal cords grafted with in situ self-assembled collagen matrices. Neurosciences 1:45–60
112. Marchant R, Hiltner A, Hamlin C, Rabinovitch A, Slobodkin R, Anderson JM (1983) In vivo biocompatibility studies. I. The cage implant system and a biodegradable hydroge. J Biomed Mater Res 17:301–325

113. Marks MG, Doillon C, Silver FH (1991) Effects of fibroblasts and basic fibroblast growth factor on facilitation of dermal wound healing by type I collagen matrices. J Biomed Mater Res 25:683–696
114. Mauch C, Adelmann-Grill B, Hatamochi A, Krieg T (1989) Collagenase gene expression in fibroblasts is regulated by a three-dimensional contact with dollagen FEBS Lett 250:301–305
115. Mauch C, Hatamochi A, Scharffetter K, Krieg T (1988) Regulation of collagen synthesis in fibroblasts within a three-dimensional collagen gel. Exp Cell Res 178:493–503
116. Meade KR, Silver FH (1990) Immunogenicity of collagenous implants. Biomaterials 11:176–180
117. Migliaresi C, Ambrosio L, Guida G, Kolarik J, Nicolais L, Ronca D (1984) A polyester fiber reinforced poly-2-hydroxyethylmethacrylate artificial tendon. Second World Congress on Biomaterials, 10th Annual Meeting of the Society for Biomaterials, Washington, April 27-May 1
118. Mikos AG, Bao Y, Cima LG, Ingber DE, Vacanti JP, Langer R (1993) Preparation of polyglycolic acid bonded fiber structures for cell attachment and transplantation. J Biomed Mater Res 27:183–189
119. Mikos AG, Sarakinos G, Leite SM, Vacanti JP, Langer R (1993) Laminated three-dimensional biodegradable foams for use in tissue engineering. Biomaterials 14:323–330
120. Mikos AG, Sarakinos G, Lyman MD, Ingber DE, Vacanti JP, Langer R (1993) Prevascularization of porous biodegradable polymers Biotechnol Bioeng 42:716–723
121. Monaham JJ (1991) Past progress and future trends in cell transplantation. J Cell Biochem 45:317–318
122. Mooney DJ, Breuer C, McNamara K et al. (1995) Fabricating tubular devices from polymers of lactic acid and glycolic acid for tissue engineering. Tissue Eng 1(2):107–119
123. Morris TH, Yuen D, Li ST, Archibald SJ, Madison R (1990) Collagen nerve guiding conduits in peripheral nerve repair. 16th Annual Meeting of the Society for Biomaterials, Charleston, USA, May 20–23
124. Mow VC, Ratcliffe A, Rosenwasser MP, Buckwalter JA (1991) Experimental studies on repair of large osteochondral defects at a high weight bearing area of the knee joint: a tissue engineering study. Transactions of the ASME. J Biomechan Eng 113:199–207
125. Murray JB, Brown L, Langer R, Klagsbrun M (1983) A micro-sustained release system for epidermal growth factor. In vitro 19:743–748
126. Nowack H, Gay S, Wick G, Becker U, Timple R (1976) Preparation and use in immunology of antibodies for type I and II collagen and procollagen. J Immunol Methods 12:117–124
127. Ohgushi H, Dohi Y, Tamai S, Tabat S (1993) Osteogenic differentiation of marrow stromal stem cells in porous hydroxyapatite ceramics. J Biomed Mater Res 27:1401–1407
128. Paige KT, Vacanti CA (1995) Engineering new tissue: formation of new cartilage. Tissue Eng (2):97–106
129. Pierce GF, Mustoe TA, ALtrock BW, Deuel TF, Thomason A (1991) Role of platelet-derived growth factor in wound healing. J Cell Biochem 45:139–326
130. Pietrucha K (1991) New collagen implant as dural substitute. Biomaterials 12:320–323
131. Poudeyhimi B (1989) Porosity of surgical mesh fabrics: new technology. J Biomed Mater Res 23:145–152
132. Prewitt JM, Homsy CA (1984) Porosity determination of a low modulus implant material. Second World Congress on Biomaterials, 10th Annual Meeting of the Society for Biomaterials. Washington, April 27-May 1
133. Puelacher WC, Kim SW, Vacanti JP, Schloo B, Mooney D, Vacanti CA (1994) Tissue-engineered growth of cartilage: the effect of varying the concentration of chondrocytes seeded onto synthetic polymer matrices. Int J Oral Maxillofac Surg 23:49–53
134. Pulapura S, Kohn J (1992) Trends in the development of bioresorbable polymers for medical applications. J Biomed Appl 6:216–250
135. Reissis N, Downes S, Bentley G (1994) Cartilage repair using a new synthetic polymer. Orthopaedic Research Society, New Orleans

136. Rich D, Johnson E, Zhou L, Grande D (1994) The use of periostal cell/polymer tissue constructs for the repair of articular cartilage defects. Orthopaedic Research Society, New Orleans
137. Rivard C-H, Chaput CJ, DesRosiers EA, Yahia LH, Selmani A (1995) Fibroblast seeding and culture in biodegradable porous substrates. J Appl Biomater 6:65–68
138. Roorda WE, de Bleyser J, Jundinger HE, Leyte JC (1990) Nuclear magnetic resonance of water in hydrogels. Biomaterials 11:17–23
139. Rosen JM, Padilla JA, Nguyen KD, Padilla MA, Sabelman EE, Pham HN (1990) Artificial nerve graft using collagen as an extracellular matrix for nerve repair compared with sutured autograft in a rat model. Ann Plast Surg 25:375–387
140. Rowling PJE, Raxworthy MJ, Wood EJ, Kearney JN, Cunliffe WJ (1990) Fabrication and reorganisation of dermal equivalents suitable for skin grafting after major cutaneous injury. Biomaterials 11:181–185
141. Roy F, DeBlois C, Doillon CJ (1993) Extracellular matrix analogs as carriers for growth factors: in vitro fibroblast behavior. J Biomed Mater Res 27:389–397
142. Rubin AL, Stenzel KH (1969) Structure and use of collagen biomaterials. Stark L, Agarwal G (eds) Plenum, New York, p 157
143. Rubin RM, Marshall JL (1975) Porous hydrophilic polymer : good and bad news in the orthopedic application of cruciate ligament substitution. J Biomed Mater Res 9:375–380
144. Sabelman EE, Atagi TA, Keeley RD, Kadlcik P (1993) In vitro model of axomal regeneration in artificial nerve grafts. 19th Annual Meeting of the Society for Biomaterials, Birmingham, USA, April 28–May 2
145. Shabana A, Ouhayoun JP, Forest N (1990) The effects of natural coral on gingival fibroblasts cultured in three-dimensional collagen gels. 16th Annual Meeting of the Society for Biomaterials. Charleston, USA, May 20–23
146. Silberman M (1990) In vitro systems for inducers of cartilage and bone development. Biomaterials (89) 12:47–49
147. Sittinger M, Bujia J, Minuth WW et al. (1994) Engineering of cartilage tissue using bioresorbable polymer carriers in perfusion culture. Biomaterials 15(6):451–456
148. Smetana K Jr, Vacik J, Souckova D, Krcova Z, Sulc J (1990) The influence of hydrogel functional groups on cell behavior. J Biomed Mater Res 24:463–470
149. Smetana K Jr, Vacik J, Souckova D, Krcova Z, Sulc J (1990) The influence of hydrogel functional groups on cell behavior. J Biomed Mater Res 24:463–470
150. Sprincl L, Kopecek J, Lim D (1971) Effect of porosity of heterogeneous poly(glycol methacrylate) gels on the healing-in of test implants. J Biomed Mater Res 5:447–458
151. Srivastava S, Gorham SD, Courtney JM (1990) The attachment and growth of an established cell line on collagen, chemically modified collagen and collagen composite surfaces. Biomaterials 11:162–168
152. Srivastava S, Gorham SD, French DA, Shivas AA, Courtney JM (1990) In vivo evaluation and comparison of collagen, acetylated collagen and collagen/glycosaminoglycan composite films and sponges as candidate biomaterials. Biomaterials 11:155–161
153. Stenzel KH, Miyata T, Rubin AL (1974) Collagen as a biomaterial. Annu Rev Biophys Bio Eng 3:231–253
154. Stone KR, Rodkey WG, Webber R, McKinney L, Steadman JR (1990) Prosthetic meniscal replacement in dogs using resorbable copolymeric collagen crostheses. 16th Annual Meeting of the Society for Biomaterials. Charleston, USA, May 20–23
155. Stone KR, Rodkey WG, Webber RJ, McKinney L, Steadman JR (1990) Future directions collagen-based prostheses for meniscal regeneration. Clin Orthop 252:129–135
156. Suzuki S, Matsuda K, Isshiki N, Tamada Y, Ikada Y (1990) Experimental study of a newly developed bilayer artificial skin. Biomaterials 11:356–360
157. Tanio T, Fukui T, Shirakura Y, Saito T, Tomita K, Kaiho T, Masamune S (1982) An extracellular poly(3-hydroxybutyrate) depolymerase from alcaligenes faecalis. Eur J Biochem 124:71–77
158. Tavis MJ, Harney JH, Thornton JW, Bartlett RH, Woodroff EA (1975) Modified collagen membrane as a skin substitute: preliminary studies. J Biomed Mater Res 9:285–301

159. Todescan R, Davies JE (1992) Reconstitued collagen produces different healing reactions in bony and soft tissue compartments. Mat Res Soc Symp Proc, vol 252. Materials Research Society pp 151–165
160. Vacanti CA, Cima LG, Ratkowski D, Upton J, Vacanti JP (1992) Tissue engineered growth of new cartilage in the shape of a human ear using synthetic polymers seeded with chondrocytes. Mat Res Soc Symp Proc, vol 252. Materials Research Society pp 367–373
161. Vacanti CA, Kim W, Mooney D, Schloo B, Vacanti JP (1993) Tissue engineered growth organized bone containing marrow, utilizing synthetic polymer templates. Orthopaedic Research Society, San-Francisco, February 15–18
162. Vacanti CA, Kim W, Upton J, Schloo B, Vacanti JP (1993) Tissue engineered composites of bone and cartilage using synthetic polymers seeded with two cell types. Orthopaedic Research Society, San-Francisco, February 15–18
163. Vacanti CA, Langer R, Schloo B, Vacanti JP (1991) Synthetic polymers seeded with chondrocytes provide a template for new cartilage formation. Plast Reconstr Surg 88:753–759
164. Vacanti JP, Morse MA, Saltzman WM, Domb AJ, Perez-Atayde A, Langer R (1988) Selective cell transplantation using biodegradable artificial polymer as matrices. J Pediatr Surg 23:3–9
165. van der Rest M, Dublet B, Champliaud M-F (1990) Fibril-associated ollagens. Biomaterials (89) 11:28–31
166. von Schroeder HP, Kwan M, Amiel D, Coutts RD (1990) The use of polylactic acid matrix and periostal grafts for the reconstruction of rabbit knee articular defects. 16th Annual Meeting of the Society for Biomaterials, Charleston, USA, May 20–23
167. von Schroeder HP, Kwan M, Amiel D, Coutts RD (1991) The use of polylactic acid matrix and periostal grafts for the reconstruction of rabbit knee articular defects. J Biomed Mater Res 25:329–339
168. Vroemen JPAM, Buurman WA, Schutte B, Maessen JG, van der Linden CJ, Kootstra G (1988) The cytokinetic behavior of donor hepatocytes after syngenic hepatocyte transplantation into the spleen. Transplantation 45:600–607
169. Wald HL, Sarakinos G, Lyman MD, Mikos AG, Vacanti JP, Langer R (1993) Cell seeding in porous transplantation devices. Biomaterials 14:270–278
170. Wasserman AJ, Kato YP, Christiansen D, Dunn MG, Silver FH (1989) Achilles tendon replacement by a collagen fiber prosthesis: morphological evaluation of neotendon formation. Scanning 3:1183–1200
171. Wesolowski SA, Fries CC, Domingo KT, Lebig WJ, Sawer PN (1963) The compound prosthetic vascular graft: a pathological survey. Surgery 58:19–44
172. Wichterle O (1960) Hydrophilic gels for biological use. Nature 185:117–118
173. Williams DF (1982) Review biodegradation of surgical polymers. J Mat Sci 17:1233–1246
174. Woerly S, Lavalée C, Marchand R (1992) Intracerebral implantation of ionic synthetic hydrogels: effect of polar substrata on astrocytis and axons. J Neural Transplant Plast 3:21–34
175. Woerly S, Marchand R (1990) Brain tissue and neural graft reactivity to the implantation in situ self-assembled collagen matrices. Restorative Neurol Neurosci 1:369–385
176. Woerly S, Marchand R (1991) Collagen chondroitin-6-Sulfate hydrogel implants in CNS lesion cavities favor glial repair, the differentiation of co-implanted neurons and the growth of axons. Restorative Neurol Neurosci 3:95–99
177. Woerly S, Marchand R, Lavallée C (1990) Interactions of copolymeric poly(glyceryl methacrylate)-collagen hydrogels with neural tissues: effects of structure and polar groups. Biointeractions '90, Oxford, August 21–23
178. Woerly S, Marchand R, Lavallée C (1990) Intercerebral implantation of synthetic polymer/biopolymer matrix: a new perspective for brain repair. Biomaterials 11:97–107
179. Woodhouse JC (1938) Esters of methacrylic acid. US patent 2 129 722

180. Yannas IV (1977) Multi-layer membrane useful as synthetic skin. US patent 4 060 081
181. Yannas IV (1988) Regeneration of skin and nerve by use of collagen templates. In: Nimni M (ed) Collagen: biochemistry, biotechnology and molecular biology, vol 3. CRC, Boca Raton, p 37
182. Yannas IV, Burke JF, Gordon PL, Huang C, Rubenstein RH (1980) Design of an artificial skin. II. Control of chemical composition. J Biomed Mater Res 14:107–131
183. Yannas IV, Burke JF, Orgill D, Skrabut EM (1984) Skin regeneration by use of a bioreplaceable polymeric template. Second World Congress on Biomaterials, 10th Annual Meeting of the Society for Biomaterials, Washington, April 27-May 1
184. Yannas IV, Burke JF, Orgill DP, Skrabut EM (1982) Wound tissue can utilize a polymeric template to synthetize a functional extension of skin. Science 215:174–176
185. EYannas IV, Orgill DP, Silver J, Norreguard NT, Zervas NT, Schoene WC (1987) Regeneration of sciatic nerve across 15 mm gap by use of a polymeric template. In: Gebelin CG (ed) Advances in biomedical polymers. Plenum, NewYork, p 1
186. El-Ghannam A, Ducheyne P, Shapiro IM (1995) Bioactive material template for in vitro synthesis of bone. J Biomed Mater Res 29:359–370
187. Dunn MG, Liesch JB, Tike ML, Zawadsky JP (1995) Development of fibroblast-seeded ligament analogs for ACL reconstruction. J Biomed Mater Res 29:1363–1371
188. Weiss R (1994) Biotech textiles with fabric makers' help, scientists are manufacturing living tissues. Health News, Washington Post, July 5, pp 7–9
189. Freed LE, Vunjak-Novakovic G, Biron RJ, Eagles DB, Lesnoy DC, Barlow SK, Langer R (1994) Biodegradable polymer scaffolds for tissue engineering. Biotechnology 12:689–693
190. Langer R, Vacanti JP (1993) Tissue engineering. Science 260:920–926

III.2 Biochemical and Mechanical Stimulation of Knee Fibroblasts for Regeneration of Ligaments In Vitro

S. Méthot, E.A. DesRosiers, C.H. Rivard, and L'H.Yahia

Introduction

The anterior cruciate ligament (ACL) is commonly believed to have a very limited intrinsic healing potential [64]. Therefore damaged ACLs are most often surgically treated, resulting in variously successful outcomes. For example, autografting damages a healthy tissue and produces a mechanically abnormal substitute. In addition, although most current synthetic prostheses are intrinsically more resistant than the natural ligament, they do not have the advantage of being persistently repaired by endogenous cells. Moreover, they lack mechanoreceptors able to provide a feedback signal to excessive motion. Therefore they usually wear and fail rapidly under the strenuous fatigue conditions prevailing in the knee. Reconstructing a living tissue with active cells now appears to be a very appealing long-term solution.

As for most soft tissues, ligament wounds usually heal through a scarring and remodeling process [33]. The ligament trauma releases collagenous debris, which is chemotactic for the various cell types involved in the healing process. Some blood vessels may also be ruptured and spill blood into the neighboring tissues. This in turn activates the coagulation sequence and leads to the formation of a cloat and to the activation of proteolytic enzymes that initiates the wound cleansing. Furthermore, platelet aggregation, the presence of white blood cells, and the exposure of endothelial cells increase the local concentration of various growth factors. A complex biochemical environment is thus instated at the wound site. The origin of the cells involved in the healing process remains uncertain, mainly because of the indistinct scar-ligament boundary. It has been observed in tendons that a fraction of the epitendineum fibroblasts are converted into active macrophages [50]. Some monocytes are also transformed into macrophages and may complete the wound cleansing. The synthesis of new matrix is accomplished by fibroblasts recruited by chemotactism, epitendineum, and especially endotendineum [49]. The scar tissue is not organized and possesses biochemical composition and mechanical properties diverging widely from those of the initial healthy tissue. The collagen fibers are deposited without preferential orientation; the matrix is hypervascularized and contains abnormally high levels of collagen type III, water and glycoaminoglycans (GAG) [41]. The scar is thereafter remodeled during the healing process. Under the combined effects of mechanical and biochemical stimulation, the healing ligament's composition and structure are slowly modified toward the normal state. The turnover rate of the matrix is high in the beginning of the remodeling process but decreases slowly to its normal level. The remodeling process is almost ended in 3–4 months, but the structure and composition of the scar still differs considerably from the initial tissue after 40 weeks [33]. Therefore the repaired ligament tissue remains very weak. The weakness of the tissue is partially compensated by its hypertrophy. However, in spite of this artifact the ultimate load reaches at most 60% of the value of a healthy tissue, and the ligament remains much less rigid [32]. This additional laxity causes pain and allows abnormal joint motions that may degenerate into articular tissue degradation (e.g., arthritis) and further ruptures.

Wound healing is the macroscopic result of the microscopic actions of legions of fibroblasts working jointly, where each cell reacts independently and individually to its local environment. This local environment can be defined almost completely by biochemical and mechanical parameters. As ligament properties are based essentially on the composition and structure of its matrix, our work focused on obtaining practical knowledge about proliferative and matrix synthesis response of ACL, PCL (posterior cruciate ligament), and PT (patellar tendon) fibroblasts to modification of their local environment. The problem is challenging, as the cell response may be influenced by a broad variety of factors. It soon became clear that the study model needed to be simplified to reduce the number of uncontrolled variables. To investigate how the cells react to a certain environment, many investigators used living animals to compare the response of tissue submitted to biochemical and mechanical stimulation. This approach yields a very good appreciation of physiological response; however, the results can be biased by a wide range of uncontrolled variables. On the other hand, in vitro studies, in spite of being remote from the clinical conditions, offer a much more controllable environment in which both the mechanical input and the cellular output can be easily quantified to allow a good reproducibility and the sequential isolation of virtually every variable that may influence the cell's behavior [51].

Biochemical Stimulation

Growth factors have a much more complex role than their name implies. They are peptidic signals ensuring intercellular communication. They may indeed stimulate cell growth, but may also inhibit it, induce cell differentiation, modify protein synthesis, activate metabolic pathways, and have many other effects. They act at different levels, through endocrine, paracrine, and autocrine pathways. Usually involved in every healing and remodeling processes, growth factors act on the recruiting of cells by their chemotactic effects and modulate cellular activity by regulating proliferation and protein synthesis. It would thus be extremely beneficial to take advantage of this potential for guiding the healing process.

Each cell type may react differently to growth factors. Furthermore, growth factors form an intricate and complex network of interactions; their effect is contextual and depends on the presence of other growth factors, the sequence of exposure, and the state of development and differentiation of the target cells. This renders investigation difficult, and gives them an apparently unpredictable effect on cell behavior.

The diversity of factors appears surprisingly limited. Up to now, in spite of considerable efforts, only about three dozen have been identified. Of these we have selected five, based on their high potential for involvement in ligament healing: insulin, insulin-like growth factor 1 (IGF-1), transforming growth factor β (TGF-β_1), platelet-derived growth factor AB (PDGF-AB) and epidermal growth factor (EGF). We explored the interactions between these growth factors at various concentrations to detect positive or negative synergistic effects [27, 29].

The first step was to select and adapt a satisfying medium. To avoid the variability associated with common serum-containing formulations we used a totally defined culture medium [24] modified from MCDB110s [11] original formulation. Canine ACL, PCL, and PT fibroblasts were transferred at the second passage in defined medium and synchronized at G_0 by nutrient depletion. The growth factors were then added alone or in two-by-two combination following a 5×5 logarithmic concentration matrix. The resultant cell proliferation was then assayed using the cellular conversion of a tetrazolium salt into a formazan-colored derivative, while collagen and proteoglycan (PG) productions were quantified using [^{14}C]glycine and $Na_2[^{35}S]O_4$. Type I/III collagen ratios and PG hydrodynamic size distributions were determined by SDS-PAGE electrophoresis and gel filtration chromatography, respectively.

Following proper block design and multifactorial statistical analysis, we observed the following. Cells from the two cruciate ligaments had a similar response to growth factors while the cells from the PT had a significantly different response [26]. For the cruciates the highest proliferation levels ranked as follows: EGF≈PDGF-AB>IGF-1>insulin≈ TGF-β_1. Neither PDGF-AB nor insulin had a significant effect by itself on collagen production. EGF slightly decreased collagen production and the type I/III collagen ratio, while both TGF-β_1 and IGF-1 increased the same parameters. EGF inhibited stimulation induced by TGF-β_1. Similarly, insulin decreased the response to IGF-1. PG production was significantly increased by all the growth factors in this study, with TGF-β_1 having the strongest effect. Small PG hydrodynamic size was correlated with high level of PG biosynthesis (Table 1).

The most remarkable response was observed in the presence of TGF-β_1. Our results show that this growth factor promotes proliferation as well as synthesis of small PG (as observed by [79]) and collagen, mostly of type I (also reported by [19]). TGF-β_1 acted in positive synergism with PDGF-AB at low TGF-β_1 concentration and inhibited the response to PDGF at a higher concentration. This could be explained by an autocrine control loop, where TGF-β_1 stimulates autocrine PDGF-AB secretion at low concentration [44, 68] and down-regulate PDGF-α receptors at a higher concentration [9, 66]. A similar feedback loop may exist between TGF-β_1 and EGF, where high concentrations of TGF-β_1 inhibits the stimulation induced by EGF.

IGF-1 stimulate cell proliferation, type I collagen synthesis, and small pro-

Table 1. Proliferative levels of canine ACL fibroblasts exposed to growth factors, alone and in combination: values are percentages of the maximal cell division

	EGF	PDGF	TGF-β	IGF-1	Insulin
	72±3	74±3	33±4	62±2	37±6
Insulin	86±6	85±9	56±8	42±6	
	+	+	+	±	
IGF-1	79±3	87±3	37±4		
	+	+	–		
TGF-β	66±6	17±3			
	–	±			
PDGF	98±3				
	+				

Considering that the 24-h test period is slightly longer than the cell cycle, these values are equivalent to the fraction of cells that divided. The plus and minus signs indicate positive or negative synergism of the effectors. A plus/minus sign means positive synergism at low concentration and mutual inhibition at high concentration. The growth factor concentrations used were the lowest ones found to have the maximal plateau effect [27]. Proportion possible divisions growth factor free control, 12% ± 10%.

teoglycan synthesis. We noted inhibition of the IGF-1 effect on proliferation by TGF-β_1, possibly through the modulation of insulin-like growth factor-binding protein 3 (IGFBP-3) synthesis. The synergism observed between insulin and IGF-1 at low concentration could be explained by ligand-binding cooperativity of insulin-IGF-1 hybrid dimer receptors [71]. Insulin has also been shown to increase production of IGFBP-3 [20] in fibroblasts, possibly leading to inhibition of the IGF-1 effects that we observed at higher insulin concentration. Insulin increased the proliferative response to EGF and PDGF.

We noted a stimulation of proliferation by EGF alone, an increase in small PG synthesis, and a specific decrease in type I collagen expression. PDGF alone stimulated proliferation but had no observed effects on collagen or PG expression. We found the strongest response with a combination of EGF and PDGF-AB.

Cells from ACL and PCL showed no significant differences between each other, while the PT cells had a consistently divergent response [27]. The proliferation curves of PT cells displayed a higher growth rate and confluent density than those shown by the cruciate cells. Without growth factor stimulation the fibroblasts isolated from PT produced less PG and more collagen – and only of type I – than the cruciate cells. These differences were approximately preserved under stimulation. We saw a general absence in type III collagen in PT cultures. Incidentally, Amiel et al. [2] reported that type III is not observed in rabbit PT. This tissue composition may in fact originate at the cellular production level.

We observed a strong relation between age and proliferation in this cell type. Based on preliminary experiments we selected an age range (3–5 years) in which this relationship is not significant. We also verified that the second-passage cells used did not show significantly lower sensitivity to growth factor than primary culture cells (Fig. 1).

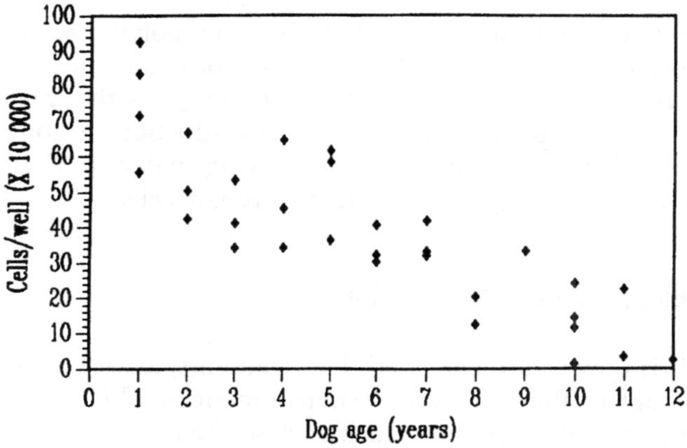

Fig. 1. Proliferative levels of canine ACL fibroblasts for dogs of different age

The physiological growth factor levels may be difficult to evaluate as the concentrations can vary rapidly in a very localized area. Furthermore, most growth factors are subjected to in vivo regulation, such as binding proteins or limiting amino acid concentration in the cell environment [35]. Because of this uncertainty we studied growth factors on a range of concentrations extending far above and below the common estimations of the physiological levels in the knee joint. The sequence of addition and precise timeframe involved in combined growth-factor stimulation may also be an important parameter.

Sequential injection of growth factor preparations could be used to stimulate cell proliferation, followed by matrix synthesis. Such a simple solution may be satisfying on the short term run. For example, it has been shown that adding TGF-β_1 to wounded medial collateral ligament stimulates collagen synthesis, mainly of type I [56]. Animal studies indicate that it is possible to improve ligament healing with growth factors in vivo [4]. Increases in biomechanical parameters, such as ultimate load, stiffness, and breaking energy have been measured in rat medial collateral ligaments treated with TGF-β_1 [21], PDGF, or two-by-two combinations of PDGF, IGF-1, and basic fibroblast growth factor [46]. However, the delivery of expensive and short-lived growth factors to a wound site is a demanding challenge. Alternative solutions to increase the local concentration of growth factors may be investigated. A promising approach consists in using growth factor gene transfer. Such transfers have been shown to increase the local growth factor concentration and accelerate the actual tissue healing through transfection of wound cells with accelerated plasmid-loaded gold particles [4]. That technique may be hardly applicable to articular ligaments. However, it is possible to imagine the injection of encapsulated recombinant cells into small cavities, such as the articular capsule. The injection or the in situ synthesis of specific inhibitors may also be very potent.

It would also be possible to stimulate platelet aggregation, and the subsequent release of effectors. The fibroblasts could be stimulated by growth factors immobilized on particles and injected in the articular capsule. Indeed, immobilization of growth factors on polymeric surfaces with bifunctionnal reagents, such as carbodiimides, does not appear to diminish the mitogen activity of the fixed growth factor [48]. As the internalization of the peptide does not appear necessary for transduction, it could be reused until the removal or degradation of the bearing material. The active site of the peptide could thus be made available to repair cells.

Biomechanical Stimulation

It is now well recognized that ligaments, as with many other biological tissues, adapt to the mechanical environment to which they are exposed [5]. Tissue-functional adaptation, such as described by Wolff's law for bony tissues, can be linked to cellular phenomena. Mobilization of the knee joint appears necessary to ligament healing. As intrinsic fibroblasts have the potential to repair, remodel the scar tissue, and play the main role in the healing process [33], the direct effects of mechanical stimulation on those deserve further investigations.

It has been shown repeatedly that immobilization of a joint deteriorates the ligament's biomechanical, biochemical, and morphological properties [77]. The rigidity and ultimate load of the ligament are decreased [63] making it more prone to ruptures by avulsion [82]. Newton et al. [60] found that immobilization of knee ligaments causes cellular changes that could be responsible for the deterioration of the tissues mechanical properties. Collagenase activity also increases and degrades collagen while lysozome hydrolase rises, destroying GAG [34].

Articulation mobilization, on the other hand, plays an important role in knee ligaments healing processes. Changes were observed in biochemical composition and biomechanical properties. Indeed, the healing rate of a scar is greatly improved and scar hypertrophy is avoided [82, 83]. On a short-term basis exercising a ligament improves collagen synthesis and increases the mean diameter and the number of fibrils in mouse tendons. The mean diameter of fibrils tends to decrease with time, but the number of fibrils remains greater than in an immobilized tendon [12]. In collagenic tissue the pressure on a scar increases the phagocytosis rate of collagen by macrophages, as well as enzymatic digestion. It is now believed that mobilization of a joint is required for scar reorganization toward a healthier and more normal structure [38, 60]. The cellular basis and biochemical pathways of such phenomena are numerous, and still obscure. The current hypotheses are reviewed elsewhere [26].

Many systems are currently used to stimulate cells in vitro. Mechanical stimulation of adherent cell cultures has long been performed, sometimes involuntarily, by the laminar flow of media in roller bottle cultures. As early as 1954 Gillespie [37] performed mechanical stimulation by poking single cells with a micropipette. Many mechanical cell stimulators are currently in use and can be classified on the basis of their mode of stimulation: uniaxial strain [1, 17, 18, 22, 25, 31, 45, 47, 52, 58, 73, 84, 85], equal biaxial strain [43, 70], radial strain having a domelike shape [3, 6, 8, 10, 15, 36, 39, 42, 61, 62, 67, 76, 80, 81], and shear stress on a cell culture created by a laminar flow [30, 45, 75]. These techniques, applied in a static or dynamic mode, are the most common for studying the mechanical environment of cells in vitro. The most common system is certainly the Flexercell, conceived by Banes et al. in 1985 [6] and commercialized since. Moore [55] designed another kind of system to simulate the mechanical behavior of arteries. This apparatus consists of a deformable tube, seeded with cells on its inner surface, which can be stimulated by a flow of fluid (applying a shear stress) and by a circumferential strain. Osmotic pressure can also be used to induce cell deformation [13]. Tripathi's research team [74] investigated the effects of vibrations transmitted to *sail-sheet* cell culture of fibroblasts.

These apparatuses usually try to simulate the in vivo mechanical conditions depending on the kind of cells under investigation. The mechanical environment of ACL fibroblasts may be simulated by two specific stress conditions: the main apparent mechanical stimuli is caused by the elongation of the ligament during physical activity, transferring a certain strain, if any, to the cells, through matrix deformation. Another stress acting on that cell type in vivo, could arise from the displacement of synovial fluid in between fibrils during ACL deformation. The first mechanical stimuli could be simulated by uniaxial loading of a substrate seeded with cells. The displacement of synovial fluid in between fibrils could be mimicked by a shear stress applied to a cell culture by a laminar flow. To investigate those two mechanical environments and expand our knowledge about cytomechanics our research group designed two apparatuses that provide the kinds of stimulation described.

Uniaxial Deformation

Mechanical Stimulator

The apparatus designed to apply a uniform uniaxial strain consists of two units: a stretching unit and its control box (Fig. 2). The stretching unit is thermostable and small enough to operate in a standard size incubator, at 37°C, for periods up to 5 days. The control box, which includes the electronic components and control cards of the motor, operates outside of the incubator.

We produced a system that could stretch an adherent cell culture up to 25% elongation of the culture surface, at frequencies as high as 4 Hz. The system is driven by an electrical gearmotor (Bodine, Ill. USA) controlled by the second unit. The motor's characteristics were chosen according to the elastic modulus of the stretchable dish described below, and to the targeted operating range of frequency/amplitude. The amplitude of deformation is given by an offset to the center of rotation of the motor, adjusted by a screw and nut system. The culture dish is fastened in a pair of specially designed nonslip grips, mounted on two low friction rails (Fig. 3). The sliding grip can be locked in any position for microscopic observation of the cells under strain in sterile conditions. Stimulation frequency can be set by a variable speed controller acting on the feedback of an optical encoder speed sensor (Red Lion, Penn., USA). The precision of the system is ±0.05% for the amplitude and ±0.02 Hz for the frequency.

Fig. 2

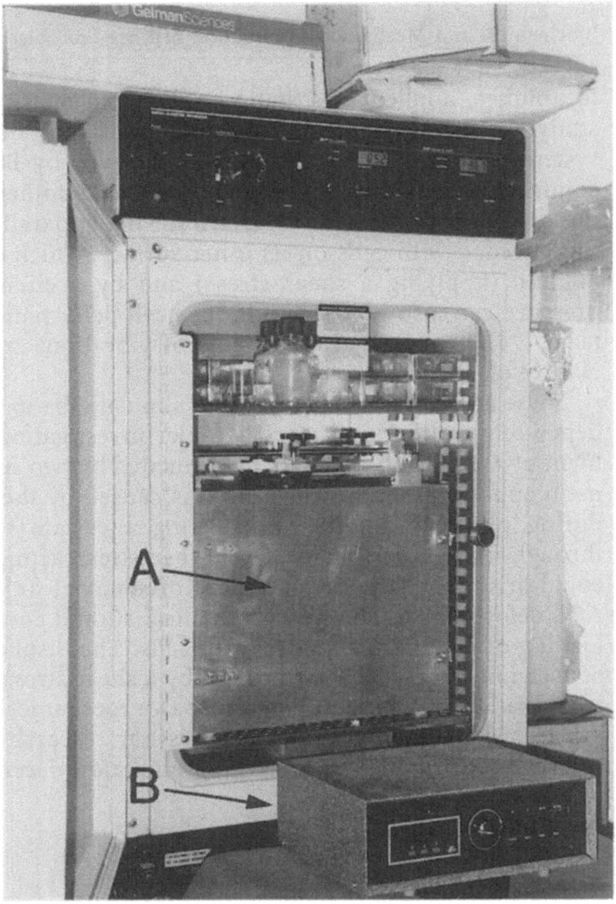

Fig. 3

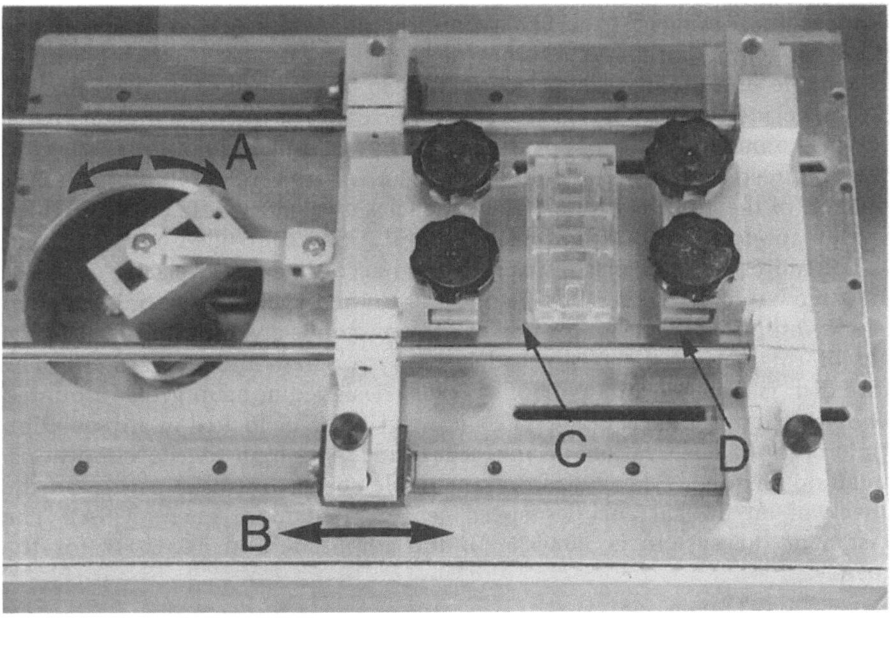

Stretchable Culture Dish

The dish provides four square culture wells of an approximate culture surface of 2 cm^2 each, similar in dimension to wells of 24-well culture plates, enabling the use of common instruments and protocols. A specially designed transparent cover allows gas exchange through air tunnels and convenient sterile handling of the cultures (Fig. 4).

Sylguard 184 silicone elastomer (Dow Corning, Ont., Canada) was retained as constitutive material since it possesses a low elastic modulus and good fatigue life, is transparent and biologically inert (cured silicone is used as negative control in standardized cytotoxicity assays), and can be easily molded.

The elastomer and curing agent are mixed by hand and degassed under vacuum. The culture dish and cover are molded by gravity and cured in an oven according to the Sylguard 184 specifications (15 min at 150°C). After full curing a waiting period of 7 days is required to achieve full mechanical strength of the silicone. The dish and cover are then thoroughly washed with mild soap and water, rinsed with deionized water, and sterilized. Autoclaving (30 min 120°C) or sanitation in ethanol 70% (30 min) can be used. Before use the culture wells are rinsed twice with culture medium under sterile conditions.

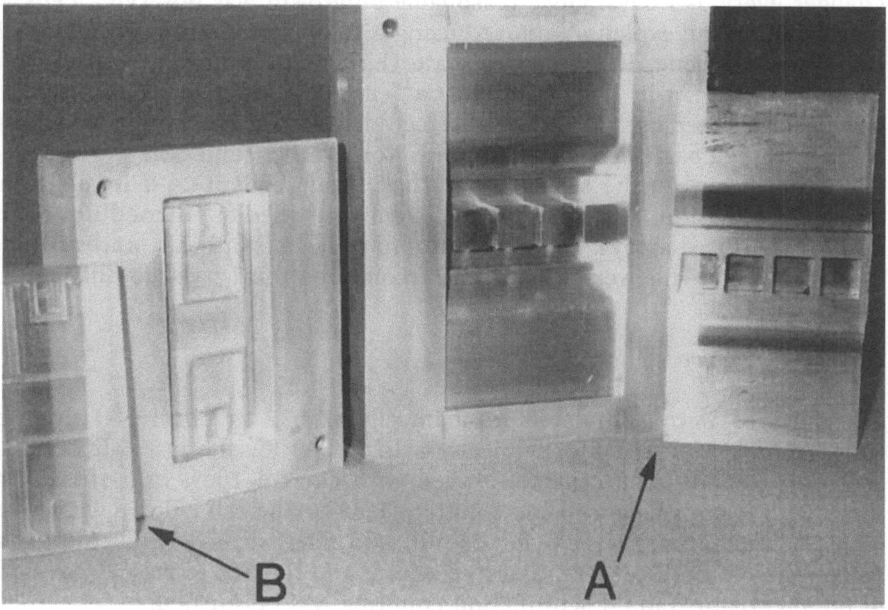

Fig. 4. The stretchable culture dish (*A*) and cover (*B*) with their respective aluminum

◄

Fig. 2. The stretching unit (*A*) is working inside a standard incubator while the control box (*B*) lies outside to prevent damaging the electronics components

Fig. 3. Top view of the stretching unit. The rotary motion (*A*) provided by the motor is transferred into a cyclic linear motion (*B*) to apply a deformation to the stretchable culture dish (*C*) held in position by special nonslip grips (*D*)

The parts that mold the four culture surfaces were polished to a rugosity less than to 0.05 μm to avoid weak spots, cell guidance, and other topography-related artifacts.

Cell Culture

ACL, PCL, and PT were aseptically removed from five 3- to 5-year-old dogs under anesthesia. All further manipulations were conducted in a laminar flow hood. The specimens were washed three in Hank's balanced salt solution (HBSS; Gibco, Grand Island, N.Y., USA) containing 10% of an antibiotics/antimycotics solution (composed of 10 kU/ml penicillin, 10 mg/ml streptomycin, and 1 ml/l fungizone; Sigma, St. Louis, Mo., USA). Loose tissues, fat, the synovial sheath, and femoral and tibial insertions were carefully dissected away. Samples were then cut in fragments of 2–5 mm^3 and placed in 75 cm^2 culture flask (Falcon Plastics, Oxnard, Calif.) coated with dried fetal bovine serum (Gibco). The fragments were then covered with 15 ml culture medium containing 10% of fetal bovine serum and 1% of the antibiotics/antimycotics solution in Dulbecco's modified Eagle's medium (Gibco). Explants were then incubated at 37°C in a humidified atmosphere of 5% CO_2, with medium changes every 3–5 days. Once reasonable outgrowth was observed, cells were detached and the explants removed using a solution containing trypsin 0.25% (w/v) and ethylene diaminatetraacetate (Fisher Scientific, Pittsburgh, Penn., USA) 0.02% (w/v) in HBSS without Ca^{2+} nor Mg^{2+} (Gibco). Cells were then frozen for further use. At time of use cell density was measured on triplicate samples (F-500 particle counter; Sysmex, Kobe, Japan), and cells were seeded at 2.5×10^4 cells/cm^2 in the stretchable dishes. A subsequent 24-h incubation allowed cell adhesion. The same batch of serum was used for all medium changes and for the three assays: cell proliferation, collagen synthesis, and proteoglycans synthesis. Cells were used at the second and third passage only.

Cellular Proliferation

The cultures were stimulated mechanically with 2 m Ci/ml [^{3}H]thymidine (60 Ci/mmol, Movarek Biochemicals, Calif., USA) for the last 4 h of the stimulation periods. The cells were fixed with ethanol 100% and rinsed three times with Hank's balanced salt solution (HBSS). The cell culture was covered with 5% trichloroacetic acid for 20 min and then washed with cold ethanol 100%. The cell culture were dissolved with 0.2 *M* NaOH/2% Na_2CO_3 overnight in the incubator. Liquid scintillation counting was performed.

Collagenic and Noncollagenic Proteins Synthesis

The cultures were mechanically stimulated with 2 m Ci/ml [^{3}H]proline (27 Ci/mmol, Amersham, UK) and then digested by 4 mg/ml pepsin (Sigma) at pH 3.0 for 24 h. The digested cultures were ultrafiltered (100 kDa PLTK,

Millipore, Bedford, Mass., USA). The filtrate was collected as noncollagenous proteins. The centrifugal ultrafilters were then thoroughly rinsed with deionized water and the retained particles incubated 6 h at 37°C in a collagenase solution: 20 m g^{-1} ml^{-1} collagenase (Worthington Biochemical, Freehold, N.J., USA) in 0.05 *M* Tris-HCl buffer at pH 7.5, supplemented by 5 m*M* $CaCl_2$ and 0.05 m*M* *N*-ethylmaleimide (Sigma) as nonspecific protease inhibitors. A second ultracentrifugation and rinse cycle allowed the collection of the collagenase-digestible fraction, that now runs through the pores of the filter. After this step each filter has been cut and placed in its respective *noncollagenous* vial. The radioactive elements in the solution which were not precipitated after 10 min in a 10% trichloroacetic acid + 0.5% tannic acid solution at 0°C following the ultracentrifugation were assumed to be collagenous and counted by scintillation (LS5000TD; Beckman, Schiller Park, Ill., USA). Collagenic and noncollagenic proteins synthesis is given in percentages relative to an unstimulated control (100%).

Proteoglycans Synthesis

The cultures were mechanically stimulated in presence of 15 m Ci/ml $Na_2[^{35}S]O_4$ (1050–1600 Ci/mmol, DuPont Canada, Ont., Canada). Incorporation of $[^{35}S]$ into macromolecules was measured as follows: The culture media and a rinse with HBSS were combined and ultrafiltered (30 kDa PLGC; Millipore). Deionised water was passed through the filters to wash out loose particles. The filters were then incubated 1 h in NaOH 1 N at 60°C and liquid scintillation counting was performed. PG synthesis is also given in percentages relative to an unstimulated control.

Experiments Planning

The software Optimum (Exact, Quebec, Canada) was used to plan some of our experiments. This software uses the Tagushi method [65] to optimize a multiparameter study, and minimizes the number of experiments based on the number of variables and their assigned weight. This method in fundamental research allows a reduction in time and costs associated with planification and realization of experiments, result analysis, modelization, and optimization of performance. Moreover, in this specific case it helps to reduce the amount of radioactivity used. Using the experimental plan and the software with our results, it is possible to optimize the variables initially entered in the system.

In our case we wanted to optimize the mechanical parameters of the system (amplitude and frequency of deformation as well as time of stimulation) in relation to cell proliferation, collagenic protein synthesis, noncollagenic protein synthesis, and proteoglycans synthesis.

Results

Uniform Strain of the Culture Surface

The square culture wells allow uniform deformation which was verified by two methods. The first method consisted in molding a culture dish with very small metal beads included on the culture surface. Each of the four culture wells were divided into five sectors (four corners and the center). A picture of each sector for every well was taken, and eight measurements between beads were performed for each picture at 0%, 5%, 10%, and 15% deformation of the culture dish (distance between the two grips). The same beads in every picture were used for measurements. We found no significant difference in strain distribution between sectors or between the four culture wells.

The second technique used to observe the uniform strain of the culture surfaces was to verify the reorientation process in those same five sectors. It is now well known that cells subjected to a mechanical stimulation reorients at 90° of the strain axis. We observed that the reorientation was constant on the whole culture surface, even in the corners, where we would have expected a distortion of the strain distribution.

Mechanical Stimulation

ACL, PCL, and PT fibroblasts were compared by measuring proliferative and metabolic response when subjected to 6% strain at 0.1 Hz for a period of 24 h.

The most obvious result was the realignment of the three cell types at 90° with the strain axis. No difference was found between ACL, PCL, and PT realignment processes. The three types of cells responded to stretch by an increased proliferation, with PT cells having the most pronounced effect. Only ACL and PCL cells responded with increased collagen synthesis. Proteoglycans synthesis was not affected by the mechanical stimulation performed (Table 2).

A study on ACL fibroblasts was performed using the experimental design described earlier. A set of nine experiments (specific plan given by the software) was performed for each measured parameters (proliferation, collagenic/noncollagenic protein synthesis and proteoglycans synthesis) and the results entered in Optimum software as an attempt to find optimum values for the input mechanical parameters (amplitude, frequency, and time of stimulation).

Table 2. Proliferative response, collagen synthesis, and proteoglycans synthesis of three types of knee ligaments fibroblasts submitted to 6% of strain at 0.1 Hz for 24 h: values are percentages of a nonstimulated control (100%)

	ACL	PCL	PT
Proliferation	140%±10%	135%±15%	210%±30%
Collagen synthesis	150%±10%	150%±10%	100%±20%
GAG synthesis	100%±20%	100%±15%	100%±10%

While no precise optimum values could be defined for the mechanical parameters, some interesting results emerged from this study. We studied the mechanical parameters above and below the physiological range of values. This can be used for other purposes than to mimic the in vivo conditions (i.e., regeneration of ligaments in vitro with the help of mechanical stimulation). Amplitude was studied up to 15% substrate deformation, with frequencies up to 3 Hz. The experiments were performed in 24-h periods. This timeframe encompassed more than a complete cell cycle and was short enough to avoid quantifying mitosis from a second cell division.

The reorientation of cell bodies occurs at low frequency and amplitude, but if one of those two parameters increases excessively, the phenomenon is inhibited. At frequencies above approx. 1.5–1.75 Hz or substrate deformations exceeding approx. 15–17% the reorientation process does not occur. This kind of stimulation may be damaging to the cells (rupture of the cell membrane, rupture of adhesion points, etc.). We observed a qualitative correlation between the number of cycles applied and the reorientation process, by visually observing the reorientation of cells at different stages of mechanical stimulation. The reorientation does not appear to depend on time: a higher alignment index was found in short-term experiments at high frequency than in others conducted at low frequency for a longer period of time.

Even at a low amplitude a frequency of 3 Hz is damaging for the cells when subjected for a long period of time (24 h). The cells became rounded, probably due to the disruption of cell adhesion points. The same mechanical stimulation was not as damaging when performed for a shorter period of time (4 h). In parallel, we observed similar results when the cells were submitted to low frequency and high amplitude. It was also found that noncollagenic protein and proteoglycans synthesis were not affected by mechanical stimulation in the range studied.

As for proliferation response and collagenic protein synthesis, it was found that ACL fibroblasts are most responsive to low amplitude and frequency of stimulation. For the amplitude of stimulation, good cell response was found around the maximal physiological level (around 6% [69]), while the range of frequencies found to have a positive effect was around and under 1 Hz (1.5–1.75 Hz being the frequency range in which the cell damage starts).

Discussion

The new uniaxial mechanical stimulator is compatible with cell culture and applies a uniform strain over the cell culture surface. The stretchable dish proved to be practical and allowed the maintanance of sterility. The uniaxial mechanical stimulator has a significant effect on fibroblast proliferative response and metabolism.

The perpendicular orientation response to cyclic deformation differs from the normal in vivo, parallel orientation. All types of cells studied responded to a "physiological" stimulation by a reorientation at 90° of the strain axis. No difference was found between ACL, PCL, and PT fibroblasts' reorientation processes. This cell reaction to a strain is well documented in the literature [7, 17, 18, 23, 28, 36, 45, 52, 57–59, 73, 76, 78], but only hypotheses exist concerning the causes of this cell reaction. Cells respond to the topography of the surface on which they are seeded. Indeed, they orient and migrate along grooves on the culture surface [16, 40, 53, 54, 72]. A grooved surface in the direction of stretch inhibits the reorientation process (at 90° of the strain axis) of skin fibroblasts when submitted to 8% deformation at 1 Hz for a period of 120 h [78]. This could explain the contradiction between in vivo (cells aligned in the stretch direction along collagen fibrils) and in vitro conditions (cells on a smooth surface reorients perpendicular to the stretch axis). The surface guidance by the collagen fibrils in vivo has a greater influence on cells than the mechanical stimulation concerning the reorientation phenomenon.

The reorientation process was found to depend on the strain magnitude [23, 57] and the number of cycles applied. Neidlinger-Wilke [59] found a relationship between the reorientation process and time of stimulation probably because the study was performed at only one frequency. The number of cycles applied was calculated for each experiment. Experiments of short time and high frequency (high number of cycles) were more effective for reorienting the cells than longer experiments at a low frequency (lower number of cycles).

The reorientation process, higher proliferative response, and collagen synthesis indicate a cell phenotype more related to scar tissue fibroblasts than to normal tissue cells. ACL and PCL have a similar proliferative response and matrix synthesis as PT fibroblasts. Cell type of these two structures differ, and this could be the main explanation for the different responses to a mechanical stimulation. PT cells are closer to a typical fibroblast while ACL and PCL cells resemble more the chondroblast type [27].

It is clear that the mechanical parameters need to be in a specific range to have a cell response (reorientation, proliferation, and matrix synthesis), otherwise the cells could be damaged when exposed to excessive stimulation inhibiting their response or the cells would not respond at all to a too small stimulation. According to Brand [14], the cells respond to a mechanical stimulation in a triggerlike manner after relatively few events or cycles of loading. Furthermore, they respond only to some window of strain magnitude, they exhibit a refractory signal, and they have a memory of previous stimuli. The complexity of the cell response to a mechanical stretch could explain the fact that no optimum values were found for the input mechanical parameters.

In vivo extracellular matrix may filter deformation to the cells modifying their orientation, proliferation, and matrix expression. It is not known how much of the strain in the ligament is transmitted to the cells, and in vitro very few studies have investigated the strain applied directly to the cells.

The cellular response (reorientation, proliferation, and matrix synthesis) to a mechanical stimulation is a result that cannot be compared among the research groups working on cytomechanics because these results are not related to the strain at the cellular level. The results of each group depend on the system used to apply the mechanical stimulation, and since groups do not all use the same apparatus, there is a need for relating the cytomechanics studies and the cellular strain. In this way cytomechanics studies could be compared among the research groups and facilitate the growth of this field of study. Furthermore, results could be used for ligament healing therapies and regeneration of ligaments in vitro.

Laminar Flow

Mechanical Stimulator

The stimulator system (Fig. 5) consists of a culture bottle (1750 cm^2; Falcon Labware, Becton-Dickinson, Mississauga, Ont., Canada) mounted on two rollers (without any slipping) running on sealed ball-bearing (General Bearing Service, Montreal, Canada) and driven by a variable-speed a.c. electric motor (Hurst, Montreal, Canada). The components are assembled in a custom-built metallic box, small enough to operate in a standard size incubator. The motor is thermally isolated from the culture bottle, and its speed is adjusted by a variable resistance. Rotation speed is measured visually by timing the cyclic apparition of a marker on the culture bottle. The incubator door is simply closed on the power supply cord, without special fitting. The whole system is thermostable and has been operated at 37°C and 100% humidity for periods up to 20 days without noticeable rotation speed variation or incubator temperature change.

Cell Culture

ACL fibroblasts were removed aseptically and cultured using the methods described above. Each tissue culture roller bottle was seeded with 1.0×10^7 cells, in 100 ml Dulbecco's modified Eagle's medium supplemented by 10% fetal

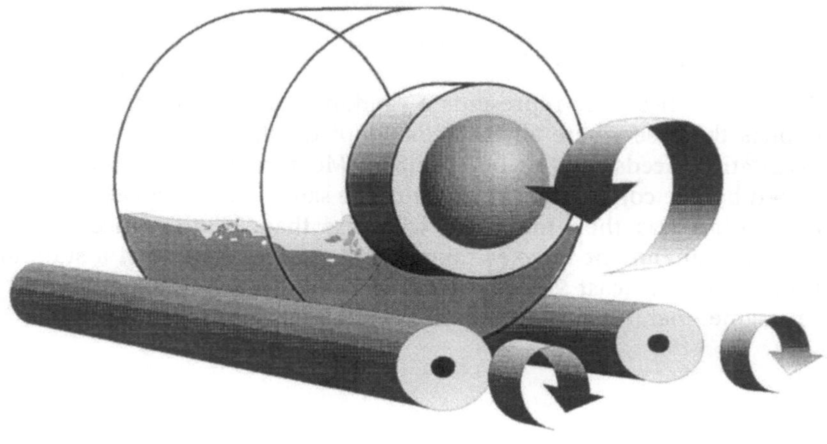

Fig. 5. A sketch of the laminar flow stimulator

bovine serum, antibiotics and antimycotics. The cells were allowed to settle at low rotation speed (1 RPM) for 24 h. The stimulation experiments were then initiated: 30 distinct trials were conducted at speeds ranging from 2 to 30 RPM in 2-RPM increments for periods of 15 and 48 h.

Image Analysis

At the end of the stimulation period the cells were stained using Diff-Quick dye kit (Baxter, Montreal, Canada). Sections of approx. 4×4 cm were then cut out of the central sections of the culture bottle walls. As these sections are concave, only a small region could be observed at a time due to the limited focal depth of the microscope. The bottle wall sections were thus flattened by an overnight exposure to chloroform vapors. Only the outer convex surface of the bottles was exposed to vapors, to avoid damaging the stained cell cultures. Microphotographs of the cells were taken (magnification: ×1000) digitized (200 DPI, 16 gray levels) and treated using a IBAS2 image analyzer. Following numeric contrast enhancement and erosion, the cell contours were defined. The longest axis of the cells could then be reckoned and highlighted by line markers. Orientation angles of the markers were then automatically determined and compiled. By convention, 0° refers to the laminar flow orientation. An alignment index (K_a) is calculated by normalizing the measured angle distribution variance by the expected dispersion of randomly oriented cells. As an example, for a sample of 90 cells a perfectly random distribution would consist in a series of cell orientation: 1°, 2°, 3°, 4°,..., 89°, 90°. The mean angle ($\bar{x}$) is then 45° and the variance of a random distribution (S_r^2) follows Eq. 1:

$$s_r^2 = \frac{\sum(x-\bar{x})^2}{(n-1)} = \frac{1}{89}\sum_1^{90}(x-45)^2 = \frac{1}{89}\left(\sum_1^{44}x^2 + \sum_1^{45}x^2\right) = 682.75 \qquad (1)$$

The alignment index, in that case, for a given sample variance (S^2) is:

$$K_a = -\frac{s^2}{s_r^2} + 1 = -\frac{s^2}{682.75} + 1 \qquad (2)$$

A K_a value of 0 thus represents a random cell orientation while a value of 1 depicts the case where all the cells adopt exactly the same orientation; mean orientation needs also to be specified. Measurement reproducibility was assessed by five consecutive analyses of the same image (45 cells). The mean cell orientation was then measured with less than 3° difference between trials. Associate alignment indexes ranged from 0.35 to 0.40, with a standard deviation of 0.04. At least 90 cells were considered in every experimental analysis. The same operator performed all the analyses.

Results

The apparatus provided reliable and smooth rotation. According to theoretical calculations, in this type of culture bottle, filled with 100 ml waterlike fluid, a rotation speed of 36 RPM can be achieved before reaching the transition from a laminar to a turbulent flow. The experimental situation is quite different, however. A zone of turbulence is noticeable near the extremities of the bottles due to the rotating end walls. The importance of these eddies increases with rotation speed and finally leads to generalized chaotic fluid movement. By filming the mixing motion of China ink in culture media at various speed and with adhering cells on the bottle wall, we determined the transition speed at 21–23 RPM.

We observed a strong alignment response of the cells in the orientation of the laminar flow, which was increased by the duration of stimulation and laminar flow speed, in a dose-dependent manner. It was noticeable even at the first observation period, 5 h of stimulation, for the higher flow speed experiments (18, 20, and 22 RPM) and after 10 h in all other trials. Figure 6 shows the orientation after 15 and 48 h. A turbulent flow does not appear to exert any influence on adherent cell orientation. It should be noted that our experiments were run on *independent* trials: increasing rotation speed until flow becomes turbulent *does not* cause aligned cells to reorient randomly. In such a case the cells maintain their aligned orientation. Similarly, an initial random cell orientation is not modified by a turbulent media flow. Cell density on the bottle wall did not increase significantly in most of our roller bottles, even though some experiments were pursued for 14 days. Numerous viable cell agglomerates could be seen in the media after these long-term experiments.

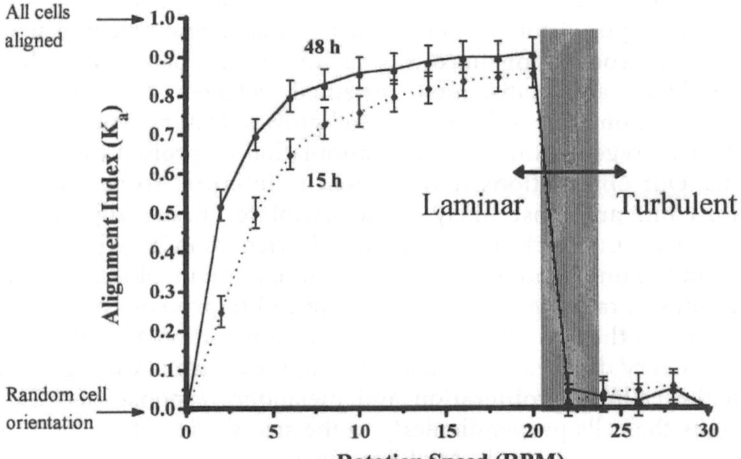

Fig. 6. Alignment of cells submitted to a flow at different speed. The transition from a laminar to a turbulent flow is well represented with a decrease in the alignment index

Discussion

Maximal laminar flow speed can be optimized by controlling parameters such as bottle dimensions or increasing media viscosity by adding dextran, for example. The relative importance of end-walls eddies can be minimized by using long bottles. Even then we considered only the central zone of the bottles for image analysis. Further hydrodynamic characterization should be pursued for specific applications of the stimulation system.

The cells appear to minimize their exposure to stress by aligning along the orientation axis offering the lowest frontal surface area to the flow, and thus lower their internal stress. This orientation ceases to be optimal in a turbulent flow, where no preferential unique flow direction is respected. From the presence of agglomerated cells suspension in the media in 14-day experiments we believe that new cells, before they attach firmly, are sheared from the bottle walls by the liquid flow. These then end up in the media, where they form agglomerates, and usually remain viable for a few days.

Future Prospects

Our results may be readily applicable to tissue engineering and other in vitro reconstruction techniques. However, we have chosen by using an in vitro system to improve reproducibility and control at the expense of loosing the physiological cell environment. The most direct application of this type of study is to optimize the in vitro reconstruction of ligament substitute through a better control of cellular activity. Even if this approach is readily approaching its clinical resolution, the combined use of growth factor and mechanical stimulation could certainly improve the current procedures. The mechanical stimulation experiments tend to show that the presence of an oriented structure is required for proper contact guidance and oriented matrix reconstruction. A grooved or fibrous structure could play that role and allow cell alignment and subsequent oriented collagen deposition and tissue reconstruction. The relevant use of growth factors would then favor a more efficient regeneration through stimulation of proliferation and matrix synthesis. Our observations may be used to interpret some physiological observations but are most likely to be useful for tissue engineering and in vitro reconstruction techniques where cell orientation may be critical.

On the one hand, growth factor stimulation invokes a wide range of cellular responses, ranging from cell cycle control to protein synthesis stimulation. On the other, the two types of mechanical stimulation acts on cell orientation and elicits very different responses. The uniaxial stimulation generates an increase in the cellular proliferation and metabolic response of ACL fibroblasts but aligns the cells perpendicularly to the stress orientation, in a manner opposed to in vivo orientation. Applying shear stress on a cell culture does not significantly increase the cellular proliferation or the metabolic response but reorients the cellular bodies along the flow.

To achieve an efficient ligament regeneration in vitro, a high number of active fibroblasts must express and deposit extracellular matrix in an oriented hierarchical assembly. Selective stimulation of proliferation and of matrix synthesis may be accomplished by the appropriate use of growth factors. The orientation of the synthesis matrix depends on the cell bodies orientation: as observed, this may be changed through mechanical stimulation. Another complementary and somewhat easier way is to provide contact guidance by the use of oriented scaffolds. The regeneration method could then be fine-tuned by optimizing the biochemical and biomechanical cell environment.

The combination of biochemical and mechanical stimulation would first stimulate a higher proliferation rate to increase the number of "working" cells. Once we have enough cells, the biochemical conditions (still combined with the mechanical stimulation) could be changed to a stage favorable for matrix synthesis. During those two processes the application of a shear stress to the cells could keep them in a good orientation, especially for matrix synthesis.

References

1. Almekinders LC (1992) The use of mechanical stimulation in in vitro models. In: Watson RR, (ed) In vitro methods of toxicology. CRC, Boca Raton, pp 157–162
2. Amiel D, Kleiner JB, Akeson WH (1986) The natural history of the anterior cruciate ligament autograft of patellar tendon origin. Am J Sports Med 14:449–462
3. Andersen KL, Norton LA (1991) A device for the application of known simulated orthodontoc forces to human cells in vitro. J Biomech 24(7):649–654
4. Andree C, Swain WF, Page CP, Macklin MD, Slam J, Hatzis D, Eriksson E (1994) In vivo transfer and expression of a human epidermal growth factor gene accelerates wound repair. Proc Natl Acad Sci USA 91:12188–12192
5. Arem AJ, Madden JW (1975) Is there a Wolff's law for connective tissue? Surg Forum 47:512–514
6. Banes AJ, Gilbert J, Taylor D, Monbureau O (1985) A new vacuum-operated stress-providing instrument that applies static or variable duration cyclic tension or compression to cells in vitro. J Cell Sci 75:35–42
7. Banes AJ, Henderson R, Donlon K, Dimeo N (1989) Migration, alignment and cell division responses of avian tendon synovial cells and internal fibroblasts subjected to increasing cyclic deformation (at 3, 10 and 40 cycles per minute) in vitro. Transactions of the 35th annual meeting of the Orthopaedic Research Society, 6–9 February, Las Vegas, p 181
8. Baskin L, Howard PS, Macarak E (1993) Effect of mechanical forces on extracellular matrix synthesis by bovine urethal fibroblasts in vitro. J Urol 150:637–641
9. Battegay EJ, Raines EW, Seifert RA, Bowen-Pope DF, Ross R (1990) TGF-β induces bimodal proliferation of connective tissue cells via complex control of an autocrine PDGF loop. Cell 63:515–524
10. Belloli DM, Williams JL, Park JY (1991) Calculated strain fields on cell stressing devices employing circular diaphragms as substrates. Trans Orthop Res Soc 16:399
11. Bettger WJ, Boyce ST, Walthall BJ, Ham RG (1981) Rapid clonal growth and serial passage of human diploid fibroblasts in a lipid-enriched synthetic medium supplemented with epidermal growth factor, insulin and dexamethasone. Proc Natl Acad Sci USA 78:5588–5592
12. Binkley JM, Peat M (1986) Effects of immobilization on the ultrastructure and mechanical properties of the medial collateral ligament of rats. Clin Orthop 203:301–309
13. Bolognani Fantin AM, Franchini A, Fuhrman Contin AM, Corbascio GC, Marchisio PC (1988) Effects of hypertonic conditions on cytoskeletal and adhesion structures of EUE cells. Eur J Cell Biol 47:327–333

14. Brand RA, Stanford CM (1994) How connective tissues temporally process mechanicall stimuli. Med Hypotheses 42:99–104
15. Brighton CT, Strafford B, Gross SB, Leatherwood DF, Williams JL, Pollack SR (1991) The proliferative and synthetic response of isolated calvarial bone cells of rats to cyclic biaxial mechanical strain. J Bone Joint Surg 73:320–331
16. Brunette DM (1986) Spreading and orientation of epithelial cells on grooved substrata. Exp Cell Res 167:203–217
17. Buck RC (1980) Reorientation response of cells to repeated stretch and recoil of the substratum. Exp Cell Res 127:470–474
18. Carver W, Nagpal ML, Nachtigal M, Borg TK, Terracio L (1991) Collagen expression in mechanically stimulated cardiac fibroblasts. Circ Res 69:116–122
19. Chua CC, Chua BH, Zhao ZY, Krebs C, Diglio C, Perrin E (1991) Effect of growth factors on collagen metabolism in cultured human heart fibroblasts. Connect Tissue Res 26:271–281
20. Conover CA (1990) Regulation of insulin-like growth factor (IGF)-Binding protein synthesis by insulin and IGF-1 in cultured bovine fibroblasts. Endocrinology 126:3139–3145
21. Conti N, Dahners L (1993) The effect of exogenous growth factors on the healing of ligaments. Trans Orthop Res Soc 18:60
22. Dartsch PC, Betz E (1989) Response of cultured endothelial cells to mechanical stimulation. Basic Res Cardiol 84:268–281
23. Dartsch PC, Betz E (1990) Cellular and cytoskeletal response of vascular cells to mechanical stimulation. In: Plank H, Dauner M, Renardy M (eds) Medical textiles for implantation. Springer, Berlin Heidelberg New York, pp 193–218
24. DesRosiers EA, Rivard C-H, Bourdeau I, Yahia L'H (1993) Culture de fibroblastes de ligaments croisés antérieurs canins en milieu défini, effets des epidermal growth factors et des platelet derived growth factors. Ann Chir 47:874–880
25. DesRosiers EA, Méthot S, Rivard C-H, Yahia L'H (1994) Metabolic response of ACL, PCL and PT fibroblasts submitted to in vitro mechanical stimulation. Proceedings of 2nd World Congress of Biomechanics, Amsterdam
26. DesRosiers EA, Rivard C-H, Yahia L'H (1995a) Mécanique cellulaire des tissus conjonctifs. Cah Rhéol 13:323–343
27. DesRosiers EA, Chaput C, Rivard C-H, Yahia L'H (1995b) Differential response of patellar tendon, anterior and posterior cruciate ligaments fibroblasts submitted to combined growth factors stimulation. Eur J Exp Musculoskel Res 4:40–48
28. DesRosiers EA, Méthot S, Yahia L'H, Rivard C-H (1995c) Réponses des fibroblastes ligamentaires à la stimulation mécanique. Ann Chir 49(8):768–774
29. DesRosiers EA, Rivard C-H, Yahia L'H (1996) Proliferative and matrix synthesis response of canine anterior cruciate ligament fibroblasts submitted to combined growth factors. J Orthop Res (in press)
30. Dewey CF Jr, Bussolari SR, Gimbrone MA Jr, Davies PF (1981) The dynamic response of vascular endothelial cells to fluid shear stress. J Biomech Eng 103:177–185
31. DeWitt MT (1984) In vitro response of chondrocytes to mechanical loading. The effect of short term mechanical tension. Connect Tiss Res 12:97–109
32. Frank C, Woo SL-Y, Amiel D, Harwood F, Gomez MA, Akeson W (1983) Medial collateral ligament healing, a multidisciplinary assessment in rabbits. Am J Sports Med 11:379–389
33. Frank C, Amiel D, Woo SL-Y, Akeson W (1985) Normal ligament properties and ligament healing. Clin Orthop 196:15–25
34. Gamble JG, Edward CC, Max SR (1984) Enzymatic adaptation in ligaments during immobilization. Am J Sports Med 12:221–228
35. Gartner MH, Shearer JD, Bereiter DF, Mills CD, Caldwell MD (1991) Wound fluid amino acid concentrations regulate the effect of epidermal growth factor on fibroblast replication. Surgery 110:448–456
36. Gilbert JA, Weinhold PS, Banes AJ, Linf GW, Jones GL (1994) Strain profiles for circular cell culture plates containing flexible surfaces employed to mechanically deform cells in vitro. J Biomech 27(9):1169–1177

37. Gillespie JA (1954) The nature of bone changes associated with nerve injuries and disease. J Bone Joint Surg Br 36:464.
38. Gomez MA, Woo SL-Y (1989) The advantages of applied tension on healing medial collateral ligament. Transactions of the 35th annual meeting of Orthopedic Research Society, 6–9 February, Las Vegas, p 184
39. Gorfein SF, Winston FK, Thibault LE, Macarak EJ (1989) Effects of biaxial deformation on pulmonary artery endothelial cells. J Cell Physiol 139:492–500
40. Green AM, Jansen JA, van der Waerden JPCM, von Recum AF (1994) Fibroblast response to microtextured silicone surfaces: texture orientation into or out of the surface. J Biomed Mater Res 28:647–653
41. Harwood F, Gomez MA, Akeson W (1983) Medial collateral ligament healing, a multidisciplinary assessment in rabbits. Am J Sports Med 11:379–389
42. Hasegawa S, Sato S, Saito S, Suzuki Y, Brunette DM (1985) Mechanical stretching increases the number of cultured bone cells synthesizing DNA and alters their pattern of protein synthesis. Calcif Tissue Int 37:431–436
43. Hung CT, Williams JL (1994) A method for inducing equi-biaxial and uniform strains in elastomeric membranes used as cell substrates. J Biomech 27(2):227–232
44. Ishikawa O, LeRoy EC, Trojanowska M (1990) Mitogenic effect of transforming growth factor-b_1 on human fibroblasts involves the induction of platelet-derived growth factor a-receptors. J Cell Phys 145:181–186
45. Ives CL, Eskin SG, McIntire LV (1986) Mechanical effects on endothelial cell morphology: in vitro assesment. In Vitro 22:500–507
46. Letson AK, Dahners LE (1994) The effect of combinations of growth factors on ligament healing. Clin Orthop 308:207–212
47. Leung DYM, Glacov S, Mathews MB (1977) A new in vitro system for studying cell response to mechanical stimulation. Exp Cell Res 109:285
48. Liu SQ, Ito Y, Imanishi Y (1992) Cell growth on immobilized cell growth factor. I. Acceleration of the growth of fibroblast cells on insulin-immobilized polymer matrix in culture medium without serum. Biomaterials 13:50–58
49. Manske PR, Gelberman RH, Vandebergh JS, Lesker PA (1984) Intrinsic flexor-tendon repair. J Bone Joint Surg Am 66:385–391
50. Mass DP, Tuel R (1989) Human flexor tendon participation in the in vitro repair process. J Hand Surg [Am] 14:64–71
51. Meikle MC, Heath JK, Reynolds JJ (1984) The use of in vitro models for investigating the response of fibrous joints to tensile mechanical stress. Am J Orthop 85:141–153
52. Méthot S, DesRosiers EA, Yahia L'H, Rivard C-H (1995) A new in vitro system for inducing cyclic deformation to adherent cell cultures. Proceedings of the 2nd Combined Meeting of the Orthopaedic Research Society, San Diego, November 5–8, p 81
53. Meyle J, Gültig K, Wolburg H, von Recum AF (1993) Fibroblast anchorage to microtextured surfaces. J Biomed Mater Res 27:1553–1557
54. Meyle J, Gültig K, Nisch W (1995) Variation in contact guidance by human cells on a microstructured surface. J Biomed Mater Res 29:81–88
55. Moore JE Jr, Bürki E, Suciu A, Zhao S, Burnier M, Brunner HR, Meister JJ (1994) A device for subjecting vascular endothelial cells to both fluid shear stress and circumferential cyclic stretch. Ann Biomed Eng 22:416–422
56. Murphy PG, Loitz BJ, Frank CB, Hart DA (1994) Influence of exogenous growth factors on the synthesis and secretion of collagen types I and III by explants of normal and healing rabbit ligaments. Biochem Cell Biol 72:403–409
57. Neidlinger-Wilke C, Holbein E, Grood M, Mörike M, Claes L (1994a) Effects of cyclic strain on proliferation, metabolic activity and alignment of human osteoblasts and fibroblasts. Proceedings of the 40th Orthopaedics Research Society, New Orleans, February 21–24, pp 101–118
58. Neidlinger-Wilke C, Wilke HJ, Claes L (1994b) Cyclic stretching of human osteoblasts affects proliferation and metabolism: a new experimental method and its application. J Orthop Res 12:70–78

59. Neidlinger-Wilke C, Brand RA, Claes L (1995) Time-dependent fibroblast orientation in response to stretch. Proceedings of the 41th Orthopaedics Research Society, Orlando, February 13–16, pp 507
60. Newton PO, Woo SL-Y, Kitabayashi LR, Lyon RM, Anderson DR, Akeson WH (1990) Ultrastructural changes in knee ligaments following immobilization. Matrix 10:314–319
61. Ngan PW, Crock B, Varghese J, Lanese R, Shanfeld J, Davidovitch Z (1988) Immunohistochemical assesment of the effect of chemical and mechanical stimuli on cAMP and prostaglandin E levels in human gingival fibroblasts in vitro. Arch Oral Biol 33:163–174
62. Ngan PW, Saito S, Saito M, Lanese R, Shanfeld J, Davidovitch Z (1990) The interactive effects of mechanical stress and interleukin-1b on prostaglandin E and cyclic AMP production in human periodontal ligament fibroblasts in vitro: comparison with cloned osteoblastic cells of mouse (MC3T3-E1). Arch Oral Biol 35:717–725
63. Noyes FR (1977) Functional properties of knee ligaments and alteration induced by immobilization: a correlative biomechanical and histological study in primates. Clin Orthop 123:210–242
64. O'Donoghue DH, Rockwood CA, Frank GR, Jack SC, Kenyon R (1966) Repair of the anterior cruciate ligament in dogs. J Bone Joint Surg Am 48:503–513
65. Phadke MS (ed) (1989) Quality engineering using robust design. Prentice Hall, Englewood Cliffs
66. Paulsson Y, Karlsson C, Heldin CH, Westermark B (1993) Density-dependent inhibitory effect of transforming growth factor-b1 on human fibroblasts involves the down-regulation of platelet-derived growth factor a-receptors. J Cell Phys 157:97–103
67. Pender N, McCulloch CA (1991) Quantification of actin polymerisation in two human fibroblast sub-types responding to mechanical stretching. J Cell Sci 100:187–193
68. Psarras S, Kletsas D, Stathakos D (1994) Restoration of down-regulated PDGF receptors by TGF-β in human embryonic fibroblasts: enhanced response during cellular in vitro aging. FEBS Lett 339:84–88
69. Rivard C-H, Yahia L'H, Brunet J, DesRosiers EA (1991) Évaluation in-vitro de la déformation maximale imposable aux ligaments collatéraux médians de lapins. Ann Chir 45:823–826
70. Schaffer JL, Rizen M, L'Italien GJ, Benbrahim A, Megerman J, Gerstenfeld LC, Gray ML (1994) Device for the application of a dynamic biaxially uniform and isotropic strain to a flexible cell culture membrane. J Bone Joint Surg 12:709–719
71. Soos MA, Whittaker J, Lammers R, Ulrich A, Siddle K (1990) Receptors for insulin and insulin-like growth factor-1 can form hybrid dimers: characterisation of hybrid receptors in transfected cells. Biochem J 270:383–390
72. Stopak D, Harris AK (1982) Connective tissue morphogenesis by fibroblast traction. Dev Biol 90:383–398
73. Terracio L, Tingström A, Peters III WH, Borg TK (1990) A potential role for mechanical stimulation in cardiac development. Ann NY Acad Sci 588:48–60
74. Tripathi SC, Kerr J (1989) Effect of mechanical stress on cellular morphology. Tissue Cell 21:747–752
75. Truskey GA, Pirone JS (1990) The effect of fluid shear stress upon cell adhesion to fibronectin-treated surfaces. J Biomed Mater Res 24:1333–1353
76. Vandenburgh HH (1988) A computerized mechanical cell stimulator for tissue culture: effects on skeletal muscle organogenesis. In Vitro 24:609–619
77. Walsh S, Frank C, Hart D (1992) Immobilization alters cell metabolism in an immature ligament. Clin Orthop 277:277–288
78. Wang H, Grood ES, Ip W, Neidlinger-Wilke C, Claes L, Gray M, Sobek D (1995) The effect of contact guidance on the realignment of fibroblasts grown on cyclically deformed substrates. Proceedings of the 2nd combined meeting of the Orthopaedic Research Society, San Diego, November 5–8, p 27
79. Westergren-Thorsson G, Antonsson P, Malmström A, Heinegard D, Oldberg A (1991) The synthesis of a family of structurally related proteoglycans in fibroblasts is differently regulated by TGF-β. Matrix 11:177–183

80. Williams JL, Chen JH, Belloli DM (1992) Strain fields on cell stressing devices employing clamped circular elastic diaphragms as substrates. J Biomech Eng 114:377–384
81. Winston FK, Macarak EJ, Gorfein SF, Thibault LE (1989) A system to reproduce and quantify the biomechanical environment of the cell. J Appl Physiol 67(1):397–405
82. Woo SL-Y, Gomez MA, Sites TJ, Newton PO, Orlando CA, Akeson WH (1987a) The biomechanical and morphological changes in the medial collateral ligament of the rabbit after immobilization and remobilization. J Bone Joint Surg 69-A:1200–1208
83. Woo SL-Y, Inoue M, McGurk-Burleson E (1987b) Treatment of the medial collateral ligament injury. Am J Sports Med 15:22–30
84. Yahia L'H, DesRosiers EA, Rivard C-H (1991) A computer controlled apparatus for in vitro mechanical stimulation and characterization of ligaments. Biomed Mater Eng 1:215–222
85. Yeh C-K, Rodan GA (1984) Tensile forces enhance prostaglandin E synthesis in osteoblastic cells grown on collagen ribbons. Calcif Tissue Int 36:67–71

III.3 Tissue-Engineered Ligament

F. Goulet, L. Germain, C. Caron, D. Rancourt,
A. Normand, and F.A. Auger

The Potential Use of Bioengineered Ligaments as New Prosthesis

Tissue bioengineering has attracted considerable interest in the biological and medical fields, notably in orthopaedics. In United States alone at least 90 000 patients per year undergo ligament reconstruction [39]. As is often reported, in contrast with the medial collateral ligament, anterior cruciate ligament (ACL) regeneration is hampered in vivo [32, 42, 48]. In the young and active population reconstruction is often the best therapeutic option when indicated. Many clinical studies have attempted to develop new therapeutic alternatives such as allografts [4, 5] and synthetic prosthesis using carbon fiber, knitted Dacron and braided polypropylene (Ligament Augmentation Device) [13]. Patellar tendon and tensor fascia have frequently been used to replace torn ACL [5]. The success of the reconstruction depends on multiple factors such as the type of tissue used for the reconstruction, surgical technique, fixation of the graft; and revascularization of the transplanted tissues, progressively ensheathed in a vascular synovial envelope [5]. The tendon should finally acquire some ligament properties; and the word "ligamentization" was given by Amiel et al. [2] to describe this physiological phenomenon in vivo postgrafting.

The use of allografts is still at an experimental stage and may lead to immunological reactions which prevent good healing of the wound [35, 49]. Synthetic material implantation was a very popular surgical technique in the 1980s; but it frequently led to implant degeneration and failure [44, 52]. Indeed, ligaments have a very particular histological organization which allows the tissue to support large mechanical stresses in vivo. An appropriate fixation is critical for the permanent implantation of bioengineered ACL in vivo, as discussed in the current literature [37, 44, 46].

Ligament General Histological Description

Little is known about physiological and pathological events that regulate the histological organization of ligaments in vivo. The cruciate ligaments appear in the human embryo by about 8 weeks of gestation from a condensation of the embryonic mesenchyme [13, 25]. One of the specific roles of ligaments is to connect bones to bones. Fascia, tendons; and ligaments share similar histological structures which are dense connective tissues consisting mainly of fibroblasts surrounded by collagen bundles [1]. Ligaments contain closely packed and thick collagen fibers (predominantly type I with a small proportion of type III collagen) but also include deposits of glycosaminoglycans, notably dermatan sulfate and hyaluronic acid [51], particularly close to the joint [1]. Extracellular matrix fibers are oriented parallel to the longitudinal axis of these tissues [1], but the fibers are less regularly arranged in ligaments than in tendons [45]. The composition of these biochemical constituents is modulated during tissue growth and development, changing also in response to functional requirements [51].

Anterior Cruciate Ligament Histological Properties

The ACL is described as one of the strongest tissues of the body. Most anatomic and histological studies on this ligament have been performed on human cadavers [3, 13, 17, 33] or on dead animals [1, 17, 51]. This ligament is composed of about 1500 fascicular collagen units measuring an average of 250 mm in diameter [3, 33].

The ACL is enveloped by synovium. A seam of synoviocytes of one to three cell layers in depth is located in a matrix covering the ACL on its anterior and posterior surface as a continuation of the synovium of the joint cavity [33]. The largest blood vessels are found in the areoloadipose synovium of the ACL, where arteries with a vascular diameter of up to 1 mm and veins of up to 2 mm are encountered [33]. Moreover, numerous nerves with diameters up to 0.1 mm lie parallel to the vessels. Occasionally, myelinated nerve fibers with a diameter of up to 10 mm can be seen among them [33].

The collagen fibrils are nonparallel but are themselves arranged into fibers which are oriented roughly in the long axis of the ligament in a wavy undulating pattern ("crimp") which slowly straightens out as loads are applied to the ligament [13]. The ACL usually supports loads of about 169 N (38 lb), and ruptures around 1730 N (about 390 lb) in vivo, [13]. An elongation of only 6% (about 2 mm) of human ACL (about 32 mm long) is reported to be the limit beyond which damage must be expected [3]. Thus, the main challenges associated with connective tissue engineering remain the production of strong and functional bioengineered ligaments.

Production of Bioengineered Ligaments

Several research groups are presently involved in the development of human bioengineered tissues to obtain ligament models and to further develop these ligament equivalents into eventual prosthesis. The development of such technology always implies the use of biocompatible and, preferably, biodegradable materials that can provide mechanical resistance to the implant and allow its progressive colonization and remodeling by living cells in situ postgrafting. Two types of ligament models, particularly ACL equivalents, are described here. The production of both models has involved the use of type I collagen; but Dunn's research group worked with a cellular composite collagen model while our bACL includes living cells seeded in an hydrated collagen matrix.

A Composite Collagen Prosthesis Used for Ligament Replacement

Dunn et al. [19–21] have reported successful replacement of ACL by cross-linked collagen prosthesis implanted in rabbits. This group, actively involved in the production of ligament biocompatible prosthesis, produced an acellular composite collagen prosthesis in order to reproduce ACL cross-linked organization of collagen fibers [50]. Their biodegradable prosthesis was produced by the alignment of 200 cross-linked collagen fibers and fixed by surgically created bone tunnels (2.8 mm diameter) in situ. Glutaraldehyde, dehydrothermal-cyanamide; or carbodiimide were used and compared as collagen cross-linking agents [20]. The cross-linked collagen fibers showed high tensile strengths (30–60 MPa) and small diameter (20–60 mm) [20], similar to fibers of the normal ACL [17]. Moreover, rapid ingrowth (within 8 weeks) of fibrous tissue and bone in the osseous tunnels as well as vascularization, cell migration, tissue remodeling; and extracellular matrix deposition (within 12 weeks) were observed in the composite collagen prosthesis after implantation in rabbit [20]. The ACL prosthesis was progressively degraded and replaced by a living and functionally ACL-like tissue in vivo. Thus, the ACL model of Dunn's group seems a promising alternative for human ligament replacement. However, living cells are not yet reported as being included in this bioengineered composite collagen prosthesis in vitro, limiting its use as a model for in vitropharmacological and biological studies.

A Living Bioengineered Ligament Model

Another type of bioengineered ligament model was developed and produced in our laboratory. Our research programs aim at producing bioengineered tissues that can serve both experimental and clinical interests [6–8, 14, 38, 40, 47]. Based on this expertise a new bioengineered ACL (bACL) model was developed. The main differences between our bACL and other ligament models [12, 19–21, 26] consist in (a) the addition of living ACL fibroblasts within its structure, (b) the absence of chemical cross-linking agents, and (c) the use of bones to anchor the bioengineered tissue. These bones are readily included to the bACL structure from the start of its production to facilitate its eventual transplantation in vivo.

Firstly, the human ligament fibroblasts (HLF) that were included in our bACL model were isolated and maintained in culture according to protocols set up in our laboratory. The cells were isolated from ACL biopsies of patients (60–80 years old) undergoing total knee arthroplasty. They were compared to dermal fibroblasts isolated from skin biopsies of healthy women (15–30 years old) undergoing breast reductive surgery. Secondly, the protocol for bACL was established, including the determination of HLF and collagen concentrations, the method of bone anchorage preparation, sterilization; manipulation and, the initial size of the model.

Figure 1 summarizes the two-step digestion method of ACL to obtain primary cultures of HLF. Briefly, the human ACL was weighted (g) and cut into small pieces following removal of the periligamentous tissue. The ACL fragments were first dissociated in 0.125% collagenase (collagenase H, Boehringer Mannheim, Germany; 1 ml enzymatic solution/mg ligament) for 16 h under gentle agitation at 37°C. A 0.1% trypsin solution (ICN biochemical, 1 ml /mg hydrated tissue) was then added to the cellular suspension for 1 h. The enzymes were dissolved in Dulbecco's modified Eagle's medium (Flow, Calif., USA), pH 7.4, containing antibiotics. HLF were cultured in this medium supplemented with 10% fetal calf serum (Gibco Labs), 100 IU/ml penicillin G; and 25 mg/ml gentamicin (Sigma, St. Louis, Mo, USA). HLF maintained their morphological properties for more than seven passages in culture.

HLF appear morphologically slightly bigger (Fig. 2a) and present a less fusiform shape than dermal fibroblasts (Fig. 2b) in culture. Immunofluorescent labeling analysis revealed that 12 different populations of HFL extracted from corresponding number of biopsies expressed vimentin, fibronectin, types I and III collagens; and elastin.

Partially demineralized chicken bones (sterilized with ethylene oxyde) were used as bACL anchorages (Fig. 3 summarizes the production steps of bACL). A solution of Dulbecco's modified Eagle's medium 2.7X containing antibiotics was mixed with a second solution containing heat-inactivated (30 min at 56°C) fetal calf serum, solubilized type I collagen (Sigma, type III) and living HLF (passages 3–5). The cells were used at a final concentration of 1×10^6 cells/ml. This mixture was quickly poured into a sterile cylinder already containing one of the bone anchorages. The second osseous anchorage was added on the top of the mixture and maintained in position using a thin thread passed through a

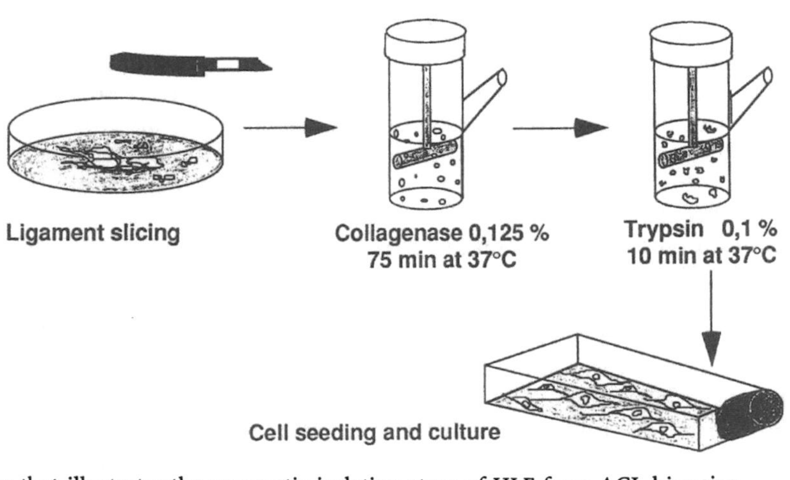

Fig. 1. Diagram that illustrates the enzymatic isolation steps of HLF from ACL biopsies

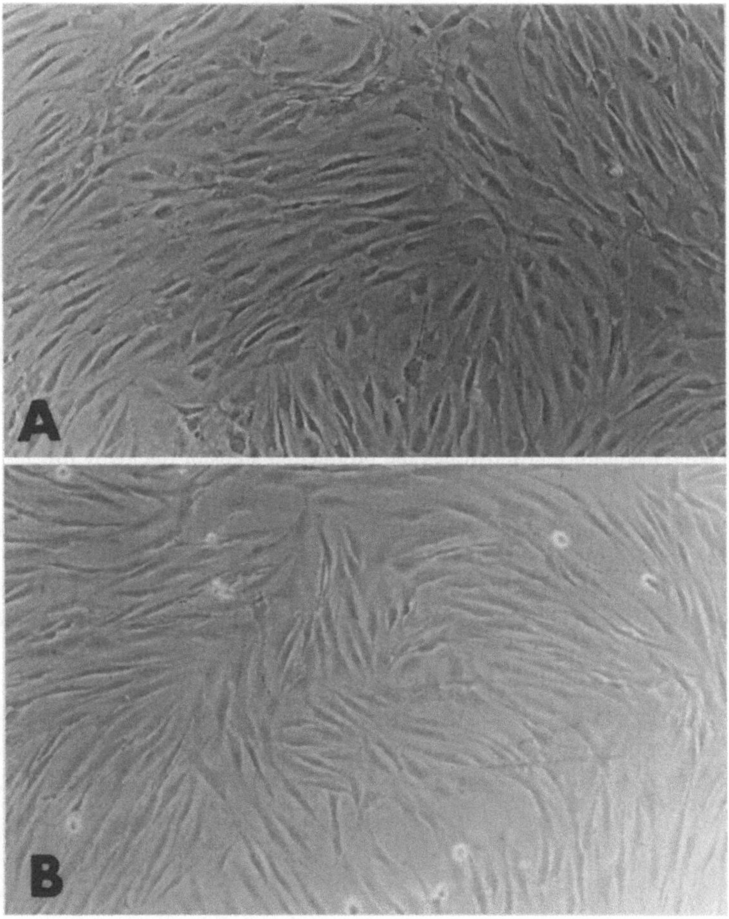

Fig. 2. Photographs of HLF (**A**) and human dermal fibroblasts (**B**) cultured for a week at passage 4. ×120

fine aperture made with a surgical needle. The final concentration of bovine type I collagen in the bACL was always 2 mg/ml. The culture medium (the same used to maintain HLF in culture) was changed daily.

The bACL hydrated collagen matrix was rapidly contracted diametrically by the HLF within the first 24 h and more slowly thereafter. Figure 4 shows one of the bACL produced according to this method and cultured for 12 h. Histological studies were performed on the bACL after various culture periods using the methods already reported [14]. The data indicate that HLF seeded in the bACL cultured in a vertical position repeatedly adopted a defined longitudinal orientation and organized the surrounding extracellular matrix in fibers parallel to the long axis of the bACL, within 2 weeks in culture (Fig. 5). Moreover, immunohistological studies showed the maintenance of vimentin, elastin; and

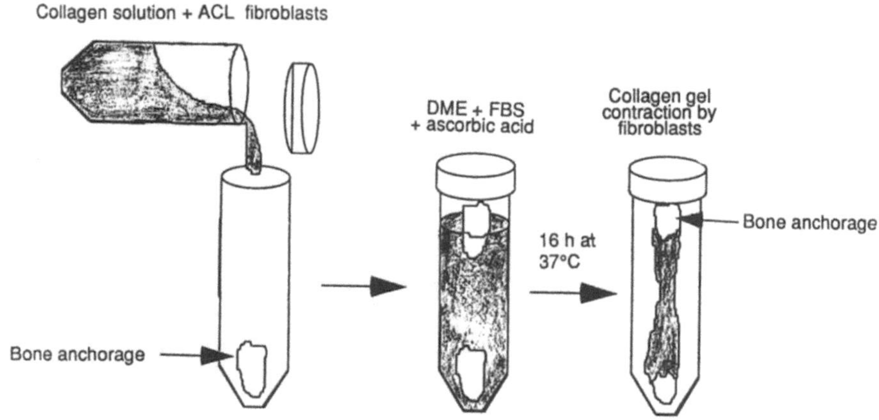

Fig. 3. Diagram that illustrates the production steps of the living bACL

type III collagen synthesis by the living HLF seeded in the bACL throughout the whole culture period (up to 1 month).

This biotechnological approach presents interesting advantages and likely could also be adapted for the production of bioengineered tendons in vitro.

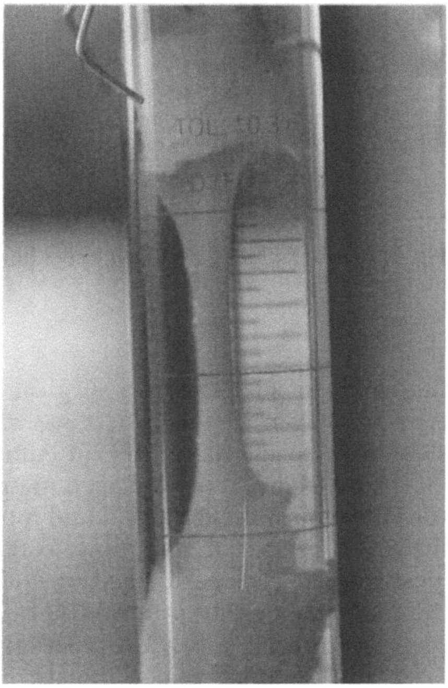

Fig. 4. Photograph of a living bACL cultured for 14 days under elongation. Note the two bone anchorages integrated to the cultured model

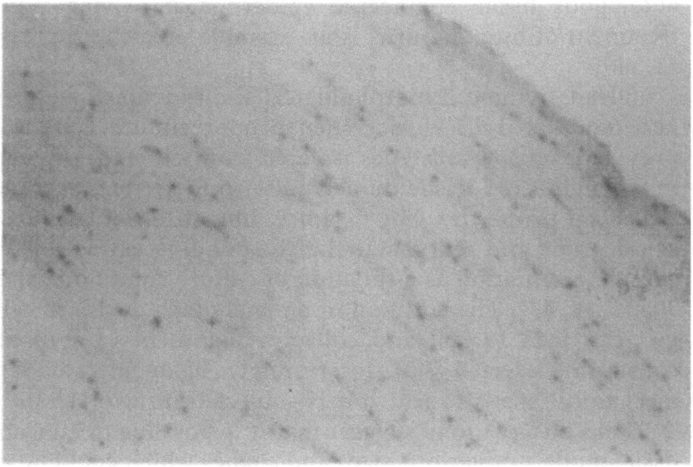

Fig. 5. Histological sections of bACL cultured for 14 days in a vertical position, under elongation. Note that a constant tension induces orientation of HLF and surrounding extracellular matrix. ×40

Since living HLF seeded in this bACL continue to secrete and reorganize extracellular matrix constituents upon culture time, this ligament model becomes a promising tool to study the role of various cellular elements involved in the maintenance or the induction of ligament organization and repair in vitro. Moreover, autologous bone fragments and HLF could be used to produce bACL for eventual ligament replacement in humans. Such possibility could greatly reduce the risks of immunological reactions and infectious risks as well as favor the permanent graft integration posttransplantation.

One of the major problems in the implantation of ACL replacement prosthesis is the permanent fixation of the implant to the bones. In our ACL model the collagen initially penetrates and polymerizes in the trabiculae structure of the spongious bone anchorages. The use of osseous anchorages seems promising since it would facilitate surgical manipulations.

The lack of strength of our bACL is presently its main drawback, less as an in vitro model than an eventual prosthesis. Thus we are actively working at testing various approaches to reinforce the bACL model, particularly its matricial scaffold. Indeed, some extruded collagenous fibers could eventually be integrated to this bACL to improve its strength. The application of various stresses could also be required in vitro in obtain comparable tissue organization between the bACL model and human ACL, although histological analyses already show very promising data.

Experimental and Clinical Use of Bioengineered Ligaments: Perspectives

The biotechnological applications of tissue engineering and cell biology have led to revolutionary advances in various medical fields. The therapeutic use of autologous tissue engineered epidermis and living skin substitutes for the treatment of human burns is an example of such progress [6, 7, 15, 16, 22–24, 31, 40].

Indeed, living cell morphological and functional features can be maintained and modulated in vitro. When proper culture conditions are defined and provided, several cell types isolated from different organs can be reintegrated in bioengineered tissue models in vitro to secrete, degrade; and reorganize its structural properties. For instance, human fibroblasts seeded in collagen gels can degrade and reorganize the surrounding extracellular matrix and adopt a specific orientation as a function of culture conditions and time [6–11, 18, 27–30, 36, 34, 47]. The addition of an anchorage device to cell-populated collagen gels can lead to cell and collagen alignment in a direction parallel to the constraints exerted [40]. Interestingly, ligament remodeling depends on the mechanical stress to which it is subjected in vivo [1] and in vitro [34, 43].

These observations suggest that it is possible to conceptualize and produce tissue models of various sizes and shapes which could be populated with living cells in vitro and/or posttransplantation in vivo. The two different bioengineered ligament models described above are good examples of such possibility, both functional and histological properties being modulated throughout culture conditions and time in vitro (our bACL) or in vivo postgrafting in animals [20].

These different bioengineered ligament equivalents must indeed be optimized prior to human transplantation. However, tissue engineering will hopefully contribute to developing and exploiting many therapeutic options. For example, gene therapy might find some interesting applications by introducing living transfected cells within various types of bioengineered tissues.

Despite the challenges inherent to organogenesis in vitro, it is quite reasonable to believe that in a near future patients will benefit from bioengineered ligament transplants initially produced in a laboratory to reach their final completion in a human body.

Acknowledgements. The authors are grateful to Mrs Julie Bergeron, Nathalie Tremblay; and Rina Guignard for skilled technical assistance, Mr. Louis-Mathieu Stevens for his helpful suggestions, Mr. Claude Marin for photographic assistance and Mrs. Jeannine Savard (Dept. of Anatomopathology) for her precious technical contribution to the histological analyses. This work was supported by The Canadian Orthopaedic Association and its Hip Hip Hip Hooray Committee. F.A.A. and L.G. are recipients of scholarships from the Fonds de la Recherche en Santé du Québec.

References

1. Amiel D, Frank C, Harwood F, Fronek J, Akeson W (1984) Tendons and ligaments: a morphological and biochemical comparison. J Orthop Res 1 257–265
2. Amiel D, Kleiner JB, Roux RD, Harwood FL, Akeson W (1986) The phenomenon of "ligamentization": anterior cruciate ligament reconstruction with autogenous patellar tendon. J Orthop Res 4(2):162–172

3. Amis AA, Dawkins GPC (1991) Functional anatomy of the anterior cruciate ligament: fibre bundle actions related to ligament replacements and injuries. J Bone Joint Surg Br 73(2):260–267
4. Andrews M, Noyes FR, Barber-Westin SD (1994) Anterior cruciate ligament allograft reconstruction in the skeletally immature athlete. Am J Sports Med 22(1):48–54
5. Arnoczky SP, Tarvin GB, Marshall JL (1982) Anterior cruciate ligament replacement using patellar tendon. J Bone Joint Surg Am 64(2):217–224
6. Auger FA (1988) The role of cultured autologous human epithelium in large burn wound treatment. Transplant Implant Today 5:21–24
7. Auger FA, Guignard R, López Valle CA, Germain L (1993) Role and inocuity of Tisseel, a tissue glue, in the grafting process and in vivo evolution of human cultured epidermis. Br J Plast Surg 46:136–142
8. Auger FA, López Valle CA, Guignard R, Tremblay N, Noël B, Goulet F, Germain L (1995) Skin equivalents produced using human collagens. In Vitro Cell Dev Biol 31:432–439
9. Bell E, Ehrlich HP, Buttle DJ, Nakatsuji T (1981) Living tissue formed in vitro and accepted as skin-equivalent tissue of full thickness. Science 211:1052–1054
10. Bell E, Ehrlich HP, Sher S, Merrill C, Sarber R, Hull B, Nakatsuji T, Church B, Buttle D (1981) Development and use of a living skin equivalent. Plast Reconstr Surg 67:386–392
11. Bell E, Ivarsson B, Merrill C (1979) Production of a tissue-like structure by contraction of collagen lattices by human fibroblasts of different proliferative potential in vitro. Proc Natl Acad Sci USA 76:1274–1278
12. Bellows CG, Melcher AH, Aubin JE (1982) Association between tension and orientation of periodontal ligament fibroblasts and exogenous collagen fibres in collagen gels in vitro. J Cell Sci 58:125–138
13. Bessette GC, Hunter RE (1990) The anterior cruciate ligament. Orthopedics 13(5) 551–562
14. Bouvard V, Germain L, Rompré P, Roy B, Auger FA (1992) Influence of dermal equivalent maturation on the development of a cultured skin equivalent. Biochem Cell Biol 70:34–42
15. Boyce ST, Foreman TJ, English KB, Stayner N, Cooper ML, Sakabu S, Hansbrough JF (1991) Skin wound closure in athymic mice with cultured human cells, biopolymers, and growth factors. Surgery 110:866–876
16. Contard P, Bartel RL, Jacobs L II, Perlish JS, MacDonald ED II, Handler L, Cone D, Fleischmajer R (1993) Culturing keratinocytes and fibroblasts in a three-dimensional mesh results in epidermal differentiation and formation of a basal lamina-anchoring zone. J Invest Dermatol 100:35–39
17. Danylchuk KD, Finlay JB, Krcek JP (1978) Microstructural organization of human and bovine cruciate ligaments. Clin Orthop 131:294–298
18. Delvoye P, Nusgens B, Lapière CM (1983) The capacity of retracting a collagen matrix is lost by dermatosparactic skin fibroblasts. J Invest Dermatol 81:267–270
19. Dunn MG, Avasarala PN, Zawadsky JP (1993) Optimization of extruded collagen fibers for ACL reconstruction. J Biomed Mater Res 27:1545–1552
20. Dunn MG, Maxian SH, Zawadsky JP (1994) Intraosseous incorporation of composite collagen prostheses designed for ligament reconstruction. J Orthop Res 12:128–137
21. Dunn MG, Tria AJ, Kato YP, Bechler JR, Ochner RS, Zawadsky JP, Silver F (1992) Anterior cruciate ligament reconstruction using a composite collagenous prosthesis: a biomechanical and histological study in rabbits. Am J Sports Med 20(5):507–515
22. Fernyhough W, Aukhil I, Link T (1987) Orientation of gingival fibroblasts in stimulated periodontal spaces in vitro containing collagen gel. J Periodontol 58:762–769
23. Germain L, Auger FA (in press) Tissue engineered biomaterials: biological and mechanical characteristics. In: Wise DL, Trantolo DJ, Altobelli DE, Yaszemski MJ, Gresser JK, Schwartz ER (eds) Encyclopedic handbook of biomaterials and bioengineering. Part B: Applications. Marcel Dekker, New York, pp 699–734
24. Germain L, Rouabhia M, Guignard R, Carrier L, Bouvard V, Auger FA (1993) Improvement of human keratinocyte isolation and culture using thermolysin. Burns 19:99–104
25. Gray DJ, Gardner E (1950) Prenatal development of the knee joint. Am J Anat 86:235–287

26. Green H, Kehinde O, Thomas J (1979) Growth of cultured human epidermal cells into multiple epithelia suitable for grafting. Proc Natl Acad Sci USA 76:5665–5668
27. Grinnell F, Lamke CR (1984) Reorganization of hydrated collagen lattices by human skin fibroblasts. J Cell Sci 66:51–63
28. Guidry C, Grinnell F (1986) Contraction of hydrated collagen gels by fibroblasts: evidence for two mechanisms by which collagen fibrils are stabilized. Collagen Rel Res 6:515–529
29. Guidry C, Grinnell F (1987) Heparin modulates the organization of hydrated collagen gels and inhibits gel contraction by fibroblasts. J Cell Biol 104:1097–1103
30. Guidry C, Grinnell F (1985) Studies on the mechanisms of hydrated collagen gel reorganization by human skin fibroblasts. J Cell Sci 79:67–81
31. Hansbrough JF, Boyce ST, Cooper ML, Foreman TJ (1989) Burn wound closure with cultured autologous keratinocytes and fibroblasts attached to a collagen-glycosaminoglycan substrate. JAMA 262:2125–2130
32. Harner CD, Paulos LE, Greenwald AE, Rosenberg TD, Cooley VC (1994) Detailed analysis of patients with bilateral anterior cruciate ligament injuries. Am J Sports Med 22:37–43
33. Haus J, Refior HJ (1987) A study of the synovial and ligamentous structure of the anterior cruciate ligament. Int Orthop 11:117–124
34. Huang D, Chang TR, Aggarwal A, Lee RC, Ehrlich HP (1993) Mechanisms and dynamics of mechanical strengthening in ligament-equivalent fibroblast-populated collagen matrices. Ann Biomed Eng 21:289–305
35. Jackson DW, Kurzeil PR (1991) Allograft in knee ligament surgery. In: Scott WN (ed) Ligament and extensor mechanism of the knee: diagnosis and treatment. Mosby, St. Louis pp 349–360
36. Kasugai, S, Susuki S, Shibata S, Yasui S, Amano H, Ogura H (1990) Measurements of the isometric contractile forces generated by dog periodontal ligament fibroblasts in vitro. Arch Oral Biol 35:597–601
37. Kurosaka M, Yoshiya S, Andrish JT (1987) A biomechanical comparison of different surgical techniques of graft fixation in anterior cruciate ligament reconstruction. Am J Sports Med 15:225–229
38. L'Heureux N, Germain L, Labbé R, Auger FA (1993) In vitro construction of a human blood vessel from cultured vascular cells: a morphological study. J Vasc Surg 17:499–509
39. Langer R, Vacanti JP (1993) Tissue engineering. Science 260:920–926
40. López Valle CA, Auger FA, Rompré P, Bouvard V, Germain L (1992) Peripheral anchorage of dermal equivalents. Br J Dermatol 127:365–371
41. López Valle CA, Glaude P, Auger FA (1992) Tie-over dressings: surgical model for in vivo evaluation of cultured epidermal sheets in mice. Plast Reconstr Surg 89:139–143
42. Lyon RM, Akeson WH, Amiel D, Kitabayashi LR, Woo SL (1991) Ultrastructural differences between the cells of the medial collateral and the anterior cruciate ligaments. Clin Orthop 272:279–286
43. Matyas J, Edwards P, Miniaci A, Shrive N, Wilson J, Bray R, Frank C (1994) Ligament tension affects nuclear shape in situ: an in vitro study. Connect Tissue Res 31(1):45–53
44. Olson EJ, Kang JD, Fu FH, Georgescu HI, Mason GC, Evans CH (1988) The biomechanical and histological effects of artificial ligament wear particles: in vitro and in vivo studies. Am J Sports Med 16(6):558–570
45. Puddu G, Ippolito E (1983) Reconstruction of the anterior cruciate ligament using the semitendinosus tendon. Histological study of a case. Am J Sports Med 11:14–16
46. Robertson DB, Daniel DM, Biden E (1986) Soft tissue fixation to bone. Am J Sports Med 14(5):398–403
47. Rompré P, Auger FA, Germain L, Bouvard V, López Valle CA, Thibault J, Le Duy A (1990) Influence of initial collagen and cellular concentrations on the final surface area of dermal and skin equivalents: a box-behnken analysis. In Vitro Cell Dev Biol 26:983–990
48. Ross SM, Joshi R, Frank CB (1990) Establishment and comparison of fibroblast cell lines from the medial collateral and anterior cruciate ligaments of the rabbit. In Vitro Cell Dev Biol 26:579–584

49. Sabiston P, Frank CY, Lam T, Shrive N (1990) Allograft ligament transplantation. A morphological and biochemical evaluation of a medial collateral ligament complex in a rabbit model. Am J Sports Med 18:160–168
50. Strocchi R, De Pasquale V, Gubellini P, Facchini A, Marcacci M, Buda R, Zaffaghini S, Ruggeri A (1992) The human anterior cruciate ligament: histological and ultrastructural observations. J Anat 180:515–519
51. Watanabe M, Nojima M, Shibata T, Hamada M (1994) Maturation-related biochemical changes in swine anterior cruciate ligament and tibialis posterior tendon. J Orthop Res 12:672–682
52. Woods GA, Indelicato PA, Prevot TJ (1991) The Gore-Tex anterior cruciate ligament prosthesis. Two versus three year results. Am J Sports Med 19:48–55

Subject Index

MIX
Papier aus verantwortungsvollen Quellen
Paper from responsible sources
FSC® C105338

If you have any concerns about our products,
you can contact us on
ProductSafety@springernature.com

In case Publisher is established outside the EU,
the EU authorized representative is:
Springer Nature Customer Service Center GmbH
Europaplatz 3, 69115 Heidelberg, Germany

Printed by Libri Plureos GmbH
in Hamburg, Germany